Mathematical Ecology

Mathematical Ecology

E. C. PIELOU

Professor of Biology
Dalhousie University
Halifax, Nova Scotia

A Wiley-Interscience Publication

JOHN WILEY & SONS, New York · London · Sydney · Toronto

Library of Congress Cataloging in Publication Data:

Pielou, E   C   1924–
 Mathematical ecology = revised edition of An introduction to mathematical ecology.

 "A Wiley-Interscience publication."
 Bibliography: p.
 Includes indexes.
 1. Ecology—Mathematical models. I. Title.
QH541.15.M3P54 1976   574.5'01'51   76-49441
ISBN 0-471-01993-3

Printed in the United States of America

10 9 8 7 6 5 4 3 2 1

Preface to the First Edition

The fact that ecology is essentially a mathematical subject is becoming ever more widely accepted. Ecologists everywhere are attempting to formulate and solve their problems by mathematical reasoning, using whatever mathematical knowledge they have acquired, usually in undergraduate courses or private study. The purpose of this book is to serve as a text for these students and to demonstrate the wide array of ecological problems that invite continued investigation.

In writing a book of this length (or, indeed, of any length) the author is always faced with the problem whether to cover a great many topics superficially or to delve deeply into a few. The compromise I have attempted has been to pick selected topics over the whole range of mathematical ecology and then to deal in detail with those aspects that seem likely to furnish good starting points for further research. The list of chapter headings shows the subjects that were chosen. Their choice, and the aspects to be pursued, has, of course, been subjective and a matter of my own judgment. It is unlikely that any other ecologist with the same objective would have selected exactly the same topics.

The book is in no sense a review. An enormous amount of interesting and valuable work has been ignored and any mathematical ecologist glancing at the bibliography is bound to wonder at what may seem to him inexplicable omissions. The decision to exclude an account of an interesting piece of work was always difficult but the exclusions were necessary to make room for a fairly full exposition of the matters dealt with. My object has been to provide a sufficiently detailed development of the topics examined for the reader to be able to consult the current literature with an understanding of what has led up to it; as far as possible nothing has been asserted without proof. For each topic I have attempted to give a connected account of the underlying theory. So that the reader will not lose the thread of an argument, methods of estimating parameters and similar practical details have not been given. Information on these matters, and numerical examples will be found in the literature cited.

v

To make the text suitable for private study, many mathematical derivations have been written out *in extenso*. I have striven to avoid the mathematician's daunting phrases "Obviously therefore . . " and "Clearly then . . ." which are so often substituted for long chains of reasoning that, for the mathematically inexperienced, are neither obvious nor clear. When these phrases are used here, a line or two at most of algebraic manipulation is all that the reader needs to interpolate.

This book is based on a one-semester graduate course in mathematical ecology that I gave in the spring of 1968 while I was Visiting Professor in the Biomathematics Program of the Department of Experimental Statistics, North Carolina State University, Raleigh. It is a pleasure to thank Dr. H. L. Lucas Jr., Director of the Program, for his hospitality.

Above all, I thank my husband D. P. Pielou for his constant help and encouragement.

E. C. PIELOU

Kingston, Ontario
July, 1969

Preface to the Second Edition

Revising *An Introduction to Mathematical Ecology* has entailed making very considerable changes to the old edition because of the explosive growth of ecology in recent years. Now, more than ever, it is impossible to cover the subject exhaustively. I have continued to focus on the same topics as before but have modernized and expanded the treatment of most of them.

The topics are compartmentalized into the same chapters as in the old edition with only two changes. The material in Chapters 5 and 6 has been rearranged, and the final section of Chapter 6 contains an (opinionated) discussion of the place of mathematical model building in ecology. One extra chapter (Chapter 17) has been added. It describes new work that I believe exemplifies the approach to ecological problems most likely to yield satisfying results: that of perceiving the many clues to ecological processes that straightforward observation of real ecosystems affords, and interpreting these clues. Much material that is not even touched upon in this book is discussed at length in my monograph on ecological diversity.* The overlap between the monograph and this text is as small as I could make it and is confined to certain parts of Chapters 17, 18, and 19 of the present book.

I hope that the enlarged and modernized version of the book will go some way toward closing the ever-widening gulf between mathematical ecology in the narrow sense (the construction and study of purely hypothetical systems models) and statistical ecology in the broad sense (the development and use of methods for interpreting observations on actual ecosystems). While it is unlikely that any research worker will be equally attracted by both fields, all serious theoretical ecologists need to be familiar with both of them.

* *Ecological Diversity*, Wiley, New York, 1975.

I am indebted to numerous people for useful comments and suggestions, especially to Dr. W. J. Ewens, Dr. N. Keyfitz, Dr. L. Tummala, and Dr. K. L. Weldon.

E. C. Pielou

Halifax, Nova Scotia,
July, 1976

Contents

x / Contents

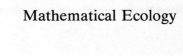

Mathematical Ecology

Introduction

Most ecological communities are made up of a vast profusion of living things. In an acre of forest, for instance, an enormous number of species is present, from trees to soil microorganisms. Not only does each species differ from every other, but also all the individuals within any one species are unique. Each individual is a complex organism that changes continuously; at any moment its behavior depends on its genetic constitution, its age, the vicissitudes of its life up to that moment, and on the conditions that prevail in its locality at that moment. Influences too numerous to list affect its physiological processes unceasingly. It can truly be said that no two of the individual units making up a community are alike and that each of them, throughout its lifetime, varies continuously in a manner peculiar to itself. In addition to all this, a community as a whole is at all times being depleted by deaths and replenished by births; often immigration and emigration are further causes of a continual turnover of individuals. Thus the components of a community are never the same on two successive occasions.

In spite of all this unending change, however, experience shows that in the absence of outside disturbance most communities either remain in a steady state for long periods or pass through an orderly progression of successional stages culminating in a steady state. The concept of a steady state is, of course, a subjective notion. It seems reasonable to say, however, that a steady state exists if the gross properties of a community persist for periods many times as long as the lifetimes of most of the species present. This cautious definition allows for the possibility that the longest lived species in a community may never attain populational stability, and if apparent stability prevails this is merely because of the comparatively short lifetimes of human observers (see Frank, 1968). A community undergoing cyclical fluctuations is also in a steady state, of course, provided the cycles are regular and no trend is superimposed on them.

In any case, that steady states should be as steady and successions as orderly, as they are found to be, is remarkable when one considers the

1

unceasing internal changes going on in a community. To say that most communities persist for generation after generation of all but a few of their constituent species is, indeed, a tautology, for if they did not do so there would be no recognizable entities that could be called communities. Although probably no two ecologists hold identical opinions over the degree to which assemblages of organisms can be regarded as autonomous and self-contained, there is no need to insist that an assemblage should have these properties for it to be defined as a community. Anyone who admits the existence of such obvious features of a landscape as forests, meadows, and swamps, realizes that the distinctness and persistence of these things deserve to be explained.

In contrast to the steadiness stressed above are the spectacular population explosions that occasionally occur. A single species, which may be a regular member of a community or a chance immigrant, rapidly increases in numbers until it has brought about profound, and sometimes irreversible, change in its surroundings. If the change is deleterious from a human point of view, the species that explodes is by definition a pest, and the latest addition to the ever-growing list of injurious pests is, of course, man himself. (Anyone who doubts this has only to ask himself: which pest is responsible for the serious modern problem of environmental pollution?) Pest control consists in preventing the population of any species from rising to a level at which it does appreciable economic damage and is one of the most important branches of applied ecology. If what is being damaged is a natural community (as opposed to a cultivated crop, stored products, or a human artefact) pest control thus consists in ensuring that the community remains in what is believed to be its natural steady state.

In very general terms, then, there are two classes of questions confronting ecologists. In the first place, what are the processes that permit maintenance of a steady state, or the gradual, orderly succession of states, that occurs in "healthy" undisturbed communities? Second, what are the causes and consequences of sudden departures from steadiness?

In a broad sense, there are three markedly different approaches to these problems: (1) ecological model building; (2) the statistical study of populations and communities; and (3) quantitative descriptive ecology. They are the topics of, respectively, Chapters 1 through 6, Chapters 7 through 18, and Chapters 19 through 22. Their distinguishing characteristics are the subject of the following paragraphs.

1. Ecological model building takes many different forms, depending on the purpose for which it is done. At one extreme are so-called "explanatory" models or systems models, whose behavior is thought to duplicate,

at least approximately, the true behavior of the populations being modeled. These models usually take the form of sets of simultaneous differential equations relating the growth rate of each species-population in a community to the sizes of all the other, interacting, species-populations. The properties (especially those having to do with the various forms of stability) of sets of differential equations is a relatively modern research field in applied mathematics (Sanchez, 1968) and mathematicians have raided the subject matter of ecology for illustrative material. In so doing, they have done a valuable service for ecologists but their results are not part of ecology itself. Their conclusions, though undeniably interesting and thought-provoking for ecologists, are no more than that; they relate to simple mathematical systems, not to complex ecosystems, as Oster (1975), for example, has emphasized. But properly applied, they do contribute to our understanding of ecological processes. It is a mistake to describe these models as "useful" if they fit actual observations (and, by extension, useless if they do not). Such a notion involves a misunderstanding of the purpose of theoretical models. A poor fit is often more informative than a good one. A good fit may imply no more than that the observations and the model reproducing them are not capable of discriminating among competing theories on ecosystem functioning. A poor fit usually permits rejection of an untenable theory and to that extent at least yields a gain in knowledge. Even a small gain is better than none.

In contrast to models designed only to *explain* the behavior of populations are models designed to *predict* their behavior in the future. The great importance of such models in all branches of applied ecology is obvious. They strive to answer the question: how will a given ecosystem continue to develop if left to itself, and how will it react if interfered with? Purely predictive models are of two kinds, regression-type models and time series models. Regression-type models summarize, in the form of sets of simultaneous equations, the observed interdependence among an ecosystem's component parts. Such models do not take account of temporal change; they merely represent the interrelationships among the components at the moment observations were made. Time series models, however, do not suffer from this defect. They are dynamic rather than static. Their use in ecology began only recently (see, for example, Poole, 1976a, 1976b) and they promise to bring about an enormous improvement in the predictive capabilities of environmental managers and the controllers of renewable resources. Their contribution is wholly to applied ecology and not at all to theoretical ecology; they are therefore not discussed in this book.

2. The kinds of ecosystems whose structure and function theoretical

ecologists seek to understand cover a very wide range in terms of size and complexity. At one end of the spectrum are laboratory microcosms in which a few species of small, short-lived, fast-breeding, active animals (for example, beetles, fruit flies, blowflies, *Daphnia*, *Paramecia*) live and interact for many generations under controlled conditions in a container. At the other end are large natural systems of baffling complexity: systems in which thousands or tens of thousands of species live together under continually changing conditions and in spatially heterogeneous environments. The more complex the ecosystem an ecologist studies, the less useful are simple explanatory "systems models" as aids to understanding and the more useful is a "statistical" approach.

The contrast is worth examining. A "modeler's" conjectures are about abstractions; he envisages possible interactions among a few idealized species-populations and deduces their consequent behavior; only then (if at all!) is the mental construct compared with reality. A "statistician's" conjectures are about concrete, observable things; contemplating real ecosystems, which initially furnish an unstructured welter of impressions, he strives to single out those observable phenomena that serve as clues to underlying processes. It is detective work. It entails discrimination, in a situation where examination of all the facts is not feasible, between facts that are, and facts that are not, relevant to the solution of particular problems. The problems themselves are usually prompted by observation.

It is worth remarking that ecosystems whose understanding demands the "statistical" approach are of greater concern to ecologists than are those that lend themselves to the "modeling" approach. Laboratory microcosms themselves are usually studied as "models" (real as opposed to conceptual) of communities in the outdoor world. But it is impossible to duplicate, in the laboratory, communities whose member species vary enormously in respect of such characteristics as size, longevity, phenotypic plasticity, generation time, fertility, morbidity, mortality, and motility. It is communities such as these, which constitute most of the biosphere, that should engage the lion's share of ecologists' collective attention.

3. The third approach to mathematical-plus-statistical ecology is closely related to the second. It consists in processing large bodies of observational data in such a way that interesting regularities, hitherto buried from sight, become apparent. These regularities can inspire the building of hypotheses, and the hypotheses can then be tested by appeal to other bodies of data. Thus approaches 2 and 3 are both statistical: both require that evidence from real ecosystems be sifted and interpreted. The difference is that testable hypotheses are derived from raw data in approach 2 and from processed data in approach 3. The processing may

be done in numerous ways. Methods of handling multivariate statistical data are constantly being improved as computers increase in sophistication; new methods are being tested in ecological contexts as soon as they are developed. The aim in all cases is to simplify and clarify unwieldy masses of "noisy" data so that their underlying structure is revealed. Only then, perhaps, can it be explained.

I

Population Dynamics

1

Birth and Death Processes

1. Introduction

All populations of organisms fluctuate in size. For any population the only assertion that can be made about it with certainty is that its size will not remain constant. The investigation of the growth and decline of populations is, historically, the oldest branch of mathematical ecology. It has been found that a fruitful way to proceed is to postulate a simple model to account for population change and then to examine the consequences of the assumptions by mathematical argument. We therefore begin by considering the simplest of all systems, the so-called pure birth process, and then deal with the simple birth and death process.

Before doing so, however, it is worthwhile to justify this approach. Field ecologists are often understandably impatient with arguments that proceed from assumptions so greatly oversimplified as to be manifestly unreasonable. But much can be said in defence of the simple, abstract models. Obviously one must study the behavior of simple models before modifying and complicating the first, simplest assumptions, and the simple models provide a basis for elaboration. Also, the way in which natural populations differ from these simple models in their behavior may suggest in what way the over-simplified assumptions are false and how they should be altered. Finally, if models are required for *ad hoc* predictive purposes, simple models are often adequate, even though we can feel no confidence that the underlying model gives a true explanation of the natural process being modeled. This is especially true when we are concerned with population changes that occur over short periods of time. The resemblance between a natural population and a simple model may then be very close.

2. The Pure Birth Process

In the pure birth process the assumptions are as follow: The organisms are assumed to be immortal and to reproduce at a rate that is the same for every individual and does not change with time. It is also assumed that

8

the individuals have no effect on one another. Since no deaths occur, a population growing in this manner can only increase or remain constant; it cannot decrease. In spite of the extreme simplicity of these assumptions they might apply, approximately at least and over a short time interval, to the growth of a population of single-celled organisms that reproduce by dividing. Possibly algal blooms in eutrophic lakes increase in this manner in spring.

If we write N_t for the size of the population at time t and λ for the rate of increase of each individual, it is clear that

$$\frac{dN_t}{dt} = \lambda N_t,$$

hence that $\ln N_t = \lambda t + C$, where C is a constant of integration. Suppose that the initial size of the population at time $t = 0$ were i. Then C may be evaluated, since at $t = 0$, $\ln i = C$. Therefore

$$\ln \frac{N_t}{i} = \lambda t \qquad \text{or} \qquad N_t = ie^{\lambda t}.$$

This is the Malthusian equation for population growth and it shows that such growth is exponential in the simple circumstances postulated.

The process just described is, of course, *deterministic*. It assumes not that an organism may reproduce but that in fact it does reproduce with absolute certainty. Clearly, however, population growth is a stochastic process. Given a population of yeast cells growing by fission, for example, one can say only that there is a certain probability that a particular cell will divide in a given time interval. We must therefore investigate the stochastic form of the pure birth process.

Suppose, then, that in any short time interval Δt the probability that a cell will divide is $\lambda \Delta t + o(\Delta t)$. Here $o(\Delta t)$ denotes a quantity of smaller order of magnitude than Δt. Thus the probability of a birth in a population of size N is $\lambda N \Delta t$, plus terms of smaller order of magnitude than Δt. Then the probability that the population is of size N at time $t + \Delta t$ is

$$p_N(t + \Delta t) = p_{N-1}(t)\lambda(N-1)\Delta t + p_N(t)(1 - \lambda N \Delta t);$$

that is, either the population was of size $N - 1$, with probability $p_{N-1}(t)$ at time t, and a division occurred in Δt or the population was of size N at time t and no division occurred in Δt. Then

$$\frac{p_N(t + \Delta t) - p_N(t)}{\Delta t} = -\lambda N p_N(t) + \lambda(N-1)p_{N-1}(t).$$

Allowing Δt to tend to zero and putting

$$\lim_{\Delta t \to 0} \frac{p_N(t+\Delta t)-p_N(t)}{\Delta t} = \frac{dp_N(t)}{dt},$$

we see that

$$\frac{dp_N(t)}{dt} = -\lambda N p_N(t) + \lambda(N-1)p_{N-1}(t). \tag{1.1}$$

Starting from this differential difference equation, we now wish to find $p_N(t)$ in terms of N, λ, t and the initial size of the population at time $t=0$, again denoted by i; that is $p_i(0)=1$ and $p_N(0)=0$ when $N \neq i$. It is also clear that since no deaths occur and consequently the population size can never be less than its initial value we must have $p_{i-1}(t)=0$. From these conditions it follows from (1.1) that

$$\frac{dp_i(t)}{dt} = -\lambda i p_i(t). \tag{1.2}$$

We shall now solve (1.2) to obtain $p_i(t)$, substitute the result in (1.1), solve for $p_{i+1}(t)$, and so on. In this manner $p_N(t)$ for all $N > i$ may be found. From (1.2) it is seen that

$$\ln p_i(t) = -\lambda i t + \text{const.}$$

Since, when $t = 0$, $p_i(t) = 1$ and $\ln p_i(t) = 0$, the constant is also 0 and therefore

$$p_i(t) = e^{-\lambda i t}.$$

This is the probability that during time t no reproduction will occur. To find $p_{i+1}(t)$ we now solve (1.1) in the form

$$\frac{dp_{i+1}(t)}{dt} + \lambda(i+1)p_{i+1}(t) = \lambda i p_i(t) = \lambda i e^{-\lambda i t}.$$

Now multiply both sides by $e^{\lambda(i+1)t}$ so that the left-hand side may be integrated:

$$e^{\lambda(i+1)t}\left\{\frac{dp_{i+1}(t)}{dt} + \lambda(i+1)p_{i+1}(t)\right\} = \lambda i e^{\lambda t}.$$

Then

$$e^{\lambda(i+1)t}p_{i+1}(t) = i e^{\lambda t} + \text{const.}$$

Since $p_{i+1}(0) = 0$, the constant is $-i$, whence

$$p_{i+1}(t) = i e^{-\lambda i t}(1 - e^{-\lambda t}).$$

Substituting this result in (1.1) so that the equation may be solved for $p_{i+2}(t)$, we have

$$\frac{dp_{i+2}(t)}{dt} + \lambda(i+2)p_{i+2}(t) = \lambda(i+1)ie^{-\lambda it}(1 - e^{-\lambda t}).$$

Multiplying both sides by $e^{\lambda(i+2)t}$ and integrating gives

$$e^{\lambda(i+2)t}p_{i+2}(t) = (i+1)i\int \lambda e^{2\lambda t}(1 - e^{-\lambda t})\, dt$$

$$= (i+1)i\frac{(e^{\lambda t} - 1)^2}{2} + \text{const.}$$

Since $p_{i+2}(0) = 0$, $\text{const.} = 0$ and

$$p_{i+2}(t) = \frac{(i+1)i}{2}e^{-\lambda it}(1 - e^{-\lambda t})^2.$$

The form of the general solution now becomes evident. It is

$$p_N(t) = \binom{N-1}{i-1}e^{-\lambda it}(1 - e^{-\lambda t})^{N-i} \tag{1.3}$$

and its correctness may be proved by induction. Assuming (1.3) to be true, $p_{N+1}(t)$ may be found by solving

$$\frac{dp_{N+1}(t)}{dt} + \lambda(N+1)p_{N+1}(t) = \lambda N\binom{N-1}{i-1}e^{-\lambda it}(1 - e^{-\lambda t})^{N-i}$$

or

$$e^{\lambda(N+1)t}p_{N+1}(t) = N\binom{N-1}{i-1}\int \lambda e^{\lambda t}(e^{\lambda t} - 1)^{N-i}\, dt$$

$$= N\binom{N-1}{i-1}\frac{(e^{\lambda t} - 1)^{N-i+1}}{N-i+1} + \text{const.}$$

The constant is 0, since $p_{N+1}(0) = 0$, and therefore

$$p_{N+1}(t) = \frac{N}{N-i+1}\binom{N-1}{i-1}e^{-\lambda(N+1)t}(e^{\lambda t} - 1)^{N-i+1}$$

$$= \binom{N}{i-1}e^{-\lambda it}(1 - e^{-\lambda t})^{N-i+1},$$

which is of the same form as (1.3).

Notice that in the formula for $p_N(t)$, λ and t cannot be separated; they occur only in the form of the product λt. It follows that the probability distribution of N, when i is given, depends only on λt. Thus a high reproductive rate acting over a short time will yield the same result as a

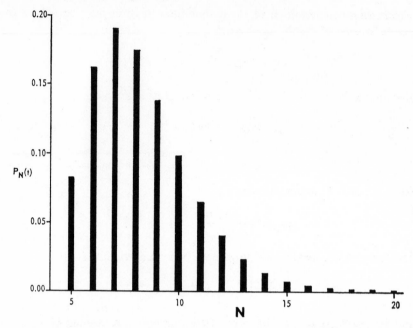

Figure 1.1. The probability distribution of the size at time t of a population undergoing a pure birth process:

$$p_N(t) = \binom{N-1}{i-1} e^{-\lambda it}(1-e^{-\lambda t})^{N-i},$$

with $\lambda t = 0.5$ and $i = 5$. The mean is $M(N\mid t) = ie^{\lambda t} = 8.24$. The variance is $\text{var}(N\mid t) = ie^{\lambda t}(e^{\lambda t}-1) = 5.35$.

low rate over a longer time if λt is the same for both. Figure 1.1 shows an example of the distribution when $\lambda t = 0.5$ and $i = 5$. The mean and variance of the distribution may be derived as follows.

Write $M(N\mid t)$ for the expected size of the population at time t; that is,

$$M(N\mid t) = \sum_{j=1}^{\infty} jp_j(t)$$

$$= e^{-\lambda it} \sum_{k=0}^{\infty} (i+k)\binom{i+k-1}{k}(1-e^{-\lambda t})^k$$

$$= ie^{-\lambda it} \sum_{k=0}^{\infty} \binom{i+k}{k}(1-e^{-\lambda t})^k$$

$$= ie^{-\lambda it}(e^{-\lambda t})^{-(i+1)}$$

$$= ie^{\lambda t}.$$

Thus the expected population size in the stochastic case is equal to that predicted with certainty for the deterministic process.

Now write $M_2(N\,|\,t)$ for the second moment about the origin; that is,

$$M_2(N\,|\,t) = \sum_{j=i}^{\infty} j^2 p_j(t)$$

$$= e^{-\lambda it} \sum_{k=0}^{\infty} (i+k)^2 \binom{i+k-1}{k}(1-e^{-\lambda t})^k$$

$$= i^2 e^{-\lambda it} \sum_{k=0}^{\infty} \binom{i+k}{k}(1-e^{-\lambda t})^k + ie^{-\lambda it} \sum_{k=0}^{\infty} k\binom{i+k}{k}(1-e^{-\lambda t})^k$$

$$= i^2 e^{\lambda t} + (i+1)ie^{-\lambda it}(1-e^{-\lambda t}) \sum_{k=0}^{\infty} \binom{i+k+1}{k}(1-e^{-\lambda t})^k$$

$$= i^2 e^{\lambda t} + (i+1)i(e^{\lambda t}-1)e^{\lambda t}.$$

Then the variance $\operatorname{var}(N\,|\,t)$ is given by

$$\operatorname{var}(N\,|\,t) = M_2(N\,|\,t) - [M(N\,|\,t)]^2$$
$$= ie^{\lambda t}(e^{\lambda t}-1).$$

As we would expect, the variance increases with t. This amounts to saying that the farther into the future we make predictions the less precise these predictions will be.

The process described by (1.1) is known as the Yule process and was first proposed by Yule (1924) in connection with evolution to describe the rate of evolution of new species within a genus. Because of the very stringent and, in most circumstances, unrealistic assumptions on which it is based, it is of limited usefulness in ecology.

The opposite process, the pure death process, is likely to be more widely applicable. In this process it is postulated that no births occur and that for each individual in the population the probability of dying in an interval of length Δt is constant and equal to $\mu\Delta t + o(\Delta t)$. The rate at which the population declines is then given by an equation analogous to (1.1), namely

$$\frac{dp_N(t)}{dt} = \mu(N+1)p_{N+1}(t) - N\mu p_N(t).$$

By arguments similar to those already given for the pure birth process we may now find the probability $p_N(t)$ that at time t the population is of size N, given that its initial size was i, $(i > N)$.

It will be found that

$$p_N(t) = \binom{i}{N}e^{-\mu it}(e^{\mu t}-1)^{i-N}. \tag{1.4}$$

The mean and variance of this distribution are

$$M(N \mid t) = ie^{-\mu t}$$

and

$$\text{var}(N \mid t) = ie^{-\mu t}(1 - e^{-\mu t}).$$

An alternative way of deriving $p_N(t)$ is as follows: Since the probability that an individual will die in any interval Δt is $\mu \Delta t + o(\Delta t)$, the probability that it will be a survivor at time t is $e^{-\mu t}$ and the probability that it will die on or before t is $1 - e^{-\mu t}$. The fate of the i individuals present at $t = 0$ may be thought of as the outcome of i Bernoulli trials for which the probability of "success" ($=$ survival) is $e^{-\mu t}$ and the probability of "failure" ($=$ death) is $1 - e^{-\mu t}$. Then $p_N(t)$ is the probability of N successes in these trials and consequently

$$p_N(t) = \binom{i}{N}(e^{-\mu t})^N (1 - e^{-\mu t})^{i-N},$$

which is the same as (1.4). The mean and variance follow immediately.

We may also find the expected time to extinction. Consider again a population whose initial size at $t = 0$ is i. The random variable T_i, the time to the first death in a population of size i, has a cumulative distribution function given by $\Pr\{T_i \le t\}$ which, from (1.4), is

$$\Pr\{T_i \le t\} = 1 - p_i(t) = 1 - e^{-\mu i t}, \qquad 0 \le t \le \infty.$$

The expectation of T_i is thus

$$E(T_i) = \int_0^\infty ite^{-\mu it}\, dt = \frac{1}{\mu i}.$$

When the first death occurs population size falls to $i - 1$, and the expected time to the next death becomes $E(T_{i-1}) = 1/\mu(i-1)$. Continuing the argument, we see that the expected time to extinction of the entire starting population, which entails a sequence of i independent deaths, is

$$\sum_{j=1}^{i} E(T_j) = \sum_{j=1}^{i} \frac{1}{\mu j}.$$

Natural circumstances in which a pure death process will occur seem quite likely to develop. If the environment of an isolated population is altered, by pollution, for instance, in such a way that all reproduction is prevented and if also the death rate of population members is independent of their ages, a pure death process will ensue. It is known (Deevy, 1947) that for many species of birds the death rate is nearly independent of age once they have become adult, and the same is possibly true of many fish.

3.   The Simple Birth and Death Process

So far we have assumed that either births or deaths occur exclusively. Now suppose that both births and deaths take place and, as before, assume that an individual's probability of giving birth or of dying is independent of its age and of the size of the population to which it belongs. We also suppose that all members of the population are capable of reproducing; either the organisms are asexual or, if the species is bisexual, we consider only the females and postulate that there is never a shortage of males.

Writing λ for the birth rate and μ for the death rate of each individual, we obtain the deterministic equation for population growth:

$$N_t = i\hat{e}^{(\lambda-\mu)t}.$$

This is formally the same as the equation for the deterministic form of the pure birth process (see page 9) but here the rate of increase $\lambda - \mu$ is the difference between a birth rate and a death rate; it may be positive or negative.

Presumably natural populations may sometimes grow exponentially in this manner, as long as the numbers are low enough for competition among population members to have no effect on the rate of increase. The growth of such a population may be treated as a simple birth and death process even when the birth and death rates are *not* independent of age, provided the population has a stable age distribution (see Chapters 3 and 4). Then, although the rates differ from one individual to another, the *mean* rate per individual remains constant because the proportions of the population in each age class do not change.

Now consider the stochastic form of the simple birth and death process. In a population of size N, and during an interval of length Δt, let the probability of a birth be $N\lambda\Delta t + o(\Delta t)$ and of a death be $N\mu\Delta t + o(\Delta t)$. The probability that more than one event (birth or death) will occur in Δt is treated as negligible. Thus, if the population is of size N at $t + \Delta t$, its size at time t must have been $N-1$, N, or $N+1$ and we see that

$$p_N(t+\Delta t) = p_{N-1}(t)\lambda(N-1)\,\Delta t + p_N(t)[1 - N\lambda\,\Delta t - N\mu\,\Delta t]$$
$$+ p_{N+1}(t)\mu(N+1)\,\Delta t.$$

Therefore

$$\frac{dp_N(t)}{dt} = -N(\lambda+\mu)p_N(t) + \lambda(N-1)p_{N-1}(t) + \mu(N+1)p_{N+1}(t). \quad (1.5)$$

From this equation we now determine the mean and variance of the population size at time t directly; that is, without first deriving the general term $p_N(t)$ of the distribution. (The same method could have been used in connection with the pure birth process.)

As before, write $M(N \mid t)$ for the expected size of the population at time t; that is,

$$M(N \mid t) = \sum_{j=1}^{\infty} j p_j(t).$$

Then

$$\frac{dM(N \mid t)}{dt} = \sum_{j=1}^{\infty} j \frac{dp_j(t)}{dt}.$$

Putting p_j in place of $p_j(t)$ for brevity and substituting from (1.5), we then see that

$$\frac{dM(N \mid t)}{dt} = \sum_{j=1}^{\infty} [-(\lambda + \mu)j^2 p_j + \lambda j(j-1)p_{j-1} + \mu j(j+1)p_{j+1}]$$

$$= \lambda \sum_{j=1}^{\infty} (-j^2 p_j + j^2 p_{j-1} - j p_{j-1}) - \mu \sum_{j=1}^{\infty} (j^2 p_j - j^2 p_{j+1} - j p_{j+1})$$

$$= \lambda \sum_{k=0}^{\infty} p_k [-k^2 + (k+1)^2 - (k+1)]$$

$$\qquad\qquad - \mu \sum_{k=1}^{\infty} p_k [k^2 - (k-1)^2 - (k-1)]$$

$$= \lambda \sum_{k=0}^{\infty} k p_k - \mu \sum_{k=1}^{\infty} k p_k = (\lambda - \mu) M(N \mid t).$$

We now have a differential equation in $M(N \mid t)$ that is easily solved. Thus

$$\frac{dM(N \mid t)}{M(N \mid t)} = (\lambda - \mu) \, dt,$$

whence $M(N \mid t) = e^{(\lambda - \mu)t} \times \text{const}$. Since at $t = 0$ the population was of size i, that is, $M(N \mid 0) = i$, we then get $M(N \mid t) = i e^{(\lambda - \mu)t}$.

Again, as was true of the pure birth process, the expected population size at time t is identical with that predicted for the deterministic process. Writing $M_2(N \mid t)$ for the second moment about the origin,

$$M_2(N \mid t) = \sum_{j=1}^{\infty} j^2 p_j(t) \quad \text{and} \quad \frac{dM_2(N \mid t)}{dt} = \sum_{j=1}^{\infty} j^2 \frac{dp_j(t)}{dt}.$$

Then, from (1.5),

$$\frac{dM_2(N \mid t)}{dt} = \lambda \sum_{j=1}^{\infty} (-j^3 p_j + j^3 p_{j-1} - j^2 p_{j-1}) - \mu \sum_{j=1}^{\infty} (j^3 p_j - j^3 p_{j+1} - j^2 p_{j+1})$$

$$= \lambda \sum_{k=0}^{\infty} p_k(2k^2+k) - \mu \sum_{k=1}^{\infty} p_k(2k^2-k)$$

$$= 2(\lambda - \mu) \sum_{k=1}^{\infty} k^2 p_k + (\lambda + \mu) \sum_{k=1}^{\infty} k p_k$$

$$= 2(\lambda - \mu)M_2(N \mid t) + (\lambda + \mu)M(N \mid t).$$

Writing M_2 in place of $M_2(N \mid t)$ for brevity and substituting for $M(N \mid t)$, we now have a differential equation in M_2:

$$\frac{dM_2}{dt} + 2(\mu - \lambda)M_2 = (\lambda + \mu)ie^{(\lambda-\mu)t},$$

which yields

$$e^{2(\mu-\lambda)t}M_2 = i(\lambda + \mu)\int e^{(\lambda-\mu)t}e^{2(\mu-\lambda)t}\, dt$$

$$= -\frac{i(\lambda + \mu)}{\lambda - \mu}e^{(\mu-\lambda)t} + c_1,$$

where c_1 is the constant of integration. If this constant were evaluated, we should then have an expression for the second moment about the origin of the distribution of N at time t. However, what we set out to determine was not $M_2(N \mid t)$ but var$(N \mid t)$, the variance. The variance may be found directly without first evaluating $M_2(N \mid t)$, since it differs from it only by a constant, the square of the mean. We may therefore write

$$e^{2(\mu-\lambda)t}\,\text{var}(N \mid t) = \frac{-i(\lambda + \mu)}{\lambda - \mu}e^{(\mu-\lambda)t} + c_2,$$

where c_2 is a different constant. To evaluate it note that at $t = 0$, var$(N \mid 0) = 0$ and thus $c_2 = i(\lambda + \mu)/(\lambda - \mu)$. Therefore

$$\text{var}(N \mid t) = e^{2(\lambda-\mu)t}\frac{i(\lambda + \mu)}{\lambda - \mu}(1 - e^{(\mu-\lambda)t})$$

$$= \frac{i(\lambda + \mu)}{\lambda - \mu}e^{(\lambda-\mu)t}(e^{(\lambda-\mu)t} - 1).$$

Clearly the variance depends not only on the difference between the birth and death rates, that is, on $\lambda - \mu$, the so-called intrinsic rate of natural increase, but also on their absolute magnitudes. This is what we should expect. For a given rate of increase predictions about the future size of the population will be less precise if births and deaths occur in rapid succession than if they are only occasional events.

An alternative way of writing var$(N \mid t)$ is obtained if we put $\lambda - \mu = r$

and $\lambda + \mu = r + 2\mu$. Thus

$$\text{var}(N \mid t) = \frac{i(r+2\mu)}{r} e^{rt}(e^{rt} - 1).$$

Suppose now that the birth and death rates are equal; that is, $r = 0$ and $M(N \mid t) = i$ for all t. The variance at time t is now given by $\lim_{r \to 0} \text{var}(N \mid t)$. This may be evaluated by using l'Hôpital's rule: differentiating both the numerator and the denominator with respect to r and then putting $r = 0$, we see that when $\mu = \lambda$:

$$\text{var}(N \mid t) = \lim_{r \to 0} \frac{i[(r+2\mu)(2te^{2rt} - te^{rt}) + e^{2rt} - e^{rt}]}{1}$$

$$= 2i\mu t.$$

It is sometimes argued that exponential population growth cannot occur in nature because it would ultimately lead to populations so large that their existence would be physically impossible. This argument is, of course, fallacious (Pielou, 1974c). It is true that for any r there exists a t such that $N_t = ie^{rt}$ exceeds the bounds of practicability. At the same time, for any t there exists an $r > 0$ sufficiently small for N_t to be easily realizable.

This raises a problem that rarely gets the attention it deserves: How far into the future is ecological prediction feasible? The answer is necessarily a matter of personal opinion and to reach a reasonable opinion one must take account of $\text{var}(N \mid t)$ as given above; and one must consider for how long a time period birth and death rates can be expected to remain sensibly constant, or to vary in a predictable manner. It should be clear that the feasibly predictable future may be disappointingly short, and that models of population growth embodying slow-acting mechanisms whose manifestations are long-delayed are intrinsically untestable.

4. The Chance of Extinction

Consider now the probability that a population governed by a simple birth and death process will die out altogether. The probability that a population will be of size 0 at some time t, given that initially there was only a single individual, is

$$p_0(t \mid i = 1) = \frac{\mu e^{(\lambda - \mu)t} - \mu}{\lambda e^{(\lambda - \mu)t} - \mu}.$$

(For a proof of this see Bailey, 1964.)

If a population initially of size i becomes extinct, it follows that each of i separate lines of descent has died out and the probability of this happening is, for any i,

$$p_0(t) = [p_0(t \mid i = 1)]^i = \left(\frac{\mu e^{(\lambda-\mu)t} - \mu}{\lambda e^{(\lambda-\mu)t} - \mu} \right)^i. \tag{1.6}$$

To find the probability of ultimate extinction of the population we must now allow t to tend to infinity.

Clearly, if $\lambda < \mu$, the exponential terms vanish as $t \to \infty$ so that $\lim_{t \to \infty} p_0(t) = 1$. Extinction is certain and an indefinitely prolonged persistence of the population is obviously impossible.

If $\lambda > \mu$, then, as $t \to \infty$,

$$p_0(t) \to \left[\frac{\mu e^{(\lambda-\mu)t}}{\lambda e^{(\lambda-\mu)t}} \right]^i = \left(\frac{\mu}{\lambda} \right)^i.$$

Continued existence is not assured, since the probability of extinction remains finite. However, as would be expected, this probability becomes smaller the more the birth rate exceeds the death rate and the larger the size of the initial population.

It remains to consider what will happen if $\lambda = \mu$. To find $\lim_{t \to \infty} p_0(t)$ in this case we may expand the exponentials in (1.6) in the form of series and put $\lambda - \mu = r$. Then

$$p_0(t) = \left[\frac{\mu(rt + r^2 t^2/2! + \cdots)}{\lambda(1 + rt + r^2 t^2/2! + \cdots) - \mu} \right]^i.$$

As $r \to 0$, terms in r^2 become negligible and

$$p_0(t) \to \left[\frac{\mu r t}{(\lambda - \mu) + \lambda r t} \right]^i \to \left(\frac{\lambda t}{1 + \lambda t} \right)^i,$$

since $\lambda - \mu = r$ and $\mu \to \lambda$. Clearly

$$\lim_{t \to \infty} \left(\frac{\lambda t}{1 + \lambda t} \right)^i = 1,$$

from which it follows that ultimate extinction is certain even though the birth and death rates are equal. Although the expected size of the population is a constant, stochastic fluctuations about this expected size will inevitably lead to extinction if a long enough period of time is allowed to pass. Only if $\lambda > \mu$, that is, only if the population has a positive rate of increase, is there a probability (not a certainty) that the population will persist indefinitely.

2

Density Dependent Population Growth

1. Logistic Population Growth

In the simple birth and death process it is assumed that the probability that an organism will reproduce or die remains constant and is independent of the size of the population of which it is a member. Obviously this can be true only for a population so small that there is no interference among its members. The growth of any population in a restricted environment must eventually be limited by a shortage of resources. Thus a stage is reached when the demands made by the existing population on these resources preclude further growth and the population is then at its "saturation level," a value determined by the "carrying capacity" of the environment.

Suppose, now, that the growth rate per individual is a function of N, the population size, for all value of N, or that

$$\frac{dN}{dt} = Nf(N).$$

Obviously, for large N at least, $df(N)/dN$ must be negative: the larger the population, the greater its inhibitory effect on further growth. The simplest assumption to make is that $f(N)$ is linear, that is, that

$$f(N) = r - sN \qquad (r, s > 0; \quad N > 0)$$

whence

$$\frac{dN}{dt} = N(r - sN), \qquad (2.1)$$

the well-known Verhulst-Pearl logistic equation.

Because of the basic importance of the logistic equation in population dynamics, it is worth mentioning two other lines of argument that lead to it.

1. Write rN for the potential rate of increase of a population of size N, that is, the rate at which the population would grow if the resources were

unlimited and the individuals did not affect one another. Here r is the intrinsic rate of natural increase. Now assume that the actual growth rate is the product of this potential rate times the proportion of the maximum attainable population size that is still unrealized and let this maximum size be K. The unrealized proportion when the population is of size N is thus $(K-N)/K$ and therefore

$$\frac{dN}{dt} = rN\left(1-\frac{N}{K}\right) = N(r-sN)$$

where $s = r/K$.

2. The argument advanced by Lotka (1925) is as follows: We require that the growth rate at any moment should be a function of population size at that moment; that is, we must have $dN/dt = F(N)$. Now suppose that by applying Taylor's theorem, $F(N)$ may be expanded as a power series in N so that

$$\frac{dN}{dt} = c_0 + c_1 N + c_2 N^2 + \cdots.$$

Obviously the growth rate is zero at $N = 0$, since at least one individual must be present for the population to grow at all. Thus $c_0 = 0$.

If we were to put $dN/dt = c_1 N$ (the equation for the simple birth and death process), we should have $dN/dt = 0$ only at $N = 0$ and $dN/dt > 0$ for all other population sizes. We now require that $F(N) = 0$ should have two roots, that is, dN/dt should be zero not only for $N = 0$ but also when N reaches its saturation level. The simplest formula for $F(N)$ to satisfy this condition is the one in which the series given above ends with the term in N^2, or

$$F(N) = \frac{dN}{dt} = c_1 N + c_2 N^2.$$

The roots of $F(N)$ are then $N = 0$ and $-c_1/c_2$. Putting $c_1 = r$ and $c_2 = -s$ yields $dN/dt = N(r-sN)$ as before.

The solution of the differential form of the logistic equation is obtained as follows. Rewriting (2.1) in the form of partial fractions, we have

$$\frac{dN}{N} + \frac{s\, dN}{r-sN} = r\, dt.$$

Integrating, we then have

$$\frac{N_t}{r-sN_t} = Ce^{rt},$$

where C is a constant of integration. Putting $N_0 = i$, the constant is found to be $C = i/(r - is)$ so that

$$N_t = \frac{K}{1 + \dfrac{K-i}{i} e^{-rt}} \qquad (2.2)$$

where as before, $K = r/s$.

An alternative form is obtained by putting $Cs = e^{-rt_0}$, where t_0 is chosen so that $N_0 = i$. Then

$$N_t = \frac{K}{1 + e^{-r(t-t_0)}} \qquad (2.3)$$

The asymptotic value of N_t, namely $\lim_{t \to \infty} N_t = r/s$, is the population's saturation level which it cannot exceed because of environmental limitations.

The form of the logistic curve is shown in Figure 2.1. There are many examples in the literature of close correspondence between the growth of an actual laboratory population and the theoretical curve. For instance, Gause (1934) describes the growth of a culture of *Paramecium caudatum*, Lotka (1925), an experimental population of *Drosophila* and a colony of bacteria, and Odum (1959), the growth of yeast in a culture. As we shall see, the apparent good fit of the theoretical curve to experimental results

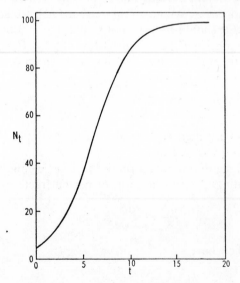

Figure 2.1. The logistic curve, $N_t = K/\{1 + \exp[-r(t - t_0)]\}$ with $r = 0.5$, $K = 100$, $i = 5$, and $t_0 = -5.89$.

may often be spurious. First, however, simple logistic growth is worth investigating in greater detail.

2. Population Growth in Discrete Time

Equations (2.2) and (2.3) express N_t as a continuous function of t. It is sometimes convenient to replace them by a difference equation expressing N_{t+1} as a function of N_t.

Put $e^r = \lambda_1$, the finite rate of natural increase, and $(K-i)/i = C'$. Then from (2.2)

$$N_t = \frac{K}{1 + C'e^{-rt}} \quad \text{and} \quad N_{t+1} = \frac{K\lambda_1}{\lambda_1 + C'e^{-rt}}.$$

But

$$C'e^{-rt} = \frac{(K - N_t)}{N_t}$$

and thus

$$N_{t+1} = \frac{\lambda_1 N_t}{1 + N_t(\lambda_1 - 1)/K} \tag{2.4}$$

This expression corresponds exactly to (2.2) and (2.3) and gives identical numerical results. It can be regarded as an alternative form of (2.2) and (2.3), to be used for predicting the size of a continuously growing population at a series of discrete instants of time, $t = 0, 1, 2, \ldots$; the length of the time interval separating these instants, that is, the unit of time, can be chosen arbitrarily. Alternatively, in a population growing discontinuously, (2.4) may be regarded as a possible model for the dependence of N_{t+1} on N_t that takes account of competition among population members. Observe that if there were no ceiling to growth set by the carrying capacity of the environment, we should have

$$\lim_{K \to \infty} N_{t+1} = \lambda_1 N_t = N_t e^r,$$

that is, exponential growth.

Population growth in nature is seldom as smoothly continuous as a classical logistic curve suggests. In species with a short annual breeding season whose members live for several breeding seasons and die at any time of the year, a continuous record of population size would undoubtedly show seasonal undulations. However, if these were disregarded, the logistic curve or some elaboration of it (see Section 5) might provide at least a reasonable approximation to the facts.

For many species of organisms, however, population growth is markedly discontinuous. These are species whose members reproduce only once in their lifetimes and die before their descendents' lives begin; that is, each generation dies before the eggs it has laid hatch (or the seeds it has set germinate) to form the succeeding generation. A continuous-time differential equation such as (2.2) is then inappropriate as a representation of population growth. A difference equation is the only suitable model and that shown in (2.4) is one possibility.

There are two other well-known difference equation models for density dependent population growth. They arise from the following considerations. Recall that exponential population growth, in difference equation form, is given by

$$N_{t+1} = N_t e^r \tag{2.5}$$

where r, the intrinsic rate of increase, is assumed to be constant. To allow for within-population competition, let us replace the constant exponent by a decreasing function of N and write

$$N_{t+1} = N_t \exp[r - sN]. \tag{2.6}$$

The parallel between (2.6) and (2.1) is obvious. An even simpler model can be obtained if we assume r to be small enough for (2.5) to be approximated by $N_{t+1} = N_t(1 + r)$ and, as in (2.6), substitute $r - sN$ for r. The result is the model

$$N_{t+1} = N_t(1 + r - sN). \tag{2.7}$$

Notice that (2.7) can only be regarded as an approximation to (2.6) when r is small. If one were free to choose the length of the time units and hence the numerical value of r, which is the per capita change in population size per unit of time, then by making the time units sufficiently short one could always arrange for (2.7) to be as close as desired to (2.6). But the duration of the time units can only be freely chosen when a discrete time model is used merely to approximate a continuous-time process. When the process itself is discontinuous the time unit is determined by the process. Thus (2.7) is useful only in exceptional circumstances and will not be considered further here.

The properties of (2.6) have been explored by May (1975). He has shown that the behavior of this model is determined by the magnitude or r and undergoes abrupt changes at critical values of r. Assuming a population whose initial size is below saturation level, that is, assuming $N_0 < K$ in each case, the various behaviors may be summarized as follows:

If $0 < r \le 1$, then $N_{t+1} - N_t > 0$ and $N_t < K$ for all t. That is, population size steadily approaches K without overshooting it.

If $1 < r \le 2$, then either $N_t < K$, $N_{t+1} > K$; or $N_t > K$, $N_{t+1} < K$. Also $|N_{t+1} - N_t| > |N_{t+2} - N_{t+1}|$. That is, the population undergoes damped oscillations as it approaches its saturation level.

If $2 < r \le 2.526$, then the population settles down to a stable two-point cycle. That is, $N_t = N^* > K$ for all even t and $N_t = N_* < K$ for all odd t (or vice versa). The relation between N^* and N_* is as given by (2.6); that is $N^* = N_* \exp[r - sN_*]$ and $N_* = N^* \exp[r - sN^*]$.

If $2.526 < r \le 2.692$, the population achieves a stable cycle with 2^n points $(n > 1)$; the value of n depends on the magnitude of r and increases as r increases.

If $r > 2.692$, the result is, as May aptly describes it, chaos. Population size varies aperiodically in a wholly unpredictable manner; different outcomes are obtained with different values of N_0. The only prediction that can be made is that the upper and lower bounds, say N^* and N_*, between which population size varies, are finite and in the ratio $N^*/N_* = \exp[\exp(r-1) - r]$; (See May, 1975).

Figure 2.2 shows examples of population behavior for selected values of r. The chaotic behavior predicted (if that is the word) when r is large is truly remarkable. We have a fully deterministic model (in the sense that no allowance is made for stochasticity) which yields a wholly arbitrary outcome. If actual populations in nature do indeed conform with the model, then their behaviors are, of course, intrinsically unpredictable.

As the foregoing discussion shows, oscillations or complete instability can occur in a discontinuously growing population. A monotonic approach to saturation is to be expected only for low r. By contrast, a continuously growing population whose behavior accords with the logistic equation [(2.1) or (2.4)] approaches saturation monotonically whatever the value of r, provided it responds to crowding without delay. If delayed response is assumed, then oscillatory behavior results as we shall demonstrate below. First it is worth discussing the parallel between discontinuous growth on the one hand, and the effect of delayed density dependence on continuous growth on the other.

A population growing at discrete moments produces, in each short breeding season, a batch of offspring to replace the current population; the result is an abrupt change in numbers that may carry population size from a level far below saturation to a level far above (or vice versa) in a single step. Likewise with a continuously growing population whose response to crowding is delayed: its growth rate at any moment is governed by its size at some earlier time. Thus a population can be increasing in size rapidly even when it is at or above saturation if its size at the relevant earlier time was well below saturation; the converse is also true. An example is given in Figure 2.3 which shows the simplest possible

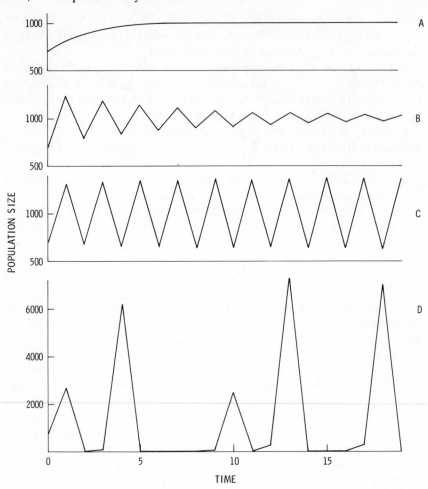

Figure 2.2. Population growth in accordance with (2.6). In all four cases $K = r/s = 1000$ and $N_0 = 700$. (a) $r = 0.5$; (b) $r = 1.9$; (c) $r = 2.1$; (d) $r = 4.5$.

version of logistic growth with delayed response; the model is arrived at by incorporating a lag term, L, in (2.4) which becomes

$$N_{t+1} = \frac{\lambda_1 N_t}{1 + N_{t-L}(\lambda_1 - 1)/K}. \tag{2.8}$$

As explained earlier, (2.4) is the exact difference equation version of continuous logistic growth. In (2.8) the very simple assumption is made

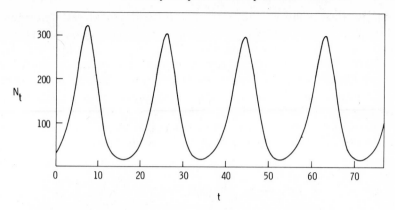

Figure 2.3. Population growth in accordance with (2.8) with $\lambda_1 =$ 1.65, $K = 100$ and $L = 4$.

that the effect of within-population competition on growth in the interval $(t,\ t+1)$ is determined solely by the population's size at time $t - L$.

3.   The Stochastic Form of Logistic Population Growth

We now return to a consideration of logistic population growth without time delays. Considering (2.1) in more detail, it will be realized that the rate of growth it gives is a net balance of births and deaths. In general, both the birth rate and the death rate will be functions of population size. Denote them by $\lambda(N)$ and $\mu(N)$, respectively. Clearly $\lambda(N)$ must decrease and $\mu(N)$ must increase as N increases; thus we may put $\lambda(N) = a_1 - b_1 N$ and $\mu(N) = a_2 + b_2 N$ with $a_1, a_2 > 0$ and $b_1, b_2 \geq 0$. Equation 2.1 now becomes

$$\frac{dN}{dt} = N[\lambda(N) - \mu(N)]$$

$$= N[(a_1 - a_2) - (b_1 + b_2)N].$$

Thus the constants r and s in (2.1) each have two components, and it is clear that for given values of r and s there are many possible values for (a_1, b_1) and (a_2, b_2). The form of the logistic curve depends only on the difference between $\lambda(N)$ and $\mu(N)$, not on their separate values. When the population reaches equilibrium at the saturation level, the birth rate and death rate are equal. Then

$$a_1 - b_1 N = a_2 + b_2 N \qquad \text{or} \qquad N = \frac{a_1 - a_2}{b_1 + b_2} = \frac{r}{s}.$$

Up to this point we have treated logistic growth as deterministic although in fact, of course, births and deaths are chance occurrences. We now consider the stochastic version of the process and how it may be simulated. The events of interest constitute a sequence of births and deaths. Disregarding for the moment the time elapsing between each event and the next, it is clear that at all times there are exactly two possibilities for the next event: it may be a birth, in which case the population will decrease from N to $N+1$; or a death, in which case the population will decrease from N to $N-1$. If all the constants a_1, a_2, b_1, and b_2 are known, we may calculate the probabilities of these events. Thus

$$\Pr(N \to N+1) \propto N\lambda(N) = a_1 N - b_1 N^2$$

and

$$\Pr(N \to N-1) \propto N\mu(N) = a_2 N + b_2 N^2.$$

The events are mutually exclusive and exhaustive, since either a birth or a death must occur next and therefore

$$\Pr(N \to N+1) = \frac{a_1 N - b_1 N^2}{(a_1 + a_2)N - (b_1 - b_2)N^2} ;$$

$$\Pr(N \to N-1) = \frac{a_2 N + b_2 N^2}{(a_1 + a_2)N - (b_1 - b_2)N^2} .$$

Consider a numerical example. Let the constants be

$$a_1 = 0.7, \qquad a_2 = 0.2,$$
$$b_1 = 0.0045, \qquad b_2 = 0.0005.$$

The deterministic equation is then

$$\frac{dN}{dt} = (a_1 - a_2)N - (b_1 + b_2)N^2 = 0.5N - 0.005N^2,$$

and the equilibrium size of the population is $r/s = 100$. (This is the curve shown in Figure 2.1.)

We shall now simulate stochastic growth; that is, we shall generate a sequence of births and deaths by assuming these values for the constants and starting with a population of size $N = 70$. First calculate $\Pr(N \to N+1)$ and $\Pr(N \to N-1)$ and denote these probabilities by α and $1 - \alpha$, respectively. Now consult a table of random numbers and pick a number in the range $(0, 1]$. If the random number is $\leq \alpha$, let the next event be a birth so that the population increases in size to 71; if the number is $> \alpha$ let the next event be a death so that the population decreases to 69. Once the event has happened and the population size has been adjusted accordingly we may calculate the new probabilities (which depend on the new population size) for the next event and proceed as before.

Table 2.1 lists a short sequence of events generated in this manner. The value of N in each row, after the first in which it is set equal to 70, is obtained by altering the value in the row above, according as the previous event chanced to be a birth (B) or a death (D).

Thus we have a sequence of events. Notice that as N increases the probability of a death increases and the probability of a birth decreases. The two probabilities become equal when $N = 100$, its equilibrium value.

Next consider *when* the events occur, assuming that simulation begins at time zero. We require the probability density function (pdf) of t, the time to the next event. The probability of an event (of either kind) in an interval of length Δt, when the population is of size N, is $N[\lambda(N) + \mu(N)]\Delta t + o(\Delta t)$ or $[(a_1 + a_2)N - (b_1 - b_2)N^2]\Delta t + o(\Delta t) \equiv \rho_N \Delta t + o(\Delta t)$, say; that is, ρ_N is the rate at which events occur in the population. It is, of course, a function of N, since events will follow one another at shorter intervals in a large population than in a small one. We assume, however, that for each individual the probability that it will reproduce or die is independent of its age. Then, so long as the population remains of size N, the probability of an event in any time interval is independent of the probability in an earlier interval. If we write $p_0(t)$ for the probability that no event will occur in an interval of length t, it then follows that

$$p_0(t + \Delta t) = p_0(t)p_0(\Delta t) = p_0(t)(1 - \rho_N \Delta t).$$

Then

$$\frac{p_0(t + \Delta t) - p_0(t)}{\Delta t} = -\rho_N p_0(t).$$

TABLE 2.1

N	$\Pr(N \rightarrow N+1)$	$\Pr(N \rightarrow N-1)$	EVENT (chosen by picking a random number)
70	0.621	0.379	B
71	0.618	0.382	D
70	0.621	0.379	B
71	0.618	0.382	B
72	0.614	0.386	B
73	0.611	0.389	B
74	0.608	0.392	D
73	$\cdots$		

Letting $\Delta t \to 0$, we have

$$\frac{dp_0(t)}{dt} = -\rho_N p_0(t),$$

whence $p_0(t) = Ke^{-\rho_N t}$ with K a constant. Since $p_0(0) = 1$, $K = 1$ and therefore

$$p_0(t) = e^{-\rho_N t}.$$

This is the probability that by time t no event will have occurred. The cumulative distribution function of the time elapsing before the next event is thus

$$F(t) = \Pr(\text{time to next event} \le t)$$
$$= 1 - p_0(t) = 1 - e^{-\rho_N t}$$

and the pdf is $f(t) = F'(t) = \rho_N e^{-\rho_N t}$.

For simulation of the process it is necessary to select at random a value of t from the population of values having this pdf. To do this, pick a number, say R, in the range $[0, 1)$ from a random numbers table and set it equal to $F(t)$. Solving for t then yields $t = -(1/\rho_N)\ln(1 - R)$, the random value of t desired. We now simulate in Table 2.2 the time intervals separating the events shown in Table 2.1.

Figure 2.4 shows the final results of the simulation process by portraying both the events and the times at which they occurred until $t = 0.72$. It is compared with the deterministic curve, which is a short segment of the curve in Figure 2.1, here enlarged.

The expected time to the next event in a population of size N is

$$E(t \mid N) = \int_0^\infty tf(t) \, dt = \frac{1}{\rho_N}.$$

TABLE 2.2

N	ρ_N	RANDOM VALUE OF t	ACCUMULATED TIME	EVENT
70	43.40	0.0005	0.0005	B
71	43.74	0.0761	0.0766	D
70	43.40	0.0033	0.0799	B
71	43.74	0.0025	0.0824	B
72	44.06	0.0176	0.1000	B
73	44.38	0.0036	0.1036	B
74	44.70	0.0400	0.1436	D
73	$\cdots$			

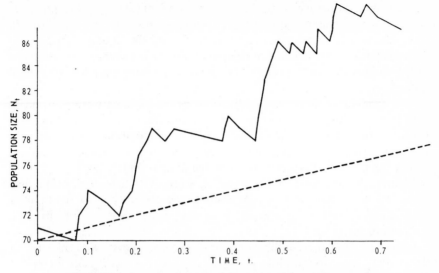

Figure 2.4. A stochastic realization of logistic population growth. The events (births and deaths) and the times of their occurrence are given, in part, in the text. The broken line shows the corresponding deterministic curve; it appears linear because only a very short segment of it is shown. The constants are $a_1 = 0.7$, $a_2 = 0.2$, $b_1 = 0.0045$, $b_2 = 0.0005$, $N_0 = 70$.

In the present example we have

$$E(t \mid N = 70) = \frac{1}{\rho_{70}} = 0.023 \text{ time units}$$

and

$$E(t \mid N = 100) = \frac{1}{\rho_{100}} = 0.020 \text{ time units}.$$

It appears that over this range of values of N the speeding up of events as the population grows is very slight; for this reason the intervals between the events shown do not become noticeably shorter as the population grows.

4. Stochastic Fluctuations at Equilibrium

If we were to allow a stochastic realization of this form of population growth to continue, the population would eventually reach a quasi-stationary state in which its size would hover around the asymptotic equilibrium value indefinitely. The state is *quasi-stationary* and not truly an ultimate stationary state, since extinction always remains possible. Because we have assumed that there is no immigration, we have $\lambda(0) = 0$. This means that when a chance deviation from the equilibrium N is of such magnitude that the size of the population falls to zero, no recovery is possible. However, the probability of this happening may be very small indeed.

If a number of stochastic realizations of the process were allowed to continue for a long time until statistical equilibrium were attained in all of them, we should not expect to find that all the populations were of exactly the same size. Instead, population size at equilibrium will be a random variate and its distribution is what we shall now find.

Clearly, $N\mu(N)P(N) = (N-1)\lambda(N-1)P(N-1)$ when equilibrium has been reached; that is, at equilibrium the probability of a death in a population of size N is the same as the probability of a birth in a population of size $N-1$.

Denoting by N the equilibrium size of the population, we now wish to determine the probabilities ... $P(N-2)$, $P(N-1)$, $P(N)$, $P(N+1)$, $P(N+2)$ ... over the range of population sizes for which the probability is appreciable. It will be found that $P(N-i) \neq P(N+i)$; that is, the distribution is unsymmetrical, although the skewness is slight. The values of $P(N\pm i)$ may be obtained recursively. Begin by choosing a convenient value for $KP(N)$; the constant K is to be evaluated later. Then $P(N+1)$ may be calculated from

$$KP(N+1) = KP(N) \cdot \frac{N\lambda(N)}{(N+1)\mu(N+1)}$$

$$= KP(N) \cdot \frac{N(a_1 - b_1 N)}{(N+1)[a_2 + b_2(N+1)]}.$$

In the same way $KP(N+2)$, $KP(N+3)$, ..., may be found in succession. Similarly,

$$KP(N-1) = KP(N) \cdot \frac{N(a_2 + b_2 N)}{(N-1)[a_1 - b_1(N-1)]},$$

and $KP(N-2)$, $KP(N-3)$, ..., may now be calculated. Each series may be prolonged until negligibly small values are reached. The sum of the calculated values is $K\sum_i P(N\pm i) \sim K$, and division of all the calculated terms by K will give the probabilities sought. An example is shown in Figure 2.5; the parameters are the same as those in the preceding example.

Consider next the properties of the distribution. We must first replace the deterministic equation

$$dN = (rN - sN^2)\, dt,$$

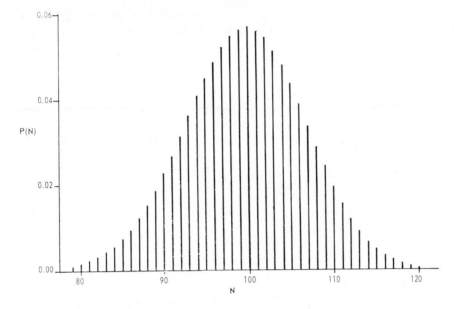

Figure 2.5. The probability distribution $P(N)$ of the size of a population that has reached stochastic equilibrium. The constants are the same as in Figure 3. The distribution has theoretical moments $\mathrm{var}(N) = \alpha\gamma/2\beta^2 = 50$ and $\hat{m} = 99.50$. The empirical moments are $s^2 = 49.13$ and $N = 99.55$.

with the stochastic equation

$$dN = (rN - sN^2)\,dt + dZ, \tag{2.9}$$

where the first term on the right represents the deterministic part of the change dN, and dZ the chance fluctuation or stochastic part (see Bartlett, 1960).

In what follows we shall need to know the mean and variance of the stochastic displacement dZ. In any small time interval Δt the population size may be displaced from equilibrium either by the occurrence of a single death with probability $N\mu(N)\,\Delta t + o(\Delta t)$ or by the occurrence of a single birth with probability $N\lambda(N)\,\Delta t + o(\Delta t)$. These are the probabilities of a displacement ΔZ of magnitudes -1 and $+1$, respectively. Since, at equilibrium, $\mu(N) = \lambda(N)$, it is seen that $E(\Delta Z) = 0$ and that

$$\mathrm{var}(\Delta Z) = (-1)^2 N\mu(N)\,\Delta t + (+1)^2 N\lambda(N)\,\Delta t.$$

Thus in the limit $E(dZ) = 0$ and

$$\mathrm{var}(dZ) = N[\lambda(N) + \mu(N)]\,dt = [(a_1 + a_2)N - (b_1 - b_2)N^2]\,dt.$$

We shall now show that the mean of the distribution of N is, because of the asymmetry, somewhat less than the equilibrium value r/s. For suppose the true mean is m and that at time t the population size N deviates from it by an amount X_t; that is, $X_t = N - m$. Then at time $t + dt$ we have

$$X_{t+dt} = X_t + (rN - sN^2)\, dt + dZ$$

Now take expectations: since the unconditional expectations of all the deviations are zero, we have

$$E(X_{t+dt}) = E(X_t) = E(dZ) = 0$$

and consequently $E[(rN - sN^2)\, dt] = 0$ also. Then, since $E(N) = m$ and $E(N^2) = m^2 + \operatorname{var}(N)$,

$$E[(rN - sN^2)\, dt] = \{rm - s[m^2 + \operatorname{var}(N)]\}\, dt = 0,$$

whence

$$rm = sm^2 + s\operatorname{var}(N) \qquad \text{or} \qquad m = \frac{r}{s} - \frac{\operatorname{var}(N)}{m}.$$

As an estimator of m we may take $\hat{m} = r/s - [(s/r)\operatorname{var}(N)]$, since $m \approx r/s$.

Next, consider $\operatorname{var}(N)$. Assume that N departs only slightly from its equilibrium value of r/s so that we may write $n = (r/s)(1 + u)$ with u small. Then $dN = (r/s)\, du$ and the stochastic equation (2.9) becomes

$$\frac{r}{s}\, du = \left[\frac{r^2}{s}(1 + u) - s\frac{r^2}{s^2}(1 + u)^2\right] dt + dZ,$$

or

$$du = -ru(1 + u)\, dt + \frac{s}{r}\, dZ \approx -ru\, dt + \frac{s}{r}\, dZ,$$

since $1 + u \approx 1$. Then

$$(u + du)^2 = u^2(1 - r\, dt)^2 + 2u\frac{s}{r}(1 - r\, dt)\, dZ + \left(\frac{s}{r}\, dZ\right)^2.$$

Take expectations, bearing in mind (1) that $E(dZ) = 0$ and $E[(dZ)^2] = \operatorname{var}(dZ)$, and (2) that since we are dealing with a state of stochastic equilibrium the quantity u has the same properties at times t and $t + dt$ or equivalently that $E(u + du)^2 = E(u^2) = \operatorname{var}(u)$.

Therefore, neglecting terms in $(dt)^2$,

$$\operatorname{var}(u) = (1 - 2r\, dt)\operatorname{var}(u) + \frac{s^2}{r^2}\operatorname{var}(dZ)$$

$$= (1 - 2r\, dt)\operatorname{var}(u) + \gamma\, dt$$

where

$$\gamma = \frac{s^2}{r^2}[(a_1+a_2)N-(b_1-b_2)N^2] \simeq \frac{s}{r}(a_1+a_2)-(b_1-b_2),$$

since $N \simeq r/s$. Solving for var(u), we find that var(u) = $\gamma/2r$ and consequently that

$$\text{var}(N) \simeq \text{var}\left(\frac{r}{s}u\right) = \frac{r^2}{s^2} \cdot \frac{\gamma}{2r} = \frac{r\gamma}{2s^2}.$$

For the numerical example already described the moments are

$$\text{var}(N) = r\gamma/2s^2 = 50$$

and

$$\hat{m} = \frac{r}{s} - \frac{s}{r}\text{var}(N) = 99.50.$$

Evaluating the moments of the calculated distribution shown in Figure 2.5 gives

$$\bar{N} = 99.55 \quad \text{and} \quad \text{var}(N) = 49.13.$$

The agreement is very good.

5. Modifications of the Logistic Equation

Although the logistic curve is often fitted to observed growth curves of natural and laboratory populations and sometimes appears to fit well, it is worth keeping in mind the highly simplified assumptions underlying it. Six assumptions that must frequently be false are the following:

1. Abiotic environmental factors are sufficiently constant not to affect the birth and death rates.
2. Crowding affects all population members equally. This is unlikely to be true if the individuals occur in clumps instead of being evenly distributed throughout the available space.
3. Birth and death rates respond instantly, without lag, to density changes.
4. Population growth rate is density-dependent even at the lowest densities. It may be more reasonable to suppose that there is some threshold density below which individuals do not interfere with one another.
5. The population has, and maintains, a stable age distribution.
6. The females in a sexually reproducing population always find mates, even when the density is low.

If assumptions (4) and (6) were incorrect, the resultant deviations would tend to cancel each other out, but it is hardly likely that the effects would ever balance precisely. In any case it is easy to modify the model so as to avoid these two unrealistic assumptions. Thus let us drop the assumption that the relationship between the per capita growth rate, $(1/N) \, dN/dt$, and N is linear for $0 < N \le K$ as in (2.1) and assume instead that the relationship takes the form shown by the curve in Figure 2.6a. That is, we suppose that for very low N the growth rate is negative because the population is too small to maintain itself; that the per capita growth rate increases with N to begin with, peaks at some intermediate N, and thereafter drops off again as increasing competition among population members more than compensates for the increase in breeding success that accompanied early population growth. We make no particular assumptions about the functional form of the curve but, instead, replace it with short line segments (shown by the dashed lines in Figure 2.6b) using as many as desired; (in the example shown, four were used). Each line segment has the same form as (2.1) and can be easily integrated.

Thus if there are z segments we have z differential equations of the form

$$\frac{1}{N}\frac{dN}{dt} = a_i + b_i N \qquad (N_i < N \le N_{i+1}) \qquad \text{for } i = 0, \dots, z-1;$$

(there is no restriction on the signs of a_i and b_i).

Put $a_i/|b_i| = \kappa_i$.

Then for

$$N_i < N \le N_{i+1} \qquad \text{and} \qquad t_i < t \le t_{i+1},$$

if $b_i > 0$,

$$N_{T_i+t} = \frac{\kappa_i}{C_i' \exp[-a_i t] - 1} \qquad \text{where} \qquad C_i' = \frac{\kappa_i + N_i}{N_i};$$

and if $b_i < 0$,

$$N_{T_i+t} = \frac{\kappa_i}{C_i'' \exp[-a_i t] + 1} \qquad \text{where} \qquad C_i'' = \frac{\kappa_i - N_i}{N_i}.$$

Here $T_i = \sum_{j=0}^{i} t_j$ is the total time spent while the population grew in size from N_0 to N_i. We first put $t_0 = 0$ and choose a value for N_0. Then the successive values of t_j are found by equating the right side of the appropriate equation to N_j for $j = 1, \dots, i$ and solving for t_j.

Table 2.3 shows the chosen constants a_i, b_i, and N_i, and the successive values of t_i and T_i, on which the example in Figure 2.6 is based.

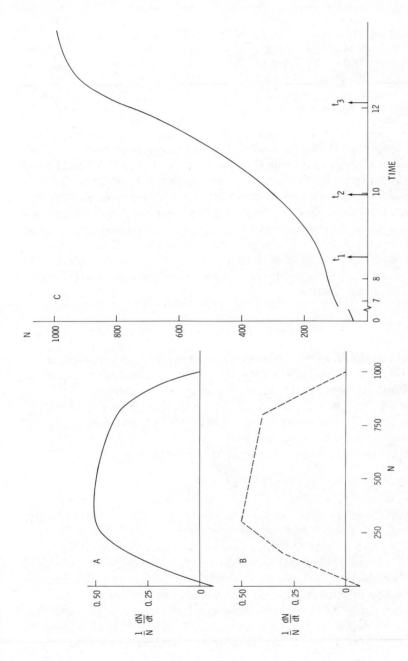

Figure 2.6. A model of sigmoid population growth. (*A*) The curvilinear relationship between per capita growth rate and population size. (*B*) Four line segments approximating the curve in *A*. (*C*) The resultant growth curve; (note the gap in the time axis).

37

TABLE 2.3

i	a_i	b_i	N_i	t_i	T_i
0	−0.06	0.0024	50	0.0	0.0
1	0.20	0.0013	150	8.51	8.51
2	0.56	−0.0002	300	1.44	9.95
3	2.00	−0.0020	800	2.15	12.10

Figure 2.6c shows the resultant growth curve. It is, as one would expect, sigmoid. Indeed, population growth curves are almost always sigmoid and it seems probable that often the good fit of the logistic curve to observed growth curves is apparent rather than real. If the only observations made on a population are determinations of its size at a succession of times and if these are the data used to estimate the parameters r and s of the logistic equation, we can scarcely avoid getting at least a fairly good fit. Therefore other data, besides those obtained by merely recording the growth of an undisturbed population, are needed before a conclusion can be drawn that the growth of a particular population is, or is not, logistic.

6. Direct Measurement of Per Capita Growth Rate

As an example of a more detailed study, consider the investigations of Smith (1963) on laboratory populations of *Daphnia magna*. An "expanding culture" technique was used in growing the populations by which a population could be maintained indefinitely at a given level of crowding. The growth rates at a succession of chosen density levels, imposed by the experimenter, were then determined.

Now consider the logistic equation in the form

$$\frac{1}{N}\frac{dN}{dt} = r\left(\frac{K-N}{K}\right). \tag{2.10}$$

Clearly, if this equation describes population growth, the growth rate per individual $(1/N)\, dN/dt$ decreases linearly with increasing N. In his experiments, however, Smith found that the relation between these quantities was not a straight line but a concave curve, as shown in Figure 2.7.

He then argued as follows. In its usual form (2.10) the logistic equation contains the assumption that growth rate is proportional to $(K - N)/K$, the proportion of maximum attainable population size still unrealized (see page 21). However, it might be truer to say that growth rate depends on the proportion of some limiting factor not yet utilized. For a food-limited

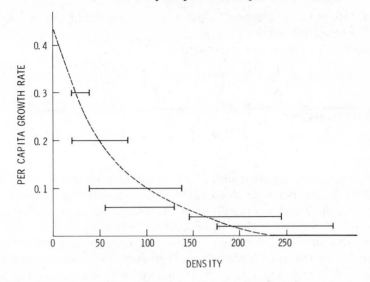

Figure 2.7. Per capita growth rates of *Daphnia magna* at different densities. The horizontal lines show the range of observed densities corresponding with each of several growth rates. The curve was independently derived. (Adapted from Figure 8 of Smith, 1963.)

population the term $(K-N)/K$ should then be replaced with a term representing the proportion of "the rate of food supply not momentarily being used by the population" (Smith, 1963). Notice that the *rate* of food supply is considered, not the amount of food. We may then write

$$\frac{1}{N}\frac{dN}{dt} = r\left(\frac{T-F}{T}\right), \tag{2.11}$$

where F is the rate at which a population of size N uses food and T is the corresponding rate when the population reaches saturation level. The ratio F/T is not the same as N/K, since a growing population will use food faster than a saturated population. When the population is growing, food is used for both maintenance and growth, whereas once the saturation level has been reached and no further growth takes place it is used for maintenance only. Thus we must have $F/T > N/K$. Now, F must depend on N (the size of the population being maintained) and dN/dt (the rate at which the population is growing), and therefore the simplest assumption to make concerning F is that

$$F = c_1 N + c_2 \frac{dN}{dt}, \qquad (c_1, c_2 > 0).$$

When saturation is reached, $dN/dt = 0$, and by definition $N = K$ and $F = T$, whence we obtain $T = c_1 K$.

Equation 2.11 may now be rewritten

$$\frac{1}{N} \cdot \frac{dN}{dt} = r\left(\frac{c_1 K - c_1 N - c_2\, dN/dt}{c_1 K}\right)$$

or, putting $c_1/c_2 = c$,

$$\frac{1}{N} \cdot \frac{dN}{dt} = r\left[\frac{K - N}{K + (r/c)N}\right] \tag{2.12}$$

and this may be contrasted with (2.10). It will be seen that if (2.12) holds the growth rate per individual will decrease rapidly with increasing N while N is small and less rapidly as N becomes larger. This is what Smith found to be true in populations of *Daphnia magna*.

If a curve showing population growth from small N_0 to near-saturation is derived in the way explained in Section 5 it is, not unexpectedly, sigmoid. A living population of *D. magna* will not necessarily show steady sigmoid growth, however, since it could happen that the response of growth rate to crowding is delayed; then an oscillatory curve would be expected. The expanding culture technique is designed to enable growth rates to be observed in populations held at constant density long enough for delays not to affect the rates finally observed.

3

Population Growth with Age-Dependent Birth and Death Rates
I: The Discrete Time Model

1. Introduction

So far we have assumed either that birth and death rates were independent of age or that population changes occurred in such a way that the age distribution remained unaltered. In arriving at the logistic equation, we did, however, allow for density dependence or a decrease in birth rate and an increase in death rate as the population became larger.

We shall now reverse these assumptions: that is, we shall assume that an individual's chances of reproducing and dying are a function of its age but that these chances are unaffected by the size of the population in which it finds itself, perhaps because population numbers never rise so high that density dependence begins to exert any influence. We shall be concerned to discover not merely the size of a population after a lapse of time, given the initial size, but also the age distribution within the population after a lapse of time, given the initial age distribution. Natural realizations of the process to be described would occur, for instance, if there were sudden colonization of a new area by a small immigrant population of arbitrary age distribution. The model considered here is deterministic. Stochastic versions of it have been described by Sykes (1969b) and Pollard (1966, 1973).

It simplifies discussion to consider only the females in a bisexal population. The same arguments would also apply if one counted members of both sexes, provided the ratio of males to females remained constant and that at all ages the death rates were the same for both sexes. In this chapter, following Lewis (1942) and Leslie (1945, 1948), we consider a discrete time model with a discrete age scale and assume that *within* an age interval birth and death rates remain constant; they differ from one interval to the next. The predictions are therefore not exact but if the time intervals are short the approximation is good. The model treating time as continuous, and with a continuous age scale, is discussed in Chapter 4.

The discrete time processes described in this chapter have been used to model the dynamics of a wide variety of ecological populations. Among the organisms concerned are brook trout (Béland, 1974), rabbits (Darwin and Williams, 1964), lice (Murray and Gordon, 1969), beetles (Lefkovitch, 1965), great tits (Pennycuick, 1969), pine trees (Usher, 1972 and references therein) and buttercups (Sarukhan and Gadgil, 1974). The model has been greatly refined and extended by students of human demography (Keyfitz, 1968; Goodman, 1969; Pollard, 1973) who find it an indispensable aid in predicting future states of human populations.

2. The Projection Matrix

Consider a population that at time $t = 0$ can be represented by the column vector

$$\mathbf{n}_0 = \begin{pmatrix} n_{00} \\ n_{10} \\ n_{20} \\ \vdots \\ n_{m0} \end{pmatrix}.$$

There are $m + 1$ elements in the vector representing $m + 1$ different age classes. In each element the first subscript denotes age and the second, time. Thus at time t there are n_{xt} individuals (females) in the age range x to $x + 1$ time units (equivalent to age x last "birthday"). Such a female is described as of age x. It is assumed that none can live to be older than m. The time intervals have the same duration as the age intervals.

Now denote by F_x the number of daughters born in one unit of time to a female of age x that will survive into the next unit of time. Until the first unit of time has passed their ages will be 0. Also, write P_x for the probability that a female aged x at time t will survive to time $t + 1$; her age will then be $x + 1$. So $P_x n_{xt} = n_{x+1,t+1}$.

Beginning, then, with a population described by the vector given above, we now ask what the population will consist of one time unit later, at $t = 1$. Clearly, in matrix notation, the change in a unit of time can be represented as follows:

$$\begin{pmatrix} F_0 & F_1 & \cdots & F_{m-1} & F_m \\ P_0 & 0 & \cdots & 0 & 0 \\ 0 & P_1 & \cdots & 0 & 0 \\ \multicolumn{5}{c}{\cdots\cdots\cdots\cdots\cdots\cdots\cdots} \\ 0 & 0 & \cdots & P_{m-1} & 0 \end{pmatrix} \begin{pmatrix} n_{00} \\ n_{10} \\ n_{20} \\ \vdots \\ n_{m0} \end{pmatrix} = \begin{pmatrix} F_0 n_{00} + \cdots + F_m n_{m0} \\ P_0 n_{00} \\ P_1 n_{10} \\ \vdots \\ P_{m-1} n_{m-1,0} \end{pmatrix} = \begin{pmatrix} n_{01} \\ n_{11} \\ n_{21} \\ \vdots \\ n_{m1} \end{pmatrix} \quad (3.1)$$

or

$$\mathbf{Mn}_0 = \mathbf{n}_1;$$

$\mathbf{M}$ may be called a "projection matrix." Similarly

$$\mathbf{n}_2 = \mathbf{Mn}_1 = \mathbf{M}^2\mathbf{n}_0; \; \ldots; \; \mathbf{n}_t = \mathbf{M}^t\mathbf{n}_0,$$

where $\mathbf{n}_t$ denotes the population structure after t time units have elapsed. Note that $\mathbf{M}$ is an $(m+1) \times (m+1)$ matrix. Unless the females are reproductive until the end of their lifespan, some of the elements in the first row of $\mathbf{M}$ (at the right-hand end when the females are in their postreproductive stage) may be zero. Then $|M| = 0$ and the matrix is singular. Thus suppose the females are sterile in the last $m - k$ age units so that $\mathbf{M}$ may be written

$$\begin{pmatrix} F_0 & F_1 & \cdots & F_{k-1} & F_k & 0 & \cdots & 0 & 0 \\ P_0 & 0 & \cdots & 0 & 0 & 0 & \cdots & 0 & 0 \\ 0 & P_1 & \cdots & 0 & 0 & 0 & \cdots & 0 & 0 \\ \multicolumn{9}{c}{\dotfill} \\ 0 & 0 & \cdots & P_{k-1} & 0 & 0 & \cdots & 0 & 0 \\ \multicolumn{9}{c}{\dotfill} \\ 0 & 0 & \cdots & 0 & 0 & 0 & \cdots & P_{m-1} & 0 \end{pmatrix}$$

$$\equiv \left(\begin{array}{c|c} \begin{matrix} \mathbf{A} \\ (k+1) \times (k+1) \end{matrix} & \begin{matrix} \mathbf{0} \\ (k+1) \times (m-k) \end{matrix} \\ \hline \begin{matrix} \mathbf{B} \\ (m-k) \times (k+1) \end{matrix} & \begin{matrix} \mathbf{C} \\ (m-k) \times (m-k) \end{matrix} \end{array} \right) \qquad (3.2)$$

Then

$$\mathbf{M}^t = \begin{pmatrix} \mathbf{A}^t & \mathbf{0} \\ f(\mathbf{ABC}) & \mathbf{C}^t \end{pmatrix}.$$

Now $\mathbf{C}$ is strictly triangular; its only nonzero elements, $P_{k+1}, \ldots, P_{m-1}$, are on the subdiagonal. Therefore, when $t \geq m - k$, $\mathbf{C}^t = \mathbf{0}$ and $\mathbf{M}^t$ has only zeroes in its last $m - k$ columns. This accords with the obvious fact that females who enter a population after they have reached postreproductive age do not contribute to its younger age classes and are themselves dead by the time $m - k$ time units (at most) have elapsed.

In what follows we treat as the population under study only that part of the total population that is of reproductive age. That is, we consider $\mathbf{A}$ only. Repeated premultiplication by $\mathbf{A}$ of the $(k+1)$-element column vector $\mathbf{n}_0 = (n_{00} \; n_{10} \; \cdots \; n_{k0})'$ permits prediction of the future growth and age distribution of this reproducing part of the population made up of females aged k or less. Let us therefore consider the mathematical properties of $\mathbf{A}$. Its only nonzero terms are the elements on the principal subdiagonal (the P's) and some or all of the elements in the top row (the F's). We must have $F_k > 0$ but for $i < k$ it is necessary only to have $F_i \geq 0$.

Observe first that $\mathbf{A}$ is square and nonsingular. Further, it is *nonnegative* (i.e., all its elements are ≥ 0); and it is *irreducible* (i.e., it is impossible to obtain from $\mathbf{A}$, however one permutes the rows with one another, and the columns with one another, a matrix that can be partitioned into square submatrices one of which is $\mathbf{0}$ of order $n > 1$).

Possession of these properties means that $\mathbf{A}$ meets the necessary conditions for the Perron-Frobenius theorem to be applicable to it. According to this theorem a nonnegative, irreducible matrix has at least one positive real latent root; the largest such root, say λ_1, is simple (i.e., of multiplicity one) and its value is at least as great as the modulus of any complex root of the matrix (i.e., $\lambda_1 \geq |\lambda_i|$ for all $i \neq 1$ where the λ's are the latent roots). The root λ_1 is known as the maximal root, or Perron root, of $\mathbf{A}$. Further, the latent vector of $\mathbf{A}$ corresponding to λ_1 has elements of the same sign (which may be taken positive) and is the only such latent vector (Sykes, 1969a).

Now consider the coefficients of the characteristic equation of $\mathbf{A}$ which is

$$|\mathbf{A} - \lambda \mathbf{I}| = 0.$$

Let us put $P_0 P_1 \cdots P_r = P_{(r)}$, and expand the determinant to give

$$\lambda^{k+1} - F_0 \lambda^k - P_{(0)} F_1 \lambda^{k-1} - \cdots - P_{(r-1)} F_r \lambda^{k-r} - \cdots - P_{(k-1)} F_k = 0 \qquad (3.3)$$

The left-hand side has only one change of sign and hence, from Descartes' rule of signs, the equation has at most one positive real root. Combining the Perron-Frobenius theorem and Descartes' rule we now see that, except for the maximal root, all the roots of $\mathbf{A}$ are negative or complex.

Now recall that by the Perron-Frobenius theorem we must have $\lambda_1 \geq |\lambda_i|$ for all $i \neq 1$. Let us consider in what circumstances $\lambda_1 = |\lambda_i|$ for some i; and in what circumstances the strict inequality $\lambda_1 > |\lambda_i|$ holds. Sykes (1969a) has proved that $\lambda_1 > |\lambda_i|$ is true except when the elements in the top row of $\mathbf{A}$ form a particular pattern. Specifically, $|\lambda_i| = \lambda_1$ for some i only if some of the F's are zero and if the greatest common divisor of the subscript j's for which $F_j > 0$ exceeds one. Equivalently, $|\lambda_i| = \lambda_1$ for some i only if some age classes do not breed and the greatest common divisor of the fertile ages exceeds one; [the unit of age measurement is, of course, the interval $(t, t+1)$].

With one exception, fertility patterns such as this are unlikely to occur in nature (unless breeding is seasonal and the time interval is a submultiple of the seasonal cycle). The exception is the fertility pattern of those species whose individuals are fertile for only a short period right at the end of their lives. (Some locust species are examples). Assume that reproduction followed by the death of all survivors occurs at age k, $F_k > 0$ and $F_j = 0$ for all $j < k$. Then, whatever the elements of $\mathbf{n}_0$, the initial age distribution, it is easy to see from (3.1) that the proportions in this initial distribution will be repeated cyclically with cycles of length $k + 1$. At the end of each cycle the population's size will be $P_{(k-1)}F_k$ times what it was at the beginning of the cycle.

Projection matrices in which the fertility pattern is not of the special kind discussed above are defined as *primitive* (see Sykes, 1969a). A projection matrix is primitive if, of those j's for which $F_j > 0$, at least two are relatively prime and hence the greatest common divisor of all the j's for which $F_j > 0$ is one. When $\mathbf{A}$ is primitive, $\lambda_1 > |\lambda_i|$ for all i. That is, the maximal root of $\mathbf{A}$ exceeds the modulus of any complex root. Further, the proportional age distribution of a population whose growth is governed by such a matrix will, whatever its initial form, approach a limiting distribution as t becomes large. This limit is known as the *stable age distribution* and it is proportional to the latent vector associated with the maximal root, λ_1, of $\mathbf{A}$. We now consider how this vector, and hence the stable age distribution, may be derived. (The root λ_1 is the population's finite rate of natural increase. If a population is of constant total size, that is, if $\lambda_1 = 1$, and if also its age distribution is stable, then the population is said to be *stationary*.)

3. The Stable Age Distribution

As shown above, we may write

$$\lim_{t \to \infty} \mathbf{A}^t \mathbf{n}_0 \propto \mathbf{n}_s;$$

The elements of $\mathbf{n}_s$ are proportional to the sizes of the age classes when the age distribution is stable. And, once stability of the age distribution has been reached, we may write

$$\mathbf{n}_{s+1} = \mathbf{A}\mathbf{n}_s = \lambda \mathbf{n}_s \tag{3.4}$$

where λ is some scalar constant. Given $\mathbf{A}$, we now wish to solve (3.4). To solve it explicitly, it is convenient to transform the matrices from their original coordinate system to a new frame of coordinates. The equation

$\mathbf{An}_s = \lambda \mathbf{n}_s$ may be difficult to solve, but if we can write an equivalent equation in transformed variables, say $\mathbf{v}_{s+1} = \mathbf{B}\mathbf{v}_s = \lambda \mathbf{v}_s$, and if $\mathbf{B}$ is of simple form, the equation may be made easily solvable. Here $\mathbf{B}$ operating on $\mathbf{v}_s$ in a new coordinate frame is equivalent to $\mathbf{A}$ acting on $\mathbf{n}_s$ in the old frame. The desired transformation may be found as follows: As before, write $P_0 P_1 \cdots P_r = P_{(r)}$. Now put $\mathbf{v} = \mathbf{H}\mathbf{n}_s$ where $\mathbf{H}$ is the diagonal matrix whose (i, i)th element is $P_{(k-1)}/P_{(i-2)}$. Then $\mathbf{v}_{s+1} = \mathbf{H}\mathbf{n}_{s+1} = (\mathbf{HAH}^{-1})\mathbf{v}_s \equiv \mathbf{B}\mathbf{v}_s$, say, where $\mathbf{B} = \mathbf{HAH}^{-1}$. Thus

$$\mathbf{B} = \begin{pmatrix} F_0 & P_{(0)}F_1 & P_{(1)}F_2 & \cdots & P_{(k-2)}F_{k-1} & P_{(k-1)}F_k \\ 1 & 0 & 0 & \cdots & 0 & 0 \\ 0 & 1 & 0 & \cdots & 0 & 0 \\ \hdotsfor{6} \\ 0 & 0 & 0 & \cdots & 1 & 0 \end{pmatrix}.$$

Next, to find λ we note that since, by definition, $\mathbf{B}\mathbf{v}_s = \lambda \mathbf{v}_s$ or $(\mathbf{B} - \lambda \mathbf{I})\mathbf{v}_s = \mathbf{0}$, λ is a solution of the characteristic equation of $\mathbf{B}$, namely $|\mathbf{B} - \lambda \mathbf{I}| = 0$. Expanding the determinant, the equation is found to be identical to (3.3), the characteristic equation of $\mathbf{A}$, that is

$$\lambda^{k+1} - F_0 \lambda^k - P_{(0)}F_1 \lambda^{k-1} - \cdots - P_{(k-1)}F_k = 0 \qquad (3.3)$$

Notice that the coefficients of λ^k, $\lambda^{k-1}, \ldots$, are the elements of the top row of $\mathbf{B}$ written with minus signs. As already discussed, the equation has only one positive real root, the maximal root λ_1.

We now find the associated latent vector. That is we solve $\mathbf{B}\mathbf{v}_s = \lambda_1 \mathbf{v}_s$ to obtain the values v_{xs} $(x = 0, 1, \ldots, k)$, which are the elements of the stable vector in the new coordinate frame. In fact, since $\mathbf{B}$ has 1's on the subdiagonal and 0's elsewhere except in its first row,

$$\begin{pmatrix} b_{00} & b_{01} & \cdots & b_{0,k-1} & b_{0k} \\ 1 & 0 & \cdots & 0 & 0 \\ 0 & 1 & \cdots & 0 & 0 \\ \hdotsfor{5} \\ 0 & 0 & \cdots & 1 & 0 \end{pmatrix} \begin{pmatrix} v_{0s} \\ v_{1s} \\ v_{2s} \\ \vdots \\ v_{ks} \end{pmatrix} = \lambda_1 \begin{pmatrix} v_{0s} \\ v_{1s} \\ v_{2s} \\ \vdots \\ v_{ks} \end{pmatrix},$$

(where we have put b_{0i} $(i = 0, \ldots, k)$ in the first row of $\mathbf{B}$) and therefore

$$v_{0s} = \lambda_1 v_{1s}, \; v_{1s} = \lambda_1 v_{2s}, \ldots, v_{k-1,s} = \lambda_1 v_{ks}.$$

Thus the stable vector in the new coordinate frame is

$$\boldsymbol{v}_s \propto \begin{pmatrix} \lambda_1^k \\ \lambda_1^{k-1} \\ \vdots \\ \lambda_1 \\ 1 \end{pmatrix}.$$

To obtain the stable vector $\mathbf{n}_s$ in the original coordinate frame it is now necessary only to note that

$$\mathbf{n}_s = \mathbf{H}^{-1}\boldsymbol{v}_s = \frac{1}{P_{(k-1)}} \begin{pmatrix} \lambda_1^k \\ \lambda_1^{k-1}P_{(0)} \\ \vdots \\ \lambda_1^{k-r}P_{(k-r-1)} \\ \vdots \\ \lambda_1 P_{(k-2)} \\ P_{(k-1)} \end{pmatrix}$$

$$\equiv \begin{pmatrix} 1 \\ \dfrac{P_{(0)}}{\lambda_1} \\ \vdots \\ \dfrac{P_{(r-1)}}{\lambda_1^r} \\ \vdots \\ \dfrac{P_{(k-2)}}{\lambda_1^{k-1}} \\ \dfrac{P_{(k-1)}}{\lambda_1^k} \end{pmatrix} \qquad (3.5)$$

Both vectors on the right are proportional to the stable age distribution. In the first of the two, the numbers of individuals in the 0th, 1st, ... $(k-1)$th age classes are shown as multiples of the number in the oldest age class; in the second, the numbers in the 1st, 2nd, ..., kth age classes are shown as multiples of the number in the youngest age class. '

Summarizing, we have done the following: using the elements of $\mathbf{A}$, we have set up (3.3) in order to obtain λ_1, the rate of increase per unit of time when stability has been achieved. Using λ_1, we have then obtained the stable age distribution from (3.5) in which the elements of $\mathbf{n}_s$ are expressed in terms of λ_1 and the elements of $\mathbf{A}$.

A numerical example (artificial) will clarify this. (The numbers are unrealistic and have been chosen to keep the arithmetic simple.)

$$\mathbf{A} = \begin{pmatrix} 1 & 3 & 4 & 12 \\ \frac{1}{2} & 0 & 0 & 0 \\ 0 & \frac{1}{4} & 0 & 0 \\ 0 & 0 & \frac{2}{3} & 0 \end{pmatrix}$$

Then

$$\mathbf{H} = \begin{pmatrix} \frac{1}{12} & 0 & 0 & 0 \\ 0 & \frac{1}{6} & 0 & 0 \\ 0 & 0 & \frac{2}{3} & 0 \\ 0 & 0 & 0 & 1 \end{pmatrix} \quad \text{and} \quad \mathbf{B} = \begin{pmatrix} 1 & \frac{3}{2} & \frac{1}{2} & 1 \\ 1 & 0 & 0 & 0 \\ 0 & 1 & 0 & 0 \\ 0 & 0 & 1 & 0 \end{pmatrix};$$

hence

$$|\mathbf{B} - \lambda\mathbf{I}| = \begin{vmatrix} 1-\lambda & \frac{3}{2} & \frac{1}{2} & 1 \\ 1 & -\lambda & 0 & 0 \\ 0 & 1 & -\lambda & 0 \\ 0 & 0 & 1 & -\lambda \end{vmatrix} = 0$$

or $\lambda^4 - \lambda^3 - \frac{3}{2}\lambda^2 - \frac{1}{2}\lambda - 1 = 0$; $\lambda_1 = 2$ is the only positive real root.

Now $\mathbf{B}\mathbf{v}_s = \lambda_1\mathbf{v}_s$ and $\lambda_1 = 2$. Therefore

$$\begin{pmatrix} 1 & \frac{3}{2} & \frac{1}{2} & 1 \\ 1 & 0 & 0 & 0 \\ 0 & 1 & 0 & 0 \\ 0 & 0 & 1 & 0 \end{pmatrix} \begin{pmatrix} v_{0s} \\ v_{1s} \\ v_{2s} \\ v_{3s} \end{pmatrix} = 2 \begin{pmatrix} v_{0s} \\ v_{1s} \\ v_{2s} \\ v_{3s} \end{pmatrix},$$

whence

$$v_{0s} = 2v_{1s}; \qquad v_{1s} = 2v_{2s}; \qquad v_{2s} = 2v_{3s}.$$

So, putting $v_{3s} = 1$, we have $v_{2s} = 2$; $v_{1s} = 4$; $v_{0s} = 8$. Thus, in transformed coordinates, the stable vector is

$$\mathbf{v}_s \propto \begin{pmatrix} 8 \\ 4 \\ 2 \\ 1 \end{pmatrix}.$$

Converting back to the original coordinate frame gives

$$\mathbf{n}_s = \mathbf{H}^{-1}\mathbf{v}_s$$

or

$$
\mathbf{n}_s = \begin{pmatrix} n_{0s} \\ n_{1s} \\ n_{2s} \\ n_{3s} \end{pmatrix} \propto \begin{pmatrix} 12 & 0 & 0 & 0 \\ 0 & 6 & 0 & 0 \\ 0 & 0 & \frac{3}{2} & 0 \\ 0 & 0 & 0 & 1 \end{pmatrix} \begin{pmatrix} 8 \\ 4 \\ 2 \\ 1 \end{pmatrix} = \begin{pmatrix} 96 \\ 24 \\ 3 \\ 1 \end{pmatrix}
$$

It will be seen that, as required, $\mathbf{A}\mathbf{n}_s = \lambda_1\mathbf{n}_s$ or $2\mathbf{n}_s$. We also see that substituting λ_1 and the P_i values given in $\mathbf{A}$ into (3.5) gives the same result.

It should now be noticed that a given value of λ_1 (which is the finite rate of natural increase of a population once its age distribution has stabilized), and a given stable age vector, do not uniquely determine a population's projection matrix. Thus two different projection matrices can have the same λ_1 and the same stable vector even though they differ in some of their nonzero elements. However, if they do so differ, they will differ also in their other latent roots, $\lambda_2, \ldots, \lambda_k$; the values of these roots affect the manner in which a population with a given initial age distribution approaches stability. Here we can give only a simple numerical example, with $k = 2$, to illustrate.

Consider two projection matrices $\mathbf{X}$ and $\mathbf{Y}$ where

$$
\mathbf{X} = \begin{pmatrix} 0.3 & 1.6 & 6.0 \\ 0.5 & 0 & 0 \\ 0 & 0.5 & 0 \end{pmatrix} \quad \text{and} \quad \mathbf{Y} = \begin{pmatrix} 0.7 & 1.9 & 1.5 \\ 0.5 & 0 & 0 \\ 0 & 0.5 & 0 \end{pmatrix}
$$

Both have the same maximal root, $\lambda_1 = 1.5$; and both have the same stable vector, $\mathbf{n}_s \propto (9 \ 3 \ 1)'$. However, their subordinate roots differ.

For $\mathbf{X}$ we have

$$
\lambda_2(\mathbf{X}) = -0.6 + 0.8i \quad \text{and} \quad \lambda_3(\mathbf{X}) = \lambda_2^*(\mathbf{X}) = -0.6 - 0.8i;
$$

(λ^* denotes the complex conjugate of λ). The modulus of these roots is $\sqrt{\lambda\lambda^*} = 1.0$.

For $\mathbf{Y}$ we have

$$
\lambda_2(\mathbf{Y}) = -0.4 + 0.3i \quad \text{and} \quad \lambda_3(\mathbf{Y}) = \lambda_2^*(\mathbf{Y}) = -0.4 - 0.3i
$$

and their modulus is 0.5.

Now imagine two populations, one with growth governed by $\mathbf{X}$ and the other by $\mathbf{Y}$, and let both populations start with the same initial size ($\sum n_{x0} = 30$) and the same initial age distribution $\mathbf{n}_0 = (10 \ 10 \ 10)'$. The populations' compositions at $t = 1, \ldots, 8$ are shown in Figure 3.1. It will be seen that the population whose projection matrix ($\mathbf{Y}$) has complex

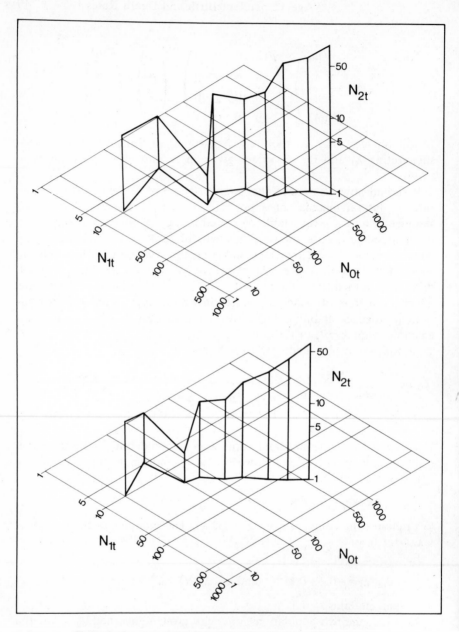

Figure 3.1a. Growth of the populations whose projection matrices are **X** (above) and **Y** (below); see text. In both cases $\mathbf{n}_0 =$ (10 10 10)'. The trajectory of each population is the sequence of points at the top of the "corners of the wall." The coordinates of these points are the numbers (logarithmically scaled) in the three age classes.

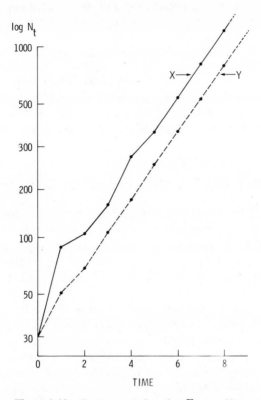

Figure 3.1*b*.   Total population size, $\sum_x n_{xt} = N_t$ at $t = 0, 1, \ldots, 8$ for the two populations. Solid line: the population governed by **X**; broken line: the population governed by **Y**.

roots of smaller modulus attains stability sooner than the other population.

Further, because of their different growth behaviors, the two populations will not be of the same total size at time t, however large t becomes, even though both started with $N_0 = \sum n_{x0} = 30$ individuals. Instead, one population gets a start on the other which it never loses.

A way of measuring this initial handicap has been given by Goodman (1969). Consider a population whose initial age distribution is $\mathbf{n}_0$ and suppose that after t time units its age distribution has "reached stability" (that is, $C - \varepsilon < n_{xt}/n_{xs} < C + \varepsilon$ for all x where C is a constant and $|\varepsilon|$ is as small as desired; the n_{xs} are the elements of the stable vector). Then we may put $n_{0t} = wn_{00}$ where w is a constant that depends on the age composition of the initial population and on the elements of the projection matrix. At later times, for example at time $t + T$, we have

$$n_{0,t+T} = \lambda_1^T n_{0t} = \lambda_1^T w n_{00} \equiv v \lambda_1^{t+T},$$

on putting $\lambda_1^{-t} w n_{00} = v$. Goodman calls v the initial population's *expected reproductive value to the 0'th age class* and describes how it may be calculated from the elements of $\mathbf{n}_0$ and $\mathbf{A}$. Clearly, if the age distribution of the population is proportional to the stable vector initially, then $w = \lambda_1^t$ and $v = n_{00}$.

4. Other Uses of Projection Matrices

Since Leslie (1945, 1948) first put their use on a firm footing, projection matrices have proved exceedingly valuable in ecology and human demography. Research continues to expand their use in many directions and in this section a few of these developments will be mentioned. They are distinct and are therefore dealt with in numbered paragraphs.

1. It should first be noticed that if a projection matrix can be found that models a population's growth and that is based on some chosen time unit (i.e., some chosen length for the discrete time interval), then it cannot, strictly speaking, be adjusted for use with some other time unit. In theory, changing the age classification by lumping or subdividing age classes is not permissible, for if a process is Markovian at some particular time interval, it will, in general, be non-Markovian at another. However, Pollard (1966) showed that predictions based on the model are not markedly affected by changes in the time interval. Thus he compared two estimates of the rate of increase per annum of Australian human females and showed that the estimates were very close, regardless of whether the age classes were one year or five years long. The error introduced by altering the time interval appears to be small.

2. When a projection matrix is used it is normally assumed that the age intervals are of equal length and that, in judging how well an actual population conforms to the model, one can tell at a glance to which age class any individual belongs. In ecological work it is often impossible to judge an individual's age, but many animals (e.g., insects) go through a sequence of easily recognizable stages that are of unequal duration. Lefkovitch (1965) has generalized Leslie's model by considering unequal stage groupings instead of equal age groupings; further, no assumptions are made about possible variation in stage duration in different individuals. Lefkovitch demonstrated the usefulness of his model in studies of the

growth of experimental populations of the cigarette beetle, *Lasioderma serricorne* (Fabricius).

3. Skellam (1967) has explored the use of projection matrices in modelling the periodic seasonal changes in populations that do not live in temporally uniform, unchanging conditions.

4. In bisexual populations with a 1:1 sex ratio and identical age distributions in the two sexes there is no need to treat the sexes separately in modeling population growth. When these assumptions do not hold, however, a modified projection matrix that keeps the sexes distinct can be used (Williamson, 1959). Thus let F_x and f_x respectively be the numbers of daughters, and of sons, born in a unit of time (and surviving into the next unit of time) to a mother in the xth age class; let P_x and p_x respectively be the survival probabilities of females and males of age x; and let f_{xt} and m_{xt} be the number of females and males in the xth age class at time t. Then, in place of (3.1), we have

$$
\begin{pmatrix}
F_0 & 0 & F_1 & 0 & \cdots & F_{k-1} & 0 & F_k & 0 \\
f_0 & 0 & f_1 & 0 & \cdots & f_{k-1} & 0 & f_k & 0 \\
P_0 & 0 & 0 & 0 & \cdots & 0 & 0 & 0 & 0 \\
0 & p_0 & 0 & 0 & \cdots & 0 & 0 & 0 & 0 \\
\multicolumn{9}{c}{\cdots\cdots\cdots\cdots\cdots\cdots\cdots\cdots} \\
0 & 0 & 0 & 0 & \cdots & P_{k-1} & 0 & 0 & 0 \\
0 & 0 & 0 & 0 & \cdots & 0 & p_{k-1} & 0 & 0
\end{pmatrix}
\begin{pmatrix}
f_{00} \\ m_{00} \\ f_{10} \\ m_{10} \\ \vdots \\ f_{k0} \\ m_{k0}
\end{pmatrix}
=
\begin{pmatrix}
f_{01} \\ m_{01} \\ f_{11} \\ m_{11} \\ \vdots \\ f_{k1} \\ m_{k1}
\end{pmatrix}
$$

5. Projection matrices are useful in determining the way in which the growth of commercially valuable plant and animal populations will be affected by harvesting. Often the harvest is not spread proportionately over all age classes but is taken chiefly from some subset of the age classes. The effect of such harvesting on future population structure can be modeled by suitably adjusting the survival probabilities of the relevant age classes; (see Williamson, 1967; Usher, 1972).

6. It is sometimes interesting to predict the contribution of some one particular age class, say those of age j at $t = 0$, to the population at some future time t. This may be done as follows. Let the population's projection matrix be $\mathbf{A}$ and put $\mathbf{A} = \{\phi_{ij}\}$ with $i, j = 0, 1, \ldots, k$. Then $\mathbf{A}^t \mathbf{n_0}$ becomes

$$
\begin{pmatrix}
\phi_{00} & \phi_{01} & \cdots & \phi_{0k} \\
\phi_{10} & \phi_{11} & \cdots & \phi_{1k} \\
\multicolumn{4}{c}{\cdots\cdots\cdots\cdots\cdots} \\
\phi_{k0} & \phi_{k1} & \cdots & \phi_{kk}
\end{pmatrix}
\begin{pmatrix}
n_{00} \\ n_{10} \\ \vdots \\ n_{k0}
\end{pmatrix}
=
\begin{pmatrix}
\cdots + \phi_{0j} n_{jo} + \cdots \\
\cdots + \phi_{1j} n_{j0} + \cdots \\
\cdots + \phi_{kj} n_{j0} + \cdots
\end{pmatrix}
=
\begin{pmatrix}
n_{0t} \\ n_{1t} \\ \vdots \\ n_{kt}
\end{pmatrix},
$$

where we have shown only those contributions to the elements in $\mathbf{n}_t$ that come from the element n_{j0} of $\mathbf{n}_0$. At time t, provided $t > k$, every element in $\mathbf{n}_t$ has some contribution from the n_{j0} individuals that were of age j at the start. At time t the total number of individuals that are survivors or descendents of this original age group is given by

$$n_{j0} \sum_{i=0}^{k} \phi_{ij};$$

that is, the product of n_{j0} (the number in the ancestral age group we are concerned with) and the sum of the elements in the $(j+1)$th column of $\mathbf{A}^t$.

7. It is often interesting to explore the consequences of "backward projection" in order to discover the age composition of a population prior to the time, say t, at which it first comes under observation. Provided we concern ourselves only with the reproducing part of the population, having a projection matrix of the form $\mathbf{A}$ in (3.2), there is no difficulty. This follows since $\mathbf{A}$ is nonsingular and therefore has an inverse given by

$$\mathbf{A}^{-1} = \begin{pmatrix} 0 & P_0^{-1} & 0 & 0 & \cdots & 0 \\ 0 & 0 & P_1^{-1} & 0 & \cdots & 0 \\ 0 & 0 & 0 & P_2^{-1} & \cdots & 0 \\ \cdot & \cdot & \cdot & \cdot & \cdots & \cdot \\ \dfrac{1}{F_k} & \dfrac{-F_0}{P_0 F_k} & \dfrac{-F_1}{P_1 F_k} & \dfrac{-F_2}{P_2 F_k} & \cdots & \dfrac{-F_{k-1}}{P_{k-1} F_k} \end{pmatrix}.$$

Clearly, we may write $\mathbf{n}_{t-1} = \mathbf{A}^{-1}\mathbf{n}_t$, $\mathbf{n}_{t-2} = \mathbf{A}^{-2}\mathbf{n}_t$, $\ldots$, $\mathbf{n}_{t-T} = \mathbf{A}^{-T}\mathbf{n}_t$, so long as T is not too large. But repeated premultiplication by $\mathbf{A}^{-1}$, which has k negative elements in the last row, may eventually give a negative value for the last element $n_{kt'}$, say, at some time t' preceding t. This is obviously impossible: it would imply a negative number of individuals in the oldest reproductive age group at some earlier time.

Therefore, although forward projection can be continued indefinitely into the future, backward projection can be performed only until the stage preceding that for which n_{kt} becomes negative. The explanation is as follows: if all our assumptions hold and if, as we have also assumed, the process is deterministic, the fate of *any* arbitrary initial age distribution can be predicted as far forward in time as we like; the starting distribution may be whatever we care to choose. But this same arbitrary distribution may be an impossible outcome for a long-continued process; that is, there may be *no* initial age distribution such that if the forward operation is repeatedly applied to it our arbitrary distribution will at some time result.

Backward projection of population composition when postreproductive age classes are considered as well as reproductive ones is more difficult. If the projection matrix has the form $\mathbf{M}$ in (3.1) with $F_i = 0$ for $i > k$ (k is the oldest age capable of reproduction), then $\mathbf{M}$ is singular and has no inverse. Thus, though forward projection is straightforward regardless of whether the projection matrix is singular or nonsingular, the same cannot be said of backward projection. Indeed, as Greville and Keyfitz (1974) write, it is "a paradox of the pure theory of population dynamics that the past is less easily accessible than the future." However, these authors show how, in certain circumstances, backward projection can be achieved by use of the generalized inverse of $\mathbf{M}$.

5. The Effect of Density Dependence on Leslie's Model

The projection matrix in its classical form makes no allowance for density dependent population growth, that is, for within-population competition. Leslie (1959) proposed a modified form of projection matrix to allow for the effect on population growth of the presence of other population members. Consider again the nonsingular matrix $\mathbf{A}$ already described. As before, let its dominant latent root be λ_1. Thus, if $\mathbf{n}_s$ is a vector proportional to the stable age distribution, $\mathbf{A}\mathbf{n}_s = \lambda_1\mathbf{n}_s$. Recall that the $(i+1)$th column of $\mathbf{A}$ gives the fertility and survival rates of the ith age group. Now let each element in this column be divided by q_{it} for $i = 0, 1, \ldots, k$, where

$$q_{it} = 1 + aN_{t-i-1} + bN_t.$$

Here a, $b > 0$ are constants that are the same for all age classes; consequently, $q_{it} > 1$. The projection matrix now becomes a function of time and is

$$\mathbf{A}_t = \mathbf{A}\mathbf{Q}_t^{-1},$$

where

$$\mathbf{Q}_t = \begin{pmatrix} q_{0t} & 0 & \cdots & 0 \\ 0 & q_{1t} & \cdots & 0 \\ & & & \\ 0 & 0 & \cdots & q_{kt} \end{pmatrix}.$$

Then $\mathbf{A}_t\mathbf{n}_t = \mathbf{A}\mathbf{Q}_t^{-1}\mathbf{n}_t = \mathbf{n}_{t+1}$.

It will be seen that the rates of production of surviving young (the elements F_i in $\mathbf{A}$) and also the survivorship of each age group (the elements P_i in $\mathbf{A}$) have been decreased in defining $\mathbf{A}_t$. Each element in $\mathbf{A}$ has been divided by an amount that depends on (a) the size of the current

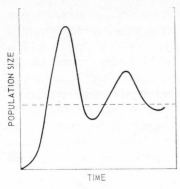

Figure 3.2. Population growth when the fertility and survival rates of an individual depend on the size of the population, both currently and at the time it was born. The broken line shows the size the population will be when stationarity is reached. (Redrawn from Leslie, 1959.)

population at time t and (b) the size of the population at time $t-i-1$, which is the beginning of the time interval in which individuals currently of age i were born. We are assuming, therefore, that the birth and death rates of all population members are influenced both by their present degree of crowding and by the crowding they experienced at the beginnings of their lifetimes which might well have a lasting effect on their future chances of reproducing and surviving.

The population will become stationary, that is, its age distribution will be stable and its total size will remain constant when, at some time τ, $q_{it} = \lambda_1$ for all i. Denoting the equilibrium size of the population by K, we then have

$$N_\tau = N_{\tau+1} = \cdots = N_{\tau+k+1} = \cdots = K.$$

Also, since $q_{i,\tau+k+1} = \lambda_1 = 1 + aN_{\tau+k-i} + bN_{\tau+k+1} \quad (i = 0, 1, \ldots, k),$

$$= 1 + (a+b)K,$$

it follows that

$$K = \frac{\lambda_1 - 1}{a + b}.$$

At this stage

$$\mathbf{A}_\tau = \mathbf{A}\mathbf{Q}_\tau^{-1} = \lambda_1^{-1}\mathbf{A}.$$

Thus, when the population has become stationary, its projection matrix is

$$\mathbf{A} = \begin{pmatrix} \dfrac{F_0}{\lambda_1} & \dfrac{F_1}{\lambda_1} & \cdots & \dfrac{F_{k-1}}{\lambda_1} & \dfrac{F_k}{\lambda_1} \\[2ex] \dfrac{P_0}{\lambda_1} & 0 & \cdots & 0 & 0 \\[2ex] 0 & \dfrac{P_1}{\lambda_1} & \cdots & 0 & 0 \\[1ex] \cdots\cdots\cdots\cdots\cdots\cdots\cdots\cdots \\[1ex] 0 & 0 & \cdots & \dfrac{P_{k-1}}{\lambda_1} & 0 \end{pmatrix}$$

The stable age distribution in this density-dependent population is seen to be the same as that in a population having the original **A** as its projection matrix, but the population's total size remains stationary.

Leslie (1959) gives a numerical example of this form of population growth and the result is shown in Figure 3.2. The population exhibits damped oscillations as it gradually approaches the stationary state.

4

Population Growth with Age-Dependent Birth and Death Rates
II: The Continuous Time Model

1. Lotka's Equations

We have already shown (page 15) that, given a simple birth and death process in which the birth and death rates are unaffected either by population size or by the ages of the individuals, the growth of a population is described by the equation $N_t = N_0 e^{rt}$. Here N_t is the size of the population at time t, and r, the intrinsic rate of natural increase, is the difference between the instantaneous birth and death rates per head of the population. To conform with established usage these rates are, in this chapter, denoted by b and d, respectively, rather than by λ and μ. The instantaneous birth rate is defined as $b = B_t/N_t$, where B_t is the number of births occurring per unit of time at time t in the whole population of size N_t. It must be emphasized that B_t and b are *rates*; they must therefore be defined in terms of a suitable unit of time but, since they vary continuously, they may have particular numerical values for only an instant of time. Further, whereas B_t is the rate for the population as a whole, b is the rate per individual. The instantaneous death rate d is defined similarly.

It was also shown (Chapter 3) that in a population in which the birth and death rates are age-dependent a stable age distribution may eventually be reached, and when this has happened

$$N_{t+1} = \lambda N_t,$$

where λ is the rate of growth per unit of time or the *finite rate of natural increase* in a population having a stable age distribution. Writing $e^r = \lambda$, this is seen to be identical to $N_{t+1} = N_t e^r$, the discrete time equivalent of $N_t = N_0 e^{rt}$. As already remarked, these formulae assume either that birth and death rates are independent of age, or more realistically, that the population has attained its stable age distribution. For it is evident that

58

when the age distribution is stable the birth and death rates per head of the population as a whole remain constant; although each age class has its own age-specific birth and death rates the proportions in which the age classes are present remain constant. So in these circumstances the two equations describing population growth are equivalent, and consequently

$$e^r = \lambda \qquad \text{or} \qquad \ln \lambda = r.$$

We now consider Lotka's original approach to the study of population growth. Time is treated as continuous and matrix methods are not used. What we wish to do is to estimate b and d (hence $r = b - d$) from observable quantities. Since two unknowns are to be derived, two equations that express b and d in terms of the observed quantities are necessary. As observables we take l_x and m_x, defined as follows:

l_x is the proportion of survivors at time x of an original cohort of individuals all born simultaneously at time 0.

m_x is the mean number of offspring born in unit time to an individual whose age is in the range $(x, x + dx)$.

It should be noticed that l_x is a pure number and m_x, a rate. Moreover, the rate m_x pertains only to those individuals in an infinitesimally small age class.

Two other quantities, to be eliminated in arriving at the final equations, are also required. These are

$c_x\, dx$, the proportion of the total population in the age range $(x, x + dx)$ when the age distribution is stable. The function c_x thus defines the stable age distribution.

$B_t\, dx$, the number of individuals born to the whole population, in an interval of length dx at time t.

Assuming the age distribution to be stable, it now follows that $N_t c_x\, dx$ is the size, at time t, of that fraction of the total population which consists of individuals whose ages are in the range $(x, x + dx)$. But these individuals are evidently the survivors of the $B_{t-x}\, dx$ individuals born at time $t - x$ in an interval of length dx, and thus

$$N_t c_x\, dx = l_x B_{t-x}\, dx. \tag{4.1}$$

Now, since the age distribution is stable, b, the instantaneous birth rate per head, is constant and equal to B_i/N_i for all i. Therefore

$$B_{t-x} = b N_{t-x}$$
$$= b N_t e^{-rx},$$

since

$$N_t = N_{t-x}e^{rx}.$$

Then, from (4.1),

$$c_x = \frac{l_x B_{t-x}}{N_t} = l_x b e^{-rx}. \tag{4.2}$$

But

$$\int_0^\infty c_x \, dx = 1 \tag{4.3}$$

and so

$$\frac{1}{b} = \int_0^\infty l_x e^{-rx} \, dx. \tag{4.3a}$$

This is the first of the two equations relating b and $d = b - r$.

Next suppose that the $N_t c_x \, dx$ individuals in the age range $(x, x + dx)$ and alive at time t produce m_x offspring per head per unit of time. Then the total number of newborn produced per unit time by all the population's age classes taken together is

$$B_t = \int_0^\infty N_t c_x m_x \, dx = \int_0^\infty b N_t l_x m_x e^{-rx} \, dx. \tag{4.4}$$

Therefore

$$\frac{1}{b} \frac{B_t}{N_t} = \int_0^\infty l_x m_x e^{-rx} \, dx.$$

But $b = B_t/N_t$ and so

$$\int_0^\infty l_x m_x e^{-rx} \, dx = 1. \tag{4.5}$$

This is the second of the required equations. When values of l_x and m_x have been found by experiment, (4.3a) and (4.5) may be solved to yield b, d, and r. These equations cannot, unfortunately, be solved explicitly, and in practice it is necessary to replace the integrals with sums and obtain approximate solutions. That is, we treat a continuous process as being approximated by a discrete time process. This is done only to facilitate computation, however. It is not assumed that the process is truly discrete. The duration of the time units is the same as the interval separating the successive observations that yielded the values of l_x and m_x required. This time unit is chosen by the observer and is without biological significance; obviously, the shorter the interval, the more precise the results.

Thus (4.5) is replaced by

$$\sum_x l_x m_x e^{-rx} = 1 \qquad (4.6)$$

and then r is obtained by successive approximation. Laughlin (1965) gives, as a rough value of r useful for starting the successive approximations,

$$r \doteqdot \frac{\sum l_x m_x \ln(\sum l_x m_x)}{\sum x l_x m_x}.$$

We may also obtain the values of c_x, the stable age distribution. From (4.2) it is seen that

$$c_x \propto l_x e^{-rx};$$

and from (4.3) that (in terms of the chosen time units which, as already remarked, may be as short as desired)

$$\sum c_x = 1.$$

Thus

$$c_x = \frac{l_x e^{-rx}}{\sum l_x e^{-rx}}$$

gives the proportion of the total population in the xth age class.

More detailed accounts of these calculations, and examples, may be found in Leslie (1945), Birch (1948), Leslie and Park (1949) and Pielou (1974).

2. The Mean Generation Time and the Net and Gross Reproductive Rates

The intrinsic rate of natural increase, r, and the finite rate, λ, both measure a population's per capita growth rate relative to some chosen time unit. For populations that grow in discrete jumps with no overlap of generations, the obvious time unit to choose is the natural interval between jumps: for instance, one year for univoltine insects and annual plants; 17 years for 17-year periodical cicadas. This is the generation time, T.

For populations in which the generations overlap, generation time is less clearcut. If births are seasonal, the interval between breeding seasons provides a natural time unit for measuring population growth rate, but it is clearly a shorter interval than generation time. However, even when generations do overlap, the concept of *mean generation time* does have an intuitive meaning though not a very precise one. Indeed, attempts to pin down the concept have led to three separate definitions. They have been fully discussed by Leslie (1966) and Caughley (1967), and are described in numbered paragraphs below.

1. First note that, for any population, the mean number of daughters born to a female in her lifetime is $\int_0^\infty l_x m_x \, dx$. Further, provided the generations do not overlap, this number is the ratio of the number of births in one generation to the number in the preceding generation; it is known as the *net reproductive rate*, R_0. Still assuming no overlap of generations, and denoting the mean generation time by T, it is seen that

$$R_0 = \frac{N_{t+T}}{N_t} = e^{rT}. \tag{4.7}$$

Hence

$$Y = \frac{\ln R_0}{r} = \frac{\ln\left(\int_0^\infty l_x m_x \, dx\right)}{r}; \tag{4.8}$$

Thus if l_x and m_x values are known, r can be determined from (4.6) and T from (4.8). T as thus defined is the classic measure of mean generation time (see Leslie, 1966) and is calculable for any population regardless of whether or not its generations overlap. However for populations in which generations do overlap, T does not represent any clearly visualizable time span as do T_c and $\bar{T}$ described below. It therefore has little to recommend it to ecologists. Also, if $r = 0$ and $R_0 = 1$, T is indeterminate.

2. A measure of generation time that has concrete meaning in all circumstances is the cohort generation time defined as

$$T_c = \frac{\int_0^\infty x l_x m_x \, dx}{\int_0^\infty l_x m_x \, dx} = \frac{1}{R_0} \int_0^\infty x l_x m_x \, dx \tag{4.9}$$

This is the mean age of mothers at the births of their daughters, as may be seen from the following argument. The expected total number of daughters born by a mother in her lifetime is $\int_0^\infty l_x m_x \, dx = R_0$ and the expected number born while she is of age x is $l_x m_x$. If we treat the proportion $(l_x m_x)/R_0$ as the probability that a female will have a daughter while her age is x, then clearly the mean age of the mothers at the times of their daughters' births is as given in (4.9).

If r is small, T and T_c are approximately equal as may be shown as follows. Recall (4.5), expand the exponential series, and at the same time put $\int_0^\infty x^j l_x m_x \, dx = R_j$. Then it is seen that

$$\sum_{j=0}^{\infty} (-1)^j \frac{r^j}{j!} R_j = 1.$$

If r is small enough for terms in r^2 and higher powers to be neglected, we then have

$$\frac{1}{R_0} \simeq 1 - r \frac{R_1}{R_0}.$$

For sufficiently small r we also have

$$\exp\left[-r \frac{R_1}{R_0} \right] \simeq 1 - r \frac{R_1}{R_0},$$

and thus

$$\frac{1}{R_0} \simeq \exp\left[-r \frac{R_1}{R_0} \right].$$

But $1/R_0 = e^{-rT}$ from (4,7), and hence

$$T \simeq \frac{R_1}{R_0} = T_c.$$

Laughlin (1965) has discussed in detail the use of T_c as a measure of mean generation time.

3. The measure regarded by Leslie (1966) and Caughley (1967) as best, both mathematically and biologically, is

$$\bar{T} = \int_0^\infty x l_x m_x e^{-rx} \, dx.$$

This is the mean age of the mothers of a cohort of newborn daughters. This follows since, from (4.4), we see that the number of newborn, B_t, produced in a unit of time by the whole population of mothers is proportional to $\int_0^\infty l_x m_x e^{-rx} \, dx$. The proportion of this total born by a mother aged x is therefore

$$\frac{l_x m_x e^{-rx}}{\displaystyle\int_0^\infty l_x m_x e^{-rx} \, dx} = l_x m_x e^{-rx},$$

taking account of (4.5). Thus the mean age of these mothers is $\int_0^\infty x l_x m_x e^{-rx} \, dx$ which is $\bar{T}$.

It is worth reiterating the distinctions among T, T_c, and $\bar{T}$. T is merely the ratio $(\ln R_0)/r$. Both T_c and $\bar{T}$ are mean ages of mothers when daughters are born but the averaging is from the mothers' points of view for T_c, and from the daughters' points of view for $\bar{T}$.

We return now to measures of population growth rate. The net reproductive rate R_0 is not strictly a rate since no measure of time enters its definition. We cannot introduce the time unit T defined in (4.8) without making the definition circular. Thus R_0 is a suitable measure of population growth rate only for species with nonoverlapping generations, in which case we can write $R_0 = N_{j+1}/N_j$ where the subscripts j and $j+1$ denote generations.

As a final measure of population growth—more precisely of "potential" growth—we have the *gross reproductive rate*, $G = \int_0^\infty m_x \, dx$. This is the expected number of offspring that would be born to a mother that survived for the whole of her reproductive period.

3. Age-Specific Fertility Tables

Sections 1 and 2 have shown how l_x and m_x values are used to determine a population's rate of increase. It must now be emphasized that equations such as (4.5) and (4.6) hold only in an environment sufficiently stable for the l_x and m_x figures to remain constant in time. Strictly, therefore, tables of l_x and m_x values should be compiled from the results of controlled experiments throughout which environmental conditions are kept constant. Values of r derived from such tables apply only to populations living in the stated conditions.

Moreover, we have assumed in the foregoing that all the individuals in a population were capable of reproduction; that is, only the female part of a bisexual population was considered. The results are easily extended to cover a whole population (both sexes combined), provided the sex ratio within each age class remains constant.

The m_x values are variously called *age-specific fertility rates*, *age-specific fecundity rates*, or *maternal frequencies*. The word *fertility* usually applies to species that bear live young. *Fecundity* is used for egg-laying species when what is counted is the numbers of eggs laid by females of different ages; in this case survival is measured from the moment an egg is laid.

Although m_x has been defined as the mean number of offspring born per unit time to a parent whose age is in the range $(x, x + dx)$, in practice parental age may be quite coarsely grouped. Thus we may take m_x equal to the mean number of offspring born per unit time to a parent in the age range $(x, x + h)$ and, provided h is small enough, the approximation will be satisfactory. This amount to treating m_x as a step function instead of as a continuous variable.

It should be observed that m_x is not the same as the element F_x in the projection matrix **M** of Chapter 3 (page 42). They are related by the identity $F_x = \lambda m_x = e^r m_x$. This ensures the equivalence of the predictions made by (4.4) concerning a continuously growing population to the predictions made when a projection matrix is used to describe the growth of the same population (assumed to have a stable age distribution) if, for convenience, we choose to treat time as discrete. It will be seen that $F_x = \lambda m_x$ is of the same form as $n_{0,s+1} = \lambda n_{0s}$ [see (3.4)]. In using a projection matrix we are, in effect, treating the youngest age class (whose ages span a finite interval) as growing in a black box. We make no stipulations about what takes place within the box during a unit of time but only about what has replaced its contents by the beginning of the next unit of time. The "replacement" group must have the same age structure as, and λ times the total size of, the group it replaces in this age class.

The relationship between l_x and the element P_x of **M** is considered in Section 4 below.

4.   Life Tables

The *life table function* l_x (sometimes called the *age-specific survival rate*) is defined as the proportion of a cohort of individuals, all born simultaneously at time zero, that still survive at time x when, of course, all are of age x. Values of l_x are often shown in a *life table* in which, besides a column of the l_x values themselves, there are usually several other columns of derived statistics.

A life table for any species is of great interest by itself, even if we do not wish to estimate the population growth rate and the stable age distribution. From a life table we may infer at what age the risk of death is greatest; if we know the principal causes of death affecting the different age classes, we can infer to which of these causes most of the deaths are due, that is, which of the many dangers a species faces is most important in keeping its numbers down. Life tables are much studied by, for instance, economic entomologists. For insect species that have nonoverlapping generations and whose eggs all hatch at roughly the same time a nautral population is itself a cohort, and l_x values may be estimated directly by obtaining sample estimates of a population's size at a sequence of times. The form of the survivorship curve (a plot of l_x versus x) may suggest when a pest species is most likely to respond to control measures and also when control is most needed. For a pest that does damage during its immature stages (e.g., caterpillars) early destruction is desirable and it is worth knowing whether natural deaths occur predominantly early or late. For species that are pests in the adult stage (e.g., mosquitoes, blackflies, tsetse flies) it does not matter whether control measures are applied early or late in the immature stages so long as they succeed; thus the time to attempt control may depend on when, as judged from a life table, a species is most vulnerable.

The data from which a life table is compiled may be gathered in one of two ways and the method used is determined by the life span of the species concerned. Therefore two different kinds of life table need to be considered. With short-lived organisms it is possible to follow the fate of a single cohort of individuals; the numbers of survivors at successive times (equivalently, at successive ages) is recorded until all members of the cohort are dead. Such data enable a *cohort life table* (sometimes called a *horizontal life table*) to be prepared.

With long-lived organisms (those whose life spans are an appreciable fraction of a human life span or longer), following the fate of a cohort is impracticable. A *current life table* (sometimes called a *vertical life table*) is compiled instead. It is based on observations of the proportion of survivors in each age class, through a single interval of time, in an all-age population. If survival rates are stationary, that is, if the probability of an individual's surviving from the xth to the $(x+1)$th age class, say p_x, is independent of time, the two kinds of life table are interconvertible; but they are not interconvertible if there is any trend of age-specific survival rates with time.

We begin by assuming that conditions are stationary and discuss how cohort and current life tables are compiled and how they are interrelated. Obviously both kinds are needed in ecological work since the life spans of different species of organisms range from a few hours for bacteria to more than a millenium for giant sequoias. Less extreme examples, of species for which life tables have been prepared, are copepods with a life span of 9 or 10 weeks (Gehrs and Robertson, 1975); spruce budworm with a life span of one year (Miller, 1963); and the Himalayan thar (an ungulate) with a life span of about 11 years (Caughley, 1967).

Cohort Life Tables

A cohort life table is prepared as follows: suppose we begin with a population of newborn individuals numbering l_0; this number is known as the radix. It may be set equal to 1, though often, to avoid small fractions, it is set equal to 1000. At a succession of times separated by equal intervals of one time unit a count is made of the number still alive. The results are tabulated thus:

Age interval	Number of survivors at age x	Number of deaths in $(x, x+1)$	Proportion of deaths in $(x, x+1)$	Number of time units lived in $(x, x+1)$	Life remaining to those aged x	Observed expectation of life at age x
$\vdots$	$\vdots$	$\vdots$	$\vdots$	$\vdots$	$\vdots$	$\vdots$
$(x, x+1)$	l_x	d_x	q_x	L_x	T_x	e_x
$\vdots$	$\vdots$	$\vdots$	$\vdots$	$\vdots$	$\vdots$	$\vdots$

The various columns of the table are interrelated as follows. Obviously

$$d_x = l_x - l_{x+1}$$

and

$$q_x = \frac{d_x}{l_x} = 1 - \frac{l_{x+1}}{l_x}.$$

The length of time lived during the interval $(x, x+1)$ by all individuals taken together is

$$L_x = \int_x^{x+1} l_x \, dx.$$

The units in which L_x is measured have the dimension individuals $\times$ time (analogous to "man-hours," say). Each individual alive at time x will contribute part of a unit if it dies during $(x, x+1)$ or one unit if it survives throughout the whole interval and is still alive at $x+1$. If the survivorship curve can be treated as approximately linear in $(x, x+1)$, we may write

$$L_x \doteqdot \tfrac{1}{2}(l_x + l_{x+1}).$$

The total lifetime remaining to all those members of the population alive at age x is the sum of their individual lifetimes; that is,

$$T_x = \sum_{j=x}^{w} L_x,$$

where w is the age at which all are dead. Therefore, putting $L_x \doteqdot \tfrac{1}{2}(l_x + l_{x+1})$,

$$T_x \doteqdot \frac{l_x}{2} + \sum_{j=x+1}^{w} l_j.$$

The element P_x in the matrix **M** used to predict population growth when time is treated as discrete (see page 42) is then

$$P_x = \frac{L_{x+1}}{L_x} \doteqdot \frac{l_{x+1} + l_{x+2}}{l_x + l_{x+1}} \qquad \text{(see Leslie, 1945)}.$$

Finally, e_x is the observed expectation of life for an individual aged x

and is given by

$$e_x = \frac{T_x}{l_x};$$

that is, e_x is the total lifetime remaining at time x to the whole of the surviving population, divided by l_x, the number of individuals by whom it is shared.

A cohort life table based on actual observations is an empirical body of data consisting of observed values of random variables. Thus the entries in the table are only estimates, subject to sampling error, of underlying population quantities. The probability distributions of the variables have been very fully discussed by Chiang (1968). As an example, we derive here the probability distribution, and expectation, of W, the age at death of the last survivor of a cohort.

The variate W has a *chain binomial distribution* generated by a mechanism that may be likened to the following urn scheme. Imagine an infinite row of urns containing mixtures of black and white balls. The proportion of white balls in the xth urn is p_x. Let l_0 balls be drawn at random from the 0th urn and suppose l_1 are white; because of this result, l_1 are drawn from the first urn of which l_2 are white; so l_2 balls are drawn from the second urn, ... ; and so on. Clearly, $l_{x+1} = l_x p_x$. In terms of the urn scheme the variate W is the ordinal number (starting from zero) of the urn that yielded black balls only and was therefore the last one drawn from.

Imagine, also, that each white ball (representing a survivor) drawn from the xth urn, say, is individually associated with one of the balls (which may turn out to be black or white) drawn from the $(x+1)$th urn. Thus an individual animal's lifetime is represented by a sequence of associated balls. It follows that the probability that a given individual survives to age $i-1$, say, which we shall denote by P_i, is given by $P_i = p_0 p_1 p_2 \cdots p_{i-1}$. And it also follows that the probability of survival to age $i-1$ of all of a given group of k individuals is P_i^k.

We can now obtain the probability that $W = w$. Thus

$$\Pr\{W = w\} = \sum_{k=1}^{l_0} [\Pr(k \text{ balls are drawn from the } w\text{th urn}) \times$$
$$\Pr(\text{all these } k \text{ balls are black})]$$

$$= \sum_{k=1}^{l_0} \binom{l_0}{k} P_w^k (1 - P_w)^{l_0 - k} (1 - p_w)^k, \qquad w = 0, 1, \ldots$$

$$= \left(\sum_{k=0}^{l_0} \binom{l_0}{k} [P_w(1 - p_w)]^k (1 - P_w)^{l_0 - k} \right) - (1 - P_w)^{l_0}$$

$$= [P_w(1 - p_w) + 1 - P_w]^{l_0} - (1 - P_w)^{l_0};$$

finally, since

$$P_w p_w = P_{w+1},$$

$$\Pr\{W = w\} = (1 - P_{w+1})^{l_0} - (1 - P_w)^{l_0}, \qquad w = 0, 1, \ldots.$$

This is the probability distribution of W. Its expectation is

$$E(W) = \sum_{w=0}^{\infty} w[(1 - P_{w+1})^{l_0} - (1 - P_w)^{l_0}].$$

For brevity, put

$$(1 - P_x)^{l_0} = Q_x.$$

Then

$$E(W) = \lim_{v \to \infty} \sum_{w=0}^{v} w(Q_{w+1} - Q_w)$$

$$= \lim_{v \to \infty} \left[v Q_{v+1} - \sum_{i=1}^{v} Q_i \right].$$

Now as $v \to \infty$, let $P_{v+1} \to 0$ and hence $Q_{v+1} \to 1$. (This amounts to assuming that an infinitely long life for any individual is impossible). Then

$$E(W) = \sum_{w=0}^{\infty} (1 - Q_w) = \sum_{w=0}^{\infty} [1 - (1 - P_w)^{l_0}].$$

In particular, if $l_0 = 1$,

$$E(W) = \sum_{w=0}^{\infty} P_w.$$

For further details, see Chiang (1968).

Current Life Tables

The fundamental difference between a cohort life table and a current life table was stated on page 66. It will be recalled that in a cohort table the raw data are the values of the life table function l_x; whereas in a current table the raw data are the numbers of individuals in the different age classes and the numbers of deaths occurring among them in one unit of time.

In practice, obtaining observed l_x values for a cohort table is often difficult except with laboratory-reared populations. In the field, not only must we be sure that a true natural cohort exists; we must also be able to recognize its members. Unless these requirements are met, l_x values have to be derived indirectly. Caughley (1966, 1967) has described various methods of obtaining the values for wild mammal populations.

Gathering data for a current life table is also difficult, except for demographers of human populations in thoroughly censused countries. The raw data for the table consist of values of Y_x, the number of individuals alive in the age class $(x, x+1)$ at the midpoint of the time interval (of unit length) to which the table relates; and values of D_x, the number of deaths in this age class in this time interval. (The difficulties of observing these values in ecological contexts are too obvious to need stressing.) The ratio $M_x = D_x/Y_x$ is the age-specific death rate.

The table* is constructed as follows.

Age class	Population in age class at midpoint of time interval	Number of deaths in age class	Age-specific death rate
.	.	.	.
.	.	.	.
.	.	.	.
$(x, x+1)$	Y_x	D_x	$M_x = \dfrac{D_x}{Y_x}$
.	.	.	.
.	.	.	.
.	.	.	.

If the population is stationary, that is, if its total size is constant and its age distribution is stable, D_x is the same as d_x in the cohort life table; that is

$$D_x = l_x - l_{x+1}.$$

Also

$$Y_x = L_x = \int_x^{x+1} l_x \, dx \doteq \tfrac{1}{2}(l_x + l_{x+1}),$$

the approximation being satisfactory provided the time interval employed is short enough.

Then the age-specific death rate is

$$M_x = \frac{D_x}{Y_x} = \frac{l_x - l_{x+1}}{L_x} \doteq \frac{2(l_x - l_{x+1})}{l_x + l_{x+1}}.$$

* Shown here is the type of table that Chiang (1968) calls an abridged life table, although here the same unit is used for age and time. Also, we here omit the column showing the mean fraction of each time interval lived by those individuals that die in the time interval. In ecological work, in which refined observations are usually not feasible, this fraction is best taken as 0.5.

Let us find the relationship between M_x, the age specific death rate and q_x, the proportion of deaths in the age class $(x, x+1)$. By definition,

$$q_x = 1 - \frac{l_{x+1}}{l_x}$$

and therefore

$$M_x = \frac{2q_x}{2-q_x}$$

or

$$q_x = \frac{M_x}{1+\tfrac{1}{2}M_x} \, .$$

Notice that q_x, which is an estimate of the probability that an individual will die when its age is in the range $(x, x+1)$, is obtained directly from a cohort life table but only indirectly from a current life table. It will be seen that the age specific death rate $M_x = D_x/Y_x$ is necessarily greater than $q_x = D_x/l_x$, since $l_{x+1} < l_x$ and therefore $Y_x \doteq \tfrac{1}{2}(l_x + l_{x+1}) < l_x$. The values of q_x may be added to the current life table as a final column.

As a final clarification of the distinction between cohort and current life tables, it is worth considering Gani's (1973) presentation. He displays the survival probabilities $p_x = 1 - q_x = l_{x+1}/l_x$ as elements in a Markov matrix, say $\mathbf{P}$, in which the states are age classes. Assuming no individuals live to be older than k and treating death as a $(k+1)$th state we have

		0	1	2	$\cdots$	$k-1$	k	$k+1$
	0	0	p_0	0	$\cdots$	0	0	q_0
	1	0	0	p_1	$\cdots$	0	0	q_1
	2	0	0	0	$\cdots$	0	0	q_2
$\mathbf{P} =$	$\vdots$							
	$k-1$	0	0	0	$\cdots$	0	p_{k-1}	q_{k-1}
	k	0	0	0	$\cdots$	0	0	1
	$k+1$	0	0	0	$\cdots$	0	0	1

Here the (i, j)th element is the probability that an individual will go from the ith to the jth state in one step. Thus the $(i, i+1)$th element for $i < k$ is the probability p_i that an individual will survive from the ith to the $(i+1)$th age class; and the $(i, k+1)$th element is the complementary probability, $q_i = 1 - p_i$, that it will not.

Now consider the relationship between the elements of $\mathbf{P}$ and the entries in the two kinds of life table. If, and only if, a population is stationary the elements of $\mathbf{P}$ are independent of time and $\mathbf{P}$'s final column gives the q_x values that appear in the cohort life table; these values may also be put in a current life table if desired.

But if the population is not stationary, there is a different Markov matrix, say $\mathbf{P}(t)$, for each time interval. The $(i, i+1)$th element in $\mathbf{P}(t)$ is $p_i(t)$, the probability that an individual in the ith age class at the end of the interval $(t-1, t)$ survives into the interval $(t, t+1)$ during which (if it lives) it will enter the $(i+1)$th age class. This individual belongs to a cohort born in the time interval $(t-i-1, t-i)$. Because the population is nonstationary, each cohort has its own unique cohort life table and to assemble the entries for $\mathbf{P}(t)$ one would need to consult $k+1$ different cohort tables constructed in $k+1$ consecutive time intervals. Conversely, to reconstruct a single cohort table from Markov matrices one requires elements from $k+1$ consecutive matrices. However, if a column of q_x values is to be appended to a current life table, the values are those in the final column of the matrix $\mathbf{P}(t)$ pertaining to the same time interval.

To summarize: The Markov matrix $\mathbf{P}(t)$ and the current life table describe events affecting individuals of all ages in the time interval $(t, t+1)$; whereas a cohort life table describes events befalling a single age class of individuals from infancy to the death of the last survivor.

5

The Dynamics of Interacting Species: I. Two-Species Interaction

1. Introduction

In Chapters 5 and 6 we consider the dynamics of systems of two or more species-populations living together and interacting. The topic is vast and because recent, authoritative, book-length treatments exist (May, 1973; Maynard Smith, 1974), the treatment here will be less general. I concentrate on aspects that seem either particularly noteworthy, or particularly capable of further useful development. An attempt will be made to steer between the scylla of a bald statement of inapplicable mathematical generalities and the charybdis of a list of narrowly specialized case studies (or if not to steer between them, to offer a little bit of both). The reader would do well to look ahead at the section headings before going farther.

Some general remarks on the place of mathematical modeling in theoretical and practical ecology appear at the end of Chapter 6. We shall not, before reaching that section, labor the fact that mathematical models must always rest on an indefinite (but large) number of simplifying assumptions whose justifiability is often debatable.

Concerning the aims and methods of analyzing a particular model (taking it for granted that the model deserves to be analyzed) the following preliminary synopsis may be helpful.

Mathematical models designed to describe the interactions of two or more species-populations usually consist of sets of differential equations, and the models are usually deterministic. [There are exceptions: difference equation models are mentioned in Sections 3 and 5; and stochastic simulation models are mentioned in Section 4.] In analyzing them, the first objective is to judge their stability.

The very meaning of the stability question depends on whether a model's differential equations are linear or nonlinear; these equations specify the growth rate of each species-population (not the per capita growth rate but that of the whole population) as a function of the sizes of the various interacting populations. The meaning of the question also depends on whether the equations are assumed to apply over all conceivable combinations of population sizes or only in the neighborhood of an equilibrium point at which all growth rates are simultaneously zero. In the former case our enquiry relates to *global stability* and in the latter to *neighborhood,* or *local, stability.*

If the difference equations are globally (alternatively, locally) linear, we may test for global (alternatively, local) stability by means of the *Routh-Hurwitz criteria.* Use of the criteria is demonstrated in the two-species case in Sections 2 and 5 of this chapter; they are stated fully for the k species case in Section 1 of Chapter 6. Their use presupposes that numerical values for the coefficients in the equations are known. In ecological contexts, detailed knowledge of an interacting-species system's properties is often unobtainable and all we can specify are the signs, not the magnitudes, of the coefficients. One can then test only for *qualitative stability*; Section 2 of Chapter 6 contains the details.

If the differential equations are globally nonlinear we may still treat them as approximately linear in a sufficiently small neighborhood of the equilibrium point (if there is one) and can then use the Routh-Hurwitz criteria to judge their local stability in that neighborhood. The result is not necessarily true globally, of course. A nonlinear model that is unstable in the neighborhood of its equilibrium point may be stable in the wider sense that it exhibits *stable limit cycles.* These are described in Section 5 of this chapter, with an example. The criterion for judging whether a model is either stable or else yields a stable limit cycle around an unstable equilibrium point may conveniently be called the *Kolmogorov criterion*; it is mentioned in Section 5 and, although a description is beyond the scope of this book references will be given to sources where full accounts may be found.

Local and global stability, qualitative stability, linear and nonlinear models, and the various diagnostic criteria are, together, the core topics of a fast-growing body of knowledge in which research is now very active (1976). However, we begin our discussion of two-species interactions with a description (in Sections 2 and 3) of the classic approach to two-species competition, of the way in which difference equations may be substituted for the classic differential equations, and of the way in which the effects of chance may be simulated.

2. Competition Between Two Species: Gause's Model

Consider two species-populations occurring together, and assume that the growth rate of each is inhibited by members both of its own and of the other species. Denoting by N_i the number of individuals in species i (for $i = 1, 2$), assume that the following pair of differential equations* describe their behavior.

$$\frac{1}{N_i}\frac{dN_i}{dt} = r_i - a_{ii}N_i - a_{ij}N_j \tag{5.1}$$

(Here $(i, j) = (1, 2)$ in the first equation of the pair and $(i, j) = (2, 1)$ in the second).

Thus we are assuming that the *per capita* growth rate of each population at any instant is a linear function of the sizes of the two competing populations at that instant. Each population would grow logistically if it were alone, with logistic parameters r_i and a_{ii} for species i; a_{ij} measures the degree to which the presence of species j affects the growth of species i.

In general, the simultaneous differential equations (5.1) cannot be explicitly solved. We consider first a particular set of circumstances in which they can be solved and return to the more general case afterward. The simplification that permits solution of the equations consists in assuming that for an individual of either species the inhibitory effect of all other individuals (of both species) is the same. Then each individual of both species behaves as if it were competing with a population of size $N = N_1 + pN_2$. The factor p allows for the fact that the members of the two species may differ from one another in their inhibitory effect. If individuals of species 2 make smaller inroads on the resources than individuals of species 1, then $p < 1$; conversely, if species 1 is the less demanding competitor, $p > 1$. In any case, N, the size of the "effective inhibiting population" is the same for both species.

Making this assumption, (5.1) may be replaced by

$$\frac{dN_1}{dt} = r_1 N_1 - (a_1 N)N_1,$$

$$\frac{dN_2}{dt} = r_2 N_2 - (a_2 N)N_2. \tag{5.2}$$

It then follows that

$$a_{11}N_1 + a_{12}N_2 = a_1 N = a_1(N_1 + pN_2),$$

and

$$a_{21}N_1 + a_{22}N_2 = a_2 N = a_2(N_1 + pN_2).$$

---

* These equations may conveniently be called Gause's competition equations. Although not originally due to Gause, it was he who studied their biological consequences in depth (Gause, 1934).

Therefore

$$(a_{11} - a_1)N_1 + (a_{12} - a_1p)N_2 = 0,$$

and

$$(a_{21} - a_2)N_1 + (a_{22} - a_2p)N_2 = 0,$$

so that

$$a_1 = a_{11}; \quad \frac{a_{12}}{a_{11}} = p,$$

$$a_2 = a_{21}; \quad \frac{a_{22}}{a_{21}} = p,$$

or

$$a_{12}a_{21} = a_{11}a_{22}.$$

This relation is therefore equivalent to the assumption already made, namely that the effective inhibiting population is of the same size for each of the two species. Equations 5.2 can now be written

$$\frac{1}{N_1}\frac{dN_1}{dt} = \frac{d}{dt}\ln N_1 = r_1 - a_1N,$$

and

$$\frac{1}{N_2}\frac{dN_2}{dt} = \frac{d}{dt}\ln N_2 = r_2 - a_2N.$$

Eliminating N gives

$$\frac{d}{dt}(a_2 \ln N_1 - a_1 \ln N_2) = \frac{d}{dt}\ln\left(\frac{N_1^{a_2}}{N_2^{a_1}}\right) = r_1a_2 - r_2a_1,$$

whence

$$\frac{N_1^{a_1}}{N_2^{a_1}} = \frac{[N_1(0)]^{a_2}}{[N_2(0)]^{a_1}}\exp[(r_1a_2 - r_2a_1)t],$$

where $N_i(0)$ for $i = 1, 2$ are the initial sizes of the two populations at $t = 0$.

Clearly, if the combined population is maintained long enough, only one of the species will persist and the other will die out. The outcome will depend only on whether r_1a_2 is greater or less than r_2a_1 and will be unaffected by the initial sizes. If $r_1a_2 > r_2a_1$, ultimately species 1 will "win" and species 2 die out. The opposite will happen if $r_1a_2 < r_2a_1$. Once the losing species has become extinct, the winner will continue to grow in accordance with the single-species logistic process.

We now drop the simplifying assumption that each species is inhibited by the same effective population. Equivalently, we assert that $a_{12}a_{21} \neq a_{11}a_{22}$. There are four possible sets of circumstances to consider and these are most clearly shown graphically (see Figure 5.1). In these *phase space* diagrams the composition of the combined two-species

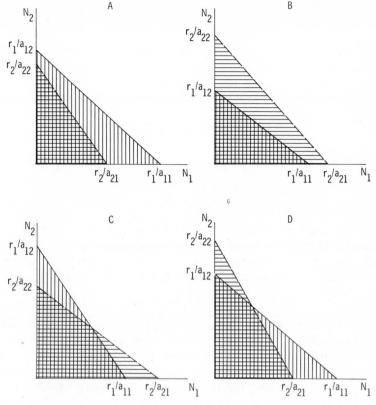

Figure 5.1 The four possible relationships between the N_1-isocline, $N_1 = (r_1 - a_{12}N_2)/a_{11}$ (solid line) and the N_2-isocline, $N_2 = (r_2 - a_{21}N_1)/a_{22}$(dashed line). In vertically hatched regions, species 1 increases; in horizontally hatched regions, species 2 increases. Thus in A species 1 wins; in B species 2 wins; in C there is stable equilibrium; in D the equilibrium is unstable.

population at any instant may be represented by a point having as coordinates the values of N_1 and N_2 at that instant.

Consider now the locus of points for which $dN_1/dt = 0$ which is known as the N_1-isocline. It is the line (shown solid in the graphs)

$$r_1 - a_{11}N_1 - a_{12}N_2 = 0 \quad \text{or} \quad N_1 = \frac{r_1 - a_{12}N_2}{a_{11}}$$

It cuts the N_1-axis at $N_1 = r_1/a_{11}$ and the N_2-axis at $N_2 = r_1/a_{12}$. If, at any instant, the point representing the combined populations falls below this line, that is, if

$$N_1 < \frac{r_1 - a_{12}N_2}{a_{11}},$$

then

$$\frac{dN_1}{dt} = N_1(r_1 - a_{11}N_1 - a_{12}N_2)$$

$$> N_1\left[r_1 - a_{11}\frac{(r_1 - a_{12}N_2)}{a_{11}} - a_{12}N_2\right] = 0$$

and the species 1 population increases in size. Conversely, if

$$N_1 > \frac{r_1 - a_{12}N_2}{a_{11}}, \qquad \text{then} \qquad \frac{dN_1}{dt} < 0$$

and the species 1 population decreases. In the graphs the region in which species 1 will increase is shown by vertical hatching and the region in which this species will decrease, by the absence of vertical hatching.

Analogous arguments show that the line (dashed in the graphs)

$$r_2 - a_{21}N_1 - a_{22}N_2 = 0$$

represents the N_2-isocline, for which $dN_2/dt = 0$. It cuts the N_1-axis at $N_1 = r_2/a_{21}$ and the N_2-axis at $N_2 = r_2/a_{22}$. Whenever the point representing the combined population falls below this line (region with horizontal hatching) $dN_2/dt > 0$ and the species 2 population increases in size. Conversely, above this line (horizontal hatching absent) species 2 must decrease.

Beginning with a combined population for which both N_1 and N_2 are small, so that the point representing the initial species composition falls within the cross-hatched area on any of the graphs, we now see that there are four possibilities, depending on the relative positions of the isoclines and hence on the values of the parameters. Thus if, as in Figure 5.1A, the N_1-isocline lies wholly above the N_2-isocline or, in other words, if $r_1a_{22} > r_2a_{12}$ and $r_1a_{21} > r_2a_{11}$, then, although species 1 is sometimes capable of increasing when species 2 is stationary or decreasing, the reverse is never true. It follows that ultimately species 1 must replace species 2 entirely.

The opposite situation, in which species 2 will win and species 1 die out is shown in Figure 5.1B in which $r_2a_{12} > r_1a_{22}$ and $r_2a_{11} > r_1a_{21}$.

In these two cases there exists no pair of values (N_1, N_2), with N_1 and N_2 both positive, that satisfies the simultaneous equations $dN_1/dt = dN_2/dt = 0$. An equilibrium condition in which the sizes of both populations remain constant is therefore impossible.

Now consider Figures 5.1C and 5.1D in which the isoclines do cross each other. They intersect at the point (N_1^*, N_2^*) where

$$N_1^* = \frac{r_1 a_{22} - r_2 a_{12}}{a_{11} a_{22} - a_{12} a_{21}} \quad \text{and} \quad N_2^* = \frac{r_2 a_{11} - r_1 a_{21}}{a_{11} a_{22} - a_{21} a_{12}}.$$

Theoretically a combined population with precisely this composition would maintain itself unchanged indefinitely. Notice, however, that in one case the equilibrium is stable and in the other, unstable.

Figure 5.1C, in which $r_1 a_{22} > r_2 a_{12}$ and $r_2 a_{11} > r_1 a_{21}$, shows the conditions for stable equilibrium. Regardless of the initial numbers of the two species, as time passes they will steadily approach N_1^* for species 1 and N_2^* for species 2. This follows since in the vertically hatched region in which $dN_1/dt \geq 0$ and $dN_2/dt \leq 0$ simultaneously $N_1 \leq N_1^*$ and $N_2 \geq N_2^*$. Correspondingly, in the horizontally hatched region in which $dN_1/dt \leq 0$ and $dN_2/dt \geq 0$ simultaneously, $N_1 \geq N_1^*$ and $N_2 \leq N_2^*$. The situation is one of negative feedback.

The conditions for unstable equilibrium, namely $r_2 a_{12} > r_1 a_{22}$ and $r_1 a_{21} > r_2 a_{11}$ are shown in Figure 5.1D. Here we have

$$\left.\begin{array}{l} \dfrac{dN_1}{dt} \leq 0 \\[2ex] \dfrac{dN_2}{dt} \geq 0 \end{array}\right\} \text{(horizontally hatched), where} \begin{cases} N_1 \leq N_1^* \\ N_2 \geq N_2^* \end{cases},$$

and conversely

$$\left.\begin{array}{l} \dfrac{dN_1}{dt} \geq 0 \\[2ex] \dfrac{dN_2}{dt} \leq 0 \end{array}\right\} \text{(vertically hatched), where} \begin{cases} N_1 \geq N_1^* \\ N_2 \leq N_2^* \end{cases}.$$

Consequently, any combined population must reach a state in which species 1 dies out and species 2 is left in sole possession of the resources or vice versa. Positive feedback occurs and therefore the equilibrium state is unstable. Whether species 1 or species 2 will be the winner depends not only on the values of the parameters but also on the sizes of the initial populations of the two species; to determine the outcome in any particular case we must trace the changes occurring in the composition of the combined populations as time passes. A way of doing this is described in Section 3.

Now recall (5.1), namely

$$\frac{1}{N_i} \frac{dN_i}{dt} = r_i - a_{ii} N_i - a_{ij} N_j, \tag{5.1}$$

with $(i, j) = (1, 2)$ in the first equation of the pair and $(2, 1)$ in the second. It is seen that $r_i/a_{ii} = K_i$ is the saturation level that species i would reach in the absence of species j; and also that $a_{ij}/a_{ii} = \alpha_{ij}$, say, can be used to measure the between-species competition experienced by species i as a proportion of the within-species competition it experiences. In terms of these coefficients, (5.1) may be rewritten as follows:

$$\frac{1}{N_i} \frac{dN_i}{dt} = \frac{r_i}{K_i} (K_i - N_i - \alpha_{ij}N_j). \tag{5.2}$$

The condition for stable equilibrium of the species, which in terms of the r's and a's was

$$r_i a_{jj} > r_j a_{ij}$$

now becomes

$$K_i > K_j \alpha_{ij}.$$

Still another way of expressing the stability conditions has been suggested by Vandermeer (1975). Let us put $a_{ij}/a_{jj} = \beta_{ij}$. Thus β_{ij} measures the between-species competition *exerted* by species j on species i, relative to the within-species competition it exerts; (note the contrast between α_{ij} which is a relative measure of the competition experienced by i and β_{ij} which is a relative measure of the competition exerted by j). The conditions for stable equilibrium now become $r_i > r_j \beta_{ij}$.

It is worth while bringing together the three ways of writing the pair of conditions that must be met for two species (governed by this model) to coexist stably. They are

$$r_i a_{jj} > r_j a_{ij} \tag{5.3a}$$

or

$$K_i > K_j \alpha_{ij} \tag{5.3b}$$

or

$$r_i > r_j \beta_{ij} \tag{5.3c}$$

where, in each pair of inequalities, we have, first, $(i, j) = (1, 2)$ and then $(i, j) = (2, 1)$.

As Vandermeer has remarked, we are free to choose which of these three pairs of inequalities we shall use to represent the conditions permitting stable coexistence. If we choose (5.3a) or (5.3c) we can say the outcome of competition does not depend on the saturation levels the populations would reach in the absence of between-species competition, but does depend on the species' intrinsic rates of increase, r_i; the converse description would be appropriate if (5.3b) were chosen as the conditions for coexistence. Clearly the choice is subjective.

Conditions (5.3b) and (5.3c) can be portrayed graphically as shown in Figure 5.2. As in Figure 5.1, the representation of conditions in which

A

B

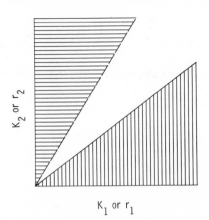

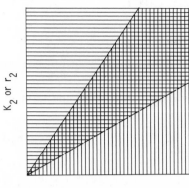

Figure 5.2. In both A and B the vertically hatched region, below the solid line $K_1 = K_2 \alpha_{12}$ (or $r_1 = r_2 \beta_{12}$) shows conditions in which species 1 increases. The horizontally hatched region, above the dashed line $K_2 = K_1 \alpha_{21}$ (or $r_2 = r_1 \beta_{21}$) shows conditions in which species 2 increases. Only if the regions overlap, as in the cross-hatched region in B, can the two species coexist stably. Under the conditions in A, they cannot.

species 1 will grow are vertically hatched and those for species 2 are horizontally hatched. Cross-hatching thus shows conditions permitting the two species to coexist in stable equilibrium.

Gause's equations, (5.1), are nonlinear; it is instructive to examine the behavior of two species-populations governed by this model in a neighborhood of their equilibrium point small enough for the growth rates to be adequately approximated by linear differential equations. To arrive at the approximating equations, we expand the right hand sides of (5.1) as Taylor series. First put

$$\frac{dN_i}{dt} = \dot{N}_i \quad \text{and} \quad N_i(r_i - a_{ii}N_i - a_{ij}N_j) = F_i(N);$$

also, denote the equilibrium point (N_1^*, N_2^*) by $(\mathbf{N}^*)$ and observe that

$$F_i(\mathbf{N}^*) = F_j(\mathbf{N}^*) = 0.$$

Now (5.1) may be written

$$\dot{N}_i = F_i(\mathbf{N})$$
$$\dot{N}_j = F_j(\mathbf{N}). \tag{5.4}$$

The Taylor series expansion around $\mathbf{N}^*$ of, for example, the first of these equations is

$$\dot{N}_i = F_i(\mathbf{N}^*) + (N_i - N_i^*)\frac{\partial \dot{N}_i}{\partial N_i}\bigg|_{\mathbf{N}^*} + (N_j - N_j^*)\frac{\partial \dot{N}_i}{\partial N_j}\bigg|_{\mathbf{N}^*}$$

$$+ \cdots \text{higher terms} \quad (5.5)$$

Now

$$\frac{\partial \dot{N}_i}{\partial N_i} = (r_i - a_{ii}N_i - a_{ij}N_j) - a_{ii}N_i,$$

whence, since at $\mathbf{N}^*$ the term in parentheses is zero,

$$\frac{\partial \dot{N}_i}{\partial N_i}\bigg|_{\mathbf{N}^*} = -a_{ii}N_i^*;$$

also,

$$\frac{\partial \dot{N}_i}{\partial N_j}\bigg|_{\mathbf{N}^*} = -a_{ij}N_i^*.$$

Next, put

$$N_i - N_i^* = n_i \quad \text{and} \quad N_j - N_j^* = n_j,$$

thus measuring population sizes as deviations from their equilibrium values. Also, consider the system's behavior in a neighborhood of the equilibrium point small enough for terms in n_i^2, $n_i n_j$, ... in (5.5) to be neglected. Then (5.4) becomes

$$\dot{n}_i = (-N_i^*)(a_{ii}n_i + a_{ij}n_j), \quad (5.6)$$

with, as always, $(i, j) = (1, 2)$ and then $(i, j) = (2, 1)$. The constant factor $-N_i^*$ may be dropped if the other coefficients are suitably adjusted.

We now have, as desired, the behavior of the two competing populations in the neighborhood of their equilibrium point represented by a pair of *linear* differential equations. One of the conditions for stability is that $a_{11} + a_{22} < 0$. (This is one of the Routh-Hurwitz criteria, to be considered in more detail subsequently; see page 100.) However, this condition, as Keyfitz (1968) has noted, is inconsistent with a_{11} and a_{22} being separately positive which they must be (when N_1 and N_2 are both small) if the species are competing so that each fares better in the absence of the other. It follows that in the neighborhood of $\mathbf{N}^*$ the rate of growth of at least one of the competing species must be negative. Whatever the initial population sizes, one of the species must overshoot its equilibrium value before decreasing towards it. That this does indeed happen is shown graphically in Figure 5.3. Equivalently the system's trajectory cannot reach (N_1^*, N_2^*) from a point in the cross-hatched region of the phase space in Figure 5.1C without first going outside that region.

The same result is reached by Rescigno and Richardson (1967) using an entirely different argument. They also show that equilibrium cannot be reached directly from the unhatched region in Figure 5.1C ; it can be reached only from one of the singly hatched regions.

3.   The Discrete Time Version of Gause's Competition Model

It has already been remarked that the simultaneous differential equations (5.1) cannot, in general, be solved explicitly. To determine the course of events in a two-species population governed by these equations, therefore, it is necessary to find difference equations that permit prediction of the sizes of the two populations at time $t+1$, given their sizes at time t; that is, we require expressions of the form

$$N_i(t+1) = f_i[N_1(t), N_2(t)] \qquad \text{for} \qquad i = 1, 2.$$

Then, given $N_1(0)$ and $N_2(0)$, we can calculate successively $N_1(1)$, $N_2(1)$, $N_1(2)$, $N_2(2)$, and so on. When the conditions are such that there is unstable equilibrium we may by this method discover which of the two competing species will be the winner for given values of $N_1(0)$ and $N_2(0)$.

It was shown on page 23 that simple logistic growth in a one-species population can be described either by the differential equation $dN/dt = N(r-sN)$ or by the difference equation

$$N(t+1) = \frac{\lambda N(t)}{1+\alpha N(t)},$$

with $\lambda = e^r$ and $\alpha = (\lambda-1)/K = s(\lambda-1)/r$; (we now attach no subscript to λ when only one species is involved).

This suggests that the differential equations (5.1) may be replaced by the pair of difference equations

$$N_i(t+1) = \frac{\lambda_i N_i(t)}{1+\alpha_i N_i(t)+\gamma_i N_j(t)}; \tag{5.7}$$

(see Leslie, 1958). As before in these pairs of equations, $(i, j) = (1, 2)$ in the first member of the pair and $(2,1)$ in the second.

We now show that (5.7) is equivalent to (5.1) provided both α_i and γ_i are proportional to $\lambda_i - 1$. Also, we shall find the parameters of the difference equations in terms of those of the differential equations.

Notice first that if species i were unaffected by within- or between-species competition we should have

$$N_i(t+1) = \lambda_i N_i(t); \qquad N_i(t+2) = \lambda_i^2 N_i(t); \ldots; N_i(t+h) = \lambda_i^h N_i(t).$$

Thus λ_i is a finite rate of natural increase. Consequently, it is a function of the time interval concerned. In writing down an expression for $N_i(t+h)$

that accords with (5.7) we must therefore put λ_i^h in place of λ_i wherever λ_i occurs.

Writing $\alpha_i = c(\lambda_i - 1)$ and $\gamma_i = c'(\lambda_i - 1)$ where c and c' are constants of proportionality, it follows that

$$N_i(t+h) = \frac{\lambda_i^h N_i(t)}{1 + c(\lambda_i^h - 1)N_i(t) + c'(\lambda_i^h - 1)N_j(t)}.$$

Then

$$\frac{N_i(t+h) - N_i(t)}{h} = N_i(t)\frac{(\lambda_i^h - 1)}{h}\left[\frac{1 - cN_i(t) - c'N_j(t)}{1 + c(\lambda_i^h - 1)N_i(t) + c'(\lambda_i^h - 1)N_j(t)}\right].$$

When $h \to 0$, the left hand side tends to dN_i/dt.

Also,

$$\lim_{h \to 0}\frac{\lambda_i^h - 1}{h} = \lim_{h \to 0}\frac{\lambda_i^h \ln \lambda_i}{1} = \ln \lambda_i;$$

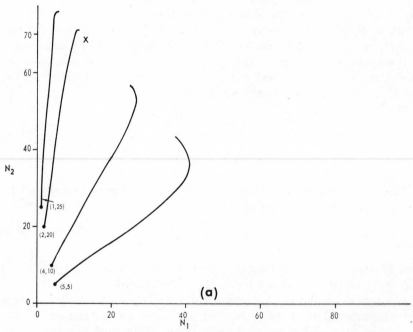

Figure 5.3a. Four curves of the family

$$dN_1/dt = N_1(0.1 - 0.0014N_1 - 0.0012N_2),$$
$$dN_2/dt = N_2(0.08 - 0.0009N_1 - 0.001N_2).$$

The stable equilibrium point is at $N_1 = 12.5$, $N_2 = 68.75$. The sizes of the initial populations $[N_1(0), N_2(0)]$ are shown at the start of each curve. All curves are shown as far as $t = 100$.

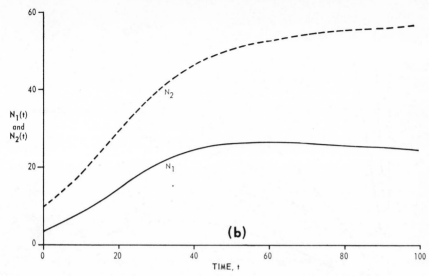

Figure 5.3b. The variation with time of N_1 and N_2, given the starting conditions $N_1(0) = 4$, $N_2(0) = 10$, up to $t = 100$: Ultimately they will level off with $N_1 = 12.5$ and $N_2 = 68.75$.

and as $h \to 0$, the denominator of the expression in square brackets tends to 1. Therefore

$$\frac{dN_i}{dt} = N_i(t) \ln \lambda_i [1 - cN_i(t) - c'N_j(t)].$$

This is equivalent to (5.1) if

(i) $\ln \lambda_i = r_i$,   whence $\lambda_i = \exp[r_i]$,

(ii) $c \ln \lambda_i = a_{ii}$,   whence $\alpha_i = \dfrac{a_{ii}(\lambda_i - 1)}{r_i}$,

(iii) $c' \ln \lambda_i = a_{ij}$,   whence $\gamma_i = \dfrac{a_{ij}(\lambda_i - 1)}{r_i}$.

Thus, if the constants in (5.1) are given, those in (5.7) may be obtained. Then for any preassigned composition of the initial population, say $[N_1(0), N_2(0)]$, the sequence $[N_1(t), N_2(t)]$ for $t = 1, 2, \ldots$, may be calculated and the trajectory of the combined population plotted. The use of different starting points leads to a whole family of such trajectories. Examples are given in Figures 5.3 and 5.4.

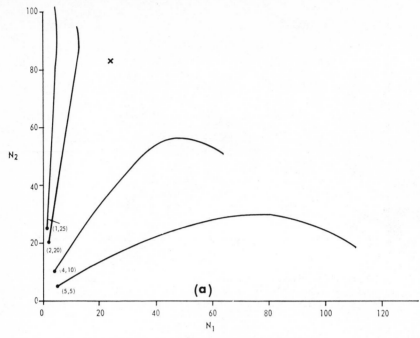

Figure 5.4a. Four curves of the family
$$dN_1/dt = N_1(0.1 - 0.0007N_1 - 0.001N_2),$$
$$dN_2/dt = N_2(0.075 - 0.0007N_1 - 0.0007N_2).$$

The equilibrium point (unstable) is at $N_1 = 23.8$, $N_2 = 83.3$. The sizes of the initial populations are shown at the start of each curve. All curves are shown as far as $t = 100$.

4. Stochastic Simulation of Population Growth with Two Competing Species

In the deterministic case described above it is necessary to know only the parameters of (5.1) and, if they are such that they give unstable equilibrium, the numbers of each species initially, to be able to predict the outcome of competition with certainty. For natural populations in which births and deaths are to some extent matters of chance this is not so. There is always a risk that a species that had been expected to succeed, either alone or in company with its competitor, will disappear because of an unlucky preponderance of deaths over births for a limited period.

It is therefore interesting to generate stochastic simulations of the

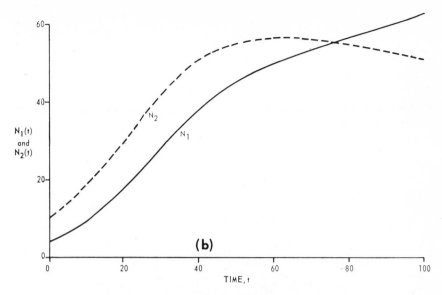

Figure 5.4b. The variation with time of N_1 and N_2 given $N_1(0) = 4$, $N_2(0) = 10$, up to $t = 100$. Ultimately species 2 will become extinct.

competition process from which we can estimate empirically the probability that the theoretically expected result will actually happen in given conditions. Before simulating the process, it is necessary to write (5.1) in more detail, remembering that they represent only a net balance of births and deaths. As in the simulation of one-species logistic growth (see page 28), so also in this case, it is necessary to treat birth rates and death rates separately. Thus, since the equation

$$\frac{dN_i}{dt} = r_i - a_{ii}N_i - a_{ij}N_j$$

represents a net result of births and deaths, we write it more fully in the form

$$\frac{dN_i}{dt} = N_i[\lambda_i(N_i \mid N_j) - \mu_i(N_i \mid N_j)]$$

where $\lambda_i(N_i \mid N_j)$ is the birth rate of species i in the presence of N_i members of its own species and N_j members of the other, and $\mu_i(N_i \mid N_j)$ is the death rate in the same circumstances. Then, if

$$\lambda_i(N_i \mid N_j) = r'_i - a'_{ii}N_i - a'_{ij}N_j,$$

and

$$\mu_i(N_i \mid N_j) = r''_i - a''_{ii}N_i - a''_{ij}N_j,$$

we have

$$\frac{dN_i}{dt} = N_i[(r'_i - r''_i) - (a'_{ii} - a''_{ii}) - (a'_{ij} - a''_{ij})N_i].$$ (5.8)

Equations (5.8) take into account the fact that both the birth rate and the death rate of each species are affected by the number of members of its own species and the number of members of the competing species present at any time.

The possible events in the population and their probabilities are given in Table 5.1, where the constant C is such that the probabilities sum to unity.

TABLE 5.1

EVENT	PROBABILITY
Species 1 birth: $N_1 \to N_1 + 1$; $N_2 \to N_2$	$CN_1\lambda_1(N_1 \mid N_2)$
Species 1 death: $N_1 \to N_1 - 1$; $N_2 \to N_2$	$CN_1\mu_1(N_1 \mid N_2)$
Species 2 birth: $N_1 \to N_1$; $N_2 \to N_2 + 1$	$CN_2\lambda_2(N_2 \mid N_1)$
Species 2 death: $N_1 \to N_1$; $N_2 \to N_2 - 1$	$CN_2\mu_2(N_2 \mid N_1)$

Simulation of the intervals between successive events can also be done if desired by the method described on page 00. Usually, however, all that is needed is a plot of N_2 versus N_1 and the time intervals are of less interest. Barnett (1962) gives a number of examples of computer simulations of the fate of two competing species. Besides plots of N_2 versus N_1, he also states for each example the number of steps (i.e., births and deaths) that occurred before one of the species became extinct.

5. Host-Parasite and Predator-Prey Interactions

We have so far considered only two-species systems in which the two populations behave similarly in that both make demands on the same limiting resource. Now consider two-species systems in which, though the first species would fare better in the absence of the second, the second is dependent on the first if it is to survive. That is, we are dealing with two-species host-parasite systems (or prey-predator systems) in which the parasite (or predator) depends for subsistence on a single species of host (or prey) and cannot turn to an alternative food source.

Because of the contrasted behaviors of the species, their numbers will be denoted by different symbols, H and P. In the host-parasite context these symbols' meanings are obvious. To remember which is which in a prey-predator context, it helps to speak instead of a "herbivore-predator" system and thus match initials with symbols. Numerals rather than letters are used for subscripts, however: 1 for hosts (or herbivores) and 2 for parasites (or predators). For brevity we speak only of hosts and parasites in what follows.

The Lotka-Volterra Model.

One of the earliest of all host-parasite models is that devised by Lotka (1925) and independently by Volterra (see Scudo, 1971). This so-called Lotka-Volterra model is summed up by the following pair of quadratic differential equations.

$$\frac{dH}{dt} = (a_1 - b_1 P)H,$$

$$\frac{dP}{dt} = (-a_2 + b_2 H)P,$$

(5.9)

with a_1, a_2, b_1, $b_2 > 0$. These equations are, of course, wholly deterministic and make no allowance for stochastic fluctuations. The parameter a_1 is the net growth rate per individual of the host species in the absence of the parasite; this rate of increase is diminished by an amount $b_1 P$ when P parasites are present. The parasite population, on the other hand, dwindles to nothing in the absence of hosts, since reproduction is then impossible; parasite increase is directly proportional to the number of hosts present.

To solve this pair of equations, note that

$$\frac{dH}{dP} = \frac{(a_1 - b_1 P)H}{(-a_2 + b_2 H)P}$$

or

$$a_2 \frac{dH}{H} - b_2 \, dH + a_1 \frac{dP}{P} - b_1 \, dP = 0.$$

On integration we then have

$$a_2 \ln H - b_2 H + a_1 \ln P - b_1 P = \text{constant}. \tag{5.10}$$

Equation (5.10) represents a family of closed curves in which each member of the family corresponds to a different value of the constant. Choice of a starting point, that is, of initial values of H and P, determines the constant. Three such curves, all with the same parameter values, are shown in Figure 5.5. Any population will continue indefinitely to follow the trajectory on which it starts, in an anticlockwise direction. There is no

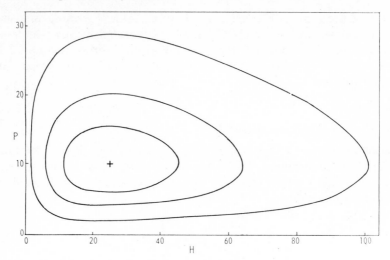

Figure 5.5. Three curves of the family $dH/dt = (a_1 - b_1 P)H$; $dP/dt = (-a_2 + b_2 H)P$, with $a_1 = 1.00$, $b_1 = 0.10$, $a_2 = 0.50$, $b_2 = 0.02$. The equilibrium point is $H = 25$, $P = 10$.

damping toward the equilibrium point at the center of the curves; this is the point $H^* = a_2/b_2$, $P^* = a_1/b_1$ at which $dH/dt = dP/dt = 0$. Graphs of H versus time, or of P versus time, would show endlessly prolonged oscillations of constant amplitude and this amplitude would be determined by the chosen initial population sizes, $H(0)$ and $P(0)$. In a word, the system has *neutral stability*.

To consider the behavior of the system in the neighborhood of the equilibrium point (H^*, P^*) one may replace (5.9) by their Taylor series approximations. Thus, rewriting (5.9) in the form

$$\dot{H} = a_1 H - b_1 PH \qquad \text{and} \qquad \dot{P} = -a_2 P + b_2 PH,$$

we see that

$$\left.\frac{\partial \dot{H}}{\partial H}\right|_{H^*,P^*} = a_1 - b_1 P^* = 0; \qquad \left.\frac{\partial \dot{H}}{\partial P}\right|_{H^*,P^*} = \frac{-b_1 a_2}{b_2};$$

$$\left.\frac{\partial \dot{P}}{\partial H}\right|_{H^*,P^*} = \frac{a_1 b_2}{b_1}; \qquad \left.\frac{\partial \dot{P}}{\partial P}\right|_{H^*,P^*} = -a_2 + b_2 H^* = 0.$$

Then, writing h and p for the deviations of H and P from equilibrium, and keeping only linear terms, we have [cf. the analogous derivation of

(5.6)]

$$\dot{h} = 0 \cdot h - \frac{b_1 a_2}{b_2} \cdot p = x_{11} h + x_{12} p,$$

say, and   (5.11)

$$\dot{p} = \frac{a_1 b_2}{b_1} \cdot h + 0 \cdot p = x_{21} h + x_{22} p,$$

say.

One of the Routh-Hurwitz criteria for stability in linear two-species systems is that $x_{11} + x_{22} < 0$. Conversely, for instability, the criterion is $x_{11} + x_{22} > 0$; and $x_{11} + x_{22} = 0$, as in (5.11), implies neutral stability, that is, endless constant cycles of the trajectory, which neither converges nor diverges. Indeed, solving this pair of equations (see, for example, Chiang, 1954) leads to the conclusion that in the neighborhood of (H^*, P^*) the trajectory becomes elliptical. Thus the salient property of the linear model (5.11) is the assumption it embodies that each species' growth rate depends only on the other species' population size; that is, neither species experiences any self limitation. And the resultant salient character of the model's behavior is its neutral stability. As already remarked, in a system governed by the model, both host and parasite populations would undergo constant oscillations whose amplitudes would bear no relation to the biology of the two species but only to the initial sizes of their populations which could be quite arbitrary.

This "unnatural" behavior of the model probably makes further study of it unprofitable. We therefore turn to a consideration of more realistic models.

Leslie and Gower's Models

Two host-parasite models that result in damped oscillations towards a stable equilibrium level in both populations were proposed by Leslie and Gower (1960). The first of these is summarized in the equations

$$\frac{dH}{dt} = (a_1 - c_1 P)H; \qquad \frac{dP}{dt} = \left(a_2 - c_2 \frac{P}{H}\right)P. \qquad (5.12)$$

The second is

$$\frac{dH}{dt} = (a_1 - b_1 H - c_1 P)H; \qquad \frac{dP}{dt} = \left(a_2 - c_2 \frac{P}{H}\right)P. \qquad (5.13)$$

(All the constants are positive.)

As may be seen, both take account of the likely effect on the parasite's per capita growth rate of the *relative* sizes of the interacting populations. Thus the larger the ratio P/H, the smaller the number of hosts per parasite and, consequently, the less rapid the growth of the parasite population. This is the reason for the term in P/H in the equation for dP/dt in both models.

The models differ in that (5.12) does not, and (5.13) does, allow for density-dependent regulation in the host population. In fact, (5.12) is a limiting form of (5.13), in which within-species competition is assumed to have negligible influence on host population growth because the parasites exert overriding control. Both models have similar outcomes, with phase-space trajectories in the form of spirals converging on an equilibrium point. Hence, each species-population undergoes dwindling oscillations with time towards its equilibrium level.

An example of the phase-space trajectory yielded by (5.12) is shown in Figure 5.6A. The spiral converges on an equilibrium point which is the intersection of the host isocline, $dH/dt = 0$, and the parasite isocline, $dP/dt = 0$. These isoclines are, respectively, the straight lines $P = a_1/c_1$ and $H = c_2 P/a_2$. The sizes of the two populations at equilibrium are thus $P^* = a_1/c_1$ and $H^* = a_1 c_2/a_2 c_1$. At every point where the trajectory crosses the H-isocline, $dH/dt = 0$ by definition and hence the tangent to the trajectory is normal to the H-axis; likewise, at points where the trajectory crosses the P-isocline, tangents to it are normal to the P-axis.

Leslie and Gower (1960) also give their models in the form of difference equations. The pair of difference equations corresponding to (5.12) is

$$H(t+1) = \frac{\lambda_1 H(t)}{1 + \gamma_1 P(t)}; \qquad P(t+1) = \frac{\lambda_2 P(t)}{1 + \gamma_2 P(t)/H(t)} \qquad (5.14)$$

with $\lambda_i = \exp[a_i]$ and $\lambda_i = c_i(\lambda_i - 1)/a_i$ for $i = 1, 2$. The equivalence of (5.14) to (5.12) may be demonstrated in exactly the same way as the equivalence of (5.7) to (5.1); (see page 83).

The Holling-Tanner Model

This model is only slightly more elaborate than the Leslie and Gower model of (5.13). It differs from the latter only in replacing the constant c_1 in (5.13) with a function of H, namely $w/(D+H)$. The model thus becomes†

$$\frac{dH}{dt} = \left(a_1 - b_1 H - \frac{wP}{D+H}\right)H; \qquad \frac{dP}{dt} = \left(a_2 - c_2 \frac{P}{H}\right)P. \qquad (5.15)$$

---

† The symbols in (5.15) differ from those used by May (1973) and Tanner (1975) who discuss the model in detail. The form given here emphasizes the close resemblance of the model to Leslie and Gower's (5.13).

Now consider the first of this pair of equations; it assumes that in the absence of the parasite, the host population would grow logistically but that in its presence the host's growth rate is reduced. However, the reduction in the host's per capita growth rate caused by the parasite is now not merely a constant multiple of P. The coefficient $w/(D+H)$ is arrived at by considering the probable effect on a parasite's (or predator's) attack rate of the density of the population of hosts (or herbivores). Holling (1965) argued that the attack rate of parasites on hosts (or of predators on herbivores), as measured by the number of hosts attacked per parasite per unit of time, say y, often takes the form $y = wH/(D+H)$. This relation allows for the fact that there must be a ceiling, w, to each parasite's attack rate that will not be exceeded however great the host's density becomes; thus when $H \gg D$, $y \simeq w$. The magnitude of the constant D varies directly with the host's ability to evade attack: the more elusive the host, the greater the value of D.

For this model the H-isocline $dH/dt = 0$ is the parabola $P = (D+H)(a_1 - b_1H)/w$; the P-isocline $dP/dt = 0$ is the straight line $H = c_2P/a_2$. Their point of intersection gives the equilibrium point (H^*, P^*). If, in the neighborhood of this point, we substitute the linear differential equations that approximate them there, and then test the local stability of the approximating linear system by the Routh-Hurwitz criteria, we find that the model is locally unstable. [The test is straightforward, though the algebraic manipulations are laborious; details are given by May (1973) and Tanner (1975)]. However the global model, as given by (5.15), is nonlinear and there is no reason to suppose that the instability in the neighborhood of (H^*, P^*) applies everywhere. On the contrary, the model exhibits *stable limit cycles*. That is to say, the system's trajectories in phase-space all converge on a single closed curve whose shape and position depend only on the coefficients in (5.15); the form of the trajectory is, ultimately, independent of its starting point.

An example is shown in Figure 5.6B in which the solid curve is the stable limit cycle; the two trajectories beginning as broken curves show how systems behave whose initial composition is represented by points in phase-space either inside or outside the limit cycle. In the former case the trajectory curves outward, and in the latter inward, to merge with the limit cycle. With respect to time, therefore, the sizes of the two species-populations oscillate perpetually with amplitudes and periods that soon tend to a limit that is independent of their initial sizes and depends only on the parameters of the model (which is to say, the coefficients of the equations). Thus continuous host-parasite (or herbivore-predator) cycles become established that closely resemble the observed behavior of many such two-species systems in nature. For example, Tanner (1975) has

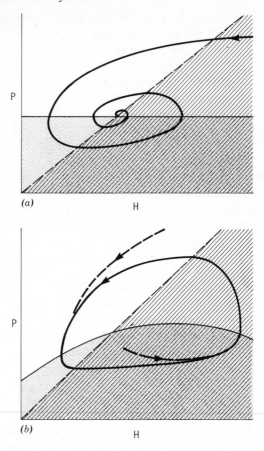

Figure 5.6. (A) Typical trajectory of an H–P system governed by model (5.12). (B) The stable limit cycle of an H–P system governed by the model (5.15). In both phase-space diagrams, the stippled region, below the H-isocline (solid line), is the region in which H increases. The hatched region, below the P-isocline (dashed line), is the region in which P increases.

compared the behavior of the model with the following natural herbivore-predator systems: house sparrows and sparrow hawks in Europe; muskrat and mink in central North America; snowshoe hare and lynx in northern Canada; mule deer and cougar in the Rockies; white-tailed deer and wolf in Ontario; moose and wolf on Isle Royale in Lake Superior; and bighorn sheep and wolf in Alaska.

The Holling-Tanner model is representative of a great many nonlinear models that produce stable limit cycles. Indeed, as May (1973) has argued, a stable equilibrium point or a stable limit cycle is yielded by practically all plausible models that embody the following four features: (1) increasing size in the parasite population reduces both its own, and the host's per capita growth rate; (2) increasing size in the host population reduces its own per capita growth rate, but increases that of the parasite population; (3) for both populations there are minimum sizes at which they have positive growth rates in all circumstances; (4) each population has a maximum size at which its growth rate sinks to zero, this limit being set either by between- or within-species interactions.

A formal statement of these conditions, and a proof that they result in stability (either point stability or a stable limit cycle) was first given by Kolmogorov in 1936. His theorem has recently been reviewed by, for example, Rescigno and Richardson (1967), Scudo (1971), and May (1973). The fact that so many models conforming to these conditions can be constructed, all of them intrinsically credible on ecological grounds, makes the problem of discriminating among them acute. We return to this point in Section 3 of Chapter 6.

6.   Harvested Populations

We now consider a two-species interaction in which one species depletes another at a constant rate; the rate is independent of the density of either of the two species. This description summarizes the simplest of all interactions between a human population and a resource, for instance between a population of fishermen and a fish stock that they harvest. The model assumes that the harvesting rate is kept constant and that the harvested population grows logistically. In spite of its unnaturalness the model is worth describing as the simplest possible example of a resource-exploitation model; it is the only one of its kind treated in this book.

Let the growth rate, $\dot{N} = dN/dt$, of the harvested population obey the logistic formula (2.1), that is, let $\dot{N} = N(r - sN)$. The relation between $\dot{N}$ and N is therefore parabolic as shown in Figure 5.7. It is easy to see that $\dot{N}$ has a maximum when $N = r/2s$, at which point its value is $\dot{N} = r^2/4s$.

Now assume that this population is harvested at a constant rate E. The effect is to reduce population growth by the constant amount E at all values of N. Thus to describe population behavior for different values of N, we must now consider the function $\dot{N} - E = y$, say; graphically this is equivalent to shifting the N-axis (the abscissa) from $\dot{N} = 0$ to $\dot{N} = E$ (see Figure 5.7). Denote the points of intersection of the parabola $\dot{N} = N(r - sN)$ with this new axis by the symbols N_* and N^* ($N_* \leq N^*$). Obviously, if $E > r^2/4s = E_c$, say, then $y < 0$ for all N. That is, if the

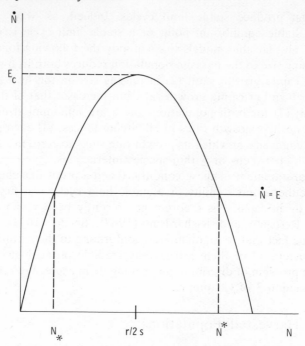

Figure 5.7. The parabola $\dot{N} \doteq N(r - sN)$ shows population growth rate versus population size for a logistically growing population. The population is harvested at a constant rate E. If $N < N_*$ ever, the population dies out. Otherwise, it stabilizes at $N = N^*$.

harvesting rate exceeds a critical level E_c, which is the population's maximum attainable growth rate, then in no circumstances can the harvested population grow and its extinction is inevitable.

Now consider a harvesting rate such that $0 < E < E_c$. It is seen that

$$y \le 0 \qquad \text{when } N \le N_* \text{ and when } N \ge N^*,$$

and

$$y > 0 \qquad \text{when } N_* < N < N^*.$$

Assuming no change whatever in E and in the functional relation between $\dot{N}$ and N, and making no allowance at all for chance (that is, assuming a situation of idealized simplicity), we can now predict what will happen to the harvested population for given values of its initial size, N_0, and chosen values of E (hence of N_* and N^*).

Clearly, if $N_0 < N_*$, the population will dwindle from the moment harvesting begins and must go extinct. If $N_* < N_0 < N^*$, the population will grow logistically to its saturation level under harvesting, N^*, and will remain at this level permanently; N^* is less than r/s, the population's saturation level when no harvest is taken. If $N_0 > N^*$, the population will decrease to begin with to the level N^* and will then remain there. To find N_* and N^* in terms of E and the population's logistic parameters r and s, observe that they are the roots of $y = 0$. Thus

$$N_* = \frac{r}{2s}(1 - \sqrt{1 - E/E_c}) \qquad \text{and} \qquad N^* = \frac{r}{2s}(1 + \sqrt{1 - E/E_c}).$$

Brauer and Sanchez (1975) have discussed the usefulness of this very simple model in predicting the effect of hunting on sandhill crane populations and find that, in spite of its simplicity, the model's predictions do not differ greatly from those of a more elaborate model. They also enlarge the simple model by taking account of the effect of time delays in population responses, and by considering what happens when the harvested population has a competing species as well as an exploiting species to contend with; here we treat constant-rate harvesting as "exploitation" to distinguish it from "predation" whose rate is always assumed to be governed by a feedback mechanism. Indeed, exploitation and predation may be regarded as the limiting extremes in the action of one species population in depressing the growth of another.

7.   Refinements in Two-Species Models

The study of population interactions is in a state of explosive growth; rapid, independent advances are being made in many different directions and these have yet to be fitted into their places within a single embracing theory. The following numbered paragraphs constitute a necessarily incomplete list of some of the interesting lines of research being followed. They are in random order with no attempt at ranking.

1. Although the old, simple models describing species interactions assumed a uniform environment, realistic models must obviously make allowances for spatial heterogeneity. Work on these lines has been done by, among others, Horn and MacArthur (1972), Levin (1974), May (1974), and Pielou (1974a, b).

2. Herbivore-predator models which take account of the age distributions of the interacting populations have been discussed by Smith and Mead (1974).

3. When a carnivore population, and the herbivore population that

supplies the bulk of its diet, both vary cyclically with the same period, it does not automatically follow that their oscillations are due to their interactions. The well-known ten-year cycle in lynx and snowshoe hare populations in the boreal forests of Canada is perhaps controlled by interaction between the hares and the vegetation, with the lynx population (which depends on the hares for food) exhibiting *driven oscillations*. Bulmer (1975) has investigated the phase relations between such cycles, in which one drives the other.

4. No theory of species interactions is convincing that does not allow for delays in the responses of each species to the other. Nearly all model builders now give due weight to this complication.

5. Host-parasite and herbivore-predator models which permit variability in the success of parasites and predators in their search for prey have been studied by many investigators, among them Macdonald and Cheng (1970) and Murdie and Hassell (1973).

6. Leòn and Tumpson (1975), in studies of competition, have emphasized the distinction between two kinds of resources. There are those that must be simultaneously present if either species is to survive (for example, fish must have oxygen and food; neither can support life without the other); other resources, however, are interchangeable since if one fails another can be substituted (for example, an owl can replace one species of rodent by another in its diet).

7. Gallopin (1971) has discussed nonautonomous models, that is, those with extrinsically controlled inputs.

8. Nonlinear models of species-interactions can exhibit local stability in the neighborhoods of a number of equilibrium points, and instability elsewhere; given a complicated model, there is no reason to assume that if its has an equilibrium state this state must be unique. Holling (1973) has discussed systems having several equilibrium states, and empirical evidence of their occurrence in nature.

9. Although this chapter has concentrated on only two kinds of two-species interaction, competition and the host-parasite (or herbivore-predator) relationship, other relationships are possible, namely: (a) Symbiosis, in which the presence of each species benefits the other. (b) Commensalism, in which the first species benefits from the second but the second is unaffected by the first. Williamson (1972) argues that many so-called predator-prey relationships could more accurately be described as commensalism since often predation does not reduce the growth rate of the prey population treated as a whole, but only of·some of its individuals members. (c) Amensalism, in which the first species is inhibited by the second but the second is unaffected by the first. Thus, what might be described as "one-sided competition," in which only one

species of "competitor" runs any risk of loss from an encounter with the other, is more properly described as amensalism than as competition (Williamson, 1972).

Formal study of these three processes and their consequences merely entails making appropriate adjustments to the signs of the interaction coefficients a_{ij} in (5.1).

6

The Dynamics of Interacting Species:
II. k-Species Interactions

1. The Routh-Hurwitz Criteria

In Chapter 5 we were concerned with interactions between only two species and this made it possible to consider competition systems and host-parasite systems separately. As soon as the number of species in a system exceeds two, of course, it is quite likely that both kinds of interactions are proceeding at the same time. Each species in a k-species system may compete with other species on its own trophic level; it may prey upon one or more species at one or more lower levels; and it may be preyed upon by one or more species at one or more higher levels. We therefore require a method for analyzing the behavior of large systems in which many pairwise interactions are going on simultaneously.

The simplest model that can be envisaged to account for k-species interactions assumes that at any instant each species has a population growth rate that is a linear function of the sizes of all k populations. Writing $dN_i/dt = \dot{N}_i$, we thus have

$$\dot{N}_i = \sum_{j=1}^{k} a_{ij}N_j, \qquad (i = 1, \ldots, k). \tag{6.1}$$

So simple a relationship is unlikely to hold except in the neighborhood of the equilibrium point (if there is one) $\mathbf{N}^* = (N_1^*, \ldots, N_k^*)$. In that neighborhood, however, it may form an adequate approximation to various more realistic models. In particular, it is an approximation to the k-species analog of Gause's competition equations (1.5) which we here† rewrite as

$$\dot{N}_i = N_i\left(r_i - \sum_{j=1}^{k} a'_{ij}N_j\right) \tag{6.2}$$

† Both (6.1) and (6.2) are commonly called Lotka-Volterra equations. Care should be taken to avoid confusion between different meanings of the overworked symbol a. To conform with custom I have used a's suitably subscripted, in both equations; but they are not the same, and the coefficients in (6.2) have been distinguished by attaching primes to them. Note also that it is *not* assumed that $a_{ii} = 1$ or $a'_{ii} = 1$.

100

Notice that in (6.1) we assume that the species' *population* growth rates are linear functions of the N_i values; whereas in (6.2) we assume that it is the *per capita* growth rates that are linear functions of the N_i's and hence that the population growth rates are quadratic functions of them.

Before exploring the properties of (6.1), it is worth showing how (6.1) is derived from (6.2) by first approximating (6.2) by a Taylor series expansion around the equilibrium point and then dropping terms that, in the neighborhood of the equilibrium, become negligibly small. (This derivation is the extension to k species of the derivation shown for two species on page 82).

Rewriting (6.2) formally as

$$\dot{N}_i = F_i(\mathbf{N}),$$

the Taylor expansion of the right-hand side around $\mathbf{n} = \mathbf{N} - \mathbf{N}^*$ is

$$\dot{N}_i = F_i(\mathbf{N}^*) + \sum_{j=1}^{k} n_j \frac{\partial \dot{N}_i}{\partial N_j}\bigg|_{\mathbf{N}^*} + \text{terms in } n_j^2, n_i n_j, \text{ etc.}$$

Now

$$\frac{\partial \dot{N}_i}{\partial N_j}\bigg|_{\mathbf{N}^*} = -a'_{ij} N_i^* \qquad \text{for} \qquad j = 1, \ldots, k;$$

and $F_i(\mathbf{N}^*) = 0$ since, by definition, $\mathbf{N}^*$ is an equilibrium point at which all species exhibit zero growth. Therefore, assuming the n_i are sufficiently small for their powers and products to be neglected, we have

$$\dot{N}_i = (-N_i^*) \sum_{j=1}^{k} a'_{ij} n_j \qquad \text{for} \qquad i = 1, \ldots, k; \qquad (6.3)$$

or, combining all k equations into a single matrix equation, and putting $\dot{\mathbf{n}}$ in place of $\dot{\mathbf{N}}$,

$$\dot{\mathbf{n}} = \mathbf{D}\mathbf{A}'\mathbf{n} \qquad (6.4)$$

where $\mathbf{D}$ is the diagonal matrix $\mathbf{D} = \text{diag}(-N_i^*)$ and $\mathbf{A}'$ is the $k \times k$ matrix $\mathbf{A}' = \{a'_{ij}\}$. [Observe that $\mathbf{D}\mathbf{A}$ is the Jacobian of (6.2)].

Now absorb the negative constants in (6.3) by putting $-N'_{ij} a'_{ij} = a_{ij}$. Then (6.4) becomes

$$\dot{\mathbf{n}} = \mathbf{A}\mathbf{n} \qquad (6.5)$$

where $\mathbf{A} = \{a_{ij}\}$.

It will be seen that if population sizes are measured as deviations from their equilibrium values, that is, if we put n_i in place of N_i in (6.1), then (6.1) and (6.5) are identical.

Recall now that (6.5) [or (6.1)] constitutes a simple model of the behavior of a k-species system in the neighborhood of its equilibrium state regardless of whether we regard it as an approximation to the model of (6.2), or as an approximation to some other, perhaps more plausible, model. Thus the properties of (6.5) deserve exploring even if we regard (6.2) as too simple to be ecologically useful. The matrix $\mathbf{A}$ has been aptly called the *community matrix*, a term coined by Levins (1968).

The fate of a k-species system whose behavior in the neighborhood of its equilibrium is governed by (6.5) depends on the values of the coefficients a_{ij} or, equivalently, on the elements of the community matrix. The solutions of the equations (6.5) are, of course, functions of time, since $\dot{\mathbf{n}}$ denotes rates with respect to time; the solution vectors may therefore be written $\mathbf{n}(t) = [n_1(t), \ldots, n_k(t)]$. If all k species are to coexist in stable equilibrium, with population sizes given by the vector $\lim_{t \to \infty} \mathbf{N}(t) = \mathbf{N}^*(t) = [N_1^*(t), \ldots, N_k^*(t)]$, we therefore require that $\mathbf{n}^*(t)$, the vector of deviations from this equilibrium, should tend to $\mathbf{0}$ as $t \to \infty$. This will indeed be the case if and only if all the latent roots of the matrix $\mathbf{A}$ have negative real parts. A matrix having this property is defined as a *stable matrix*, and we now require a method for recognizing a stable matrix, in other words a method of judging whether a given matrix is stable or not.

This may be done by means of the *Routh-Hurwitz criteria*. To apply these, we first expand the characteristic equation of the matrix, namely $\det(\mathbf{A} - \lambda \mathbf{I}) = 0$ as a polynomial equation in λ, say†

$$f(\lambda) = \lambda^k + c_1 \lambda^{k-1} + c_2 \lambda^{k-2} + \cdots + c_k = 0 \tag{6.6}$$

We next write down the so-called *Hurwitz matrices* of which there are k. The jth of them, $\mathbf{H}_j$, is defined as

$$\mathbf{H}_j = \begin{pmatrix} c_1 & 1 & 0 & 0 & \cdots & 0 \\ c_3 & c_2 & c_1 & 1 & \cdots & 0 \\ c_5 & c_4 & c_3 & c_2 & \cdots & 0 \\ \cdot & \cdot & \cdot & \cdot & \cdots & \cdot \\ c_{2j-1} & c_{2j-2} & c_{2j-3} & c_{2j-4} & & c_j \end{pmatrix}$$

The general term in this matrix, say the (p, q)th, is seen to be c_{2p-q}. If

$$2p - q = 0, \qquad c_{2p-q} = c_0 = 1;$$

also, if

$$2p - q < 0 \quad \text{or} \quad 2p - q > k, \quad \text{then} \quad c_{2p-q} = 0.$$

It can be proved that all the zeroes of $f(\lambda)$ (equivalently, all the latent roots of $\mathbf{A}$) have negative real parts if and only if the determinants of all k

---

† The coefficients c_i in (6.6) are often written as a_i. One then has a's with two subscripts as elements of $\mathbf{A}$; and a's with one subscript as the coefficients of its characteristic equation. To avoid possible confusion I have therefore raided the alphabet for another letter. See also the footnote on page 100.

Hurwitz matrices associated with $\mathbf{A}$ are positive. Or, more concisely, $\mathbf{A}$ is stable if and only if

$$\det \mathbf{H}_j > 0 \qquad \text{for} \qquad j = 1, \ldots, k.$$

A proof of this theorem (the Routh-Hurwitz theorem) will be found in, for example, Lancaster (1969). To illustrate its use in judging the stability of a k-species system, we consider only $k = 2$ and $k = 3$. This will suffice to show that application of the theorem, apart from being manipulatively laborious for large k, is straightforward. Thus suppose $k = 2$ so that

$$\mathbf{A} = \begin{pmatrix} a_{11} & a_{12} \\ a_{21} & a_{22} \end{pmatrix}.$$

The characteristic polynomial, $f(\lambda)$, of $\mathbf{A}$ is then

$$f(\lambda) = \det(\mathbf{A} - \lambda \mathbf{I}) = \lambda^2 - (a_{11} + a_{22})\lambda + (a_{11}a_{22} - a_{12}a_{21}),$$

whence

$$c_1 = -(a_{11} + a_{22}); \qquad c_2 = a_{11}a_{22} - a_{12}a_{21}.$$

When $k = 2$ the Hurwitz matrices are

$$\mathbf{H}_1 = (c_1) \qquad \text{and} \qquad \mathbf{H}_2 = \begin{pmatrix} c_1 & 1 \\ 0 & c_2 \end{pmatrix}.$$

The conditions for stability are therefore

$$\det \mathbf{H}_1 = c_1 = -(a_{11} + a_{22}) > 0 \qquad \text{or} \qquad a_{11} + a_{22} < 0;$$

and

$$\det \mathbf{H}_2 = c_1 c_2 = -(a_{11} + a_{22})(a_{11}a_{22} - a_{12}a_{21}) > 0.$$

Assuming the first condition is met, the second requires that we should have as well $a_{11}a_{22} > a_{12}a_{21}$.

Next, let $k = 3$ and write $f(\lambda) = \lambda^3 + c_1\lambda^2 + c_2\lambda + c_3$. (We shall not here go to the labor of writing out the coefficients c_i in terms of the elements of $\mathbf{A}$; as always they are found by expanding $\det(\mathbf{A} - \lambda \mathbf{I})$ as a power series in λ.) For $k = 3$ the Hurwitz matrices are

$$\mathbf{H}_1 = (c_1), \qquad \mathbf{H}_2 = \begin{pmatrix} c_1 & 1 \\ c_3 & c_2 \end{pmatrix}, \qquad \text{and} \qquad \mathbf{H}_3 = \begin{pmatrix} c_1 & 1 & 0 \\ c_3 & c_2 & c_1 \\ 0 & 0 & c_3 \end{pmatrix}.$$

The conditions for stability are therefore

$$\det \mathbf{H}_1 = c_1 > 0,$$

$$\det \mathbf{H}_2 = c_1 c_2 - c_3 > 0,$$

and

$$\det \mathbf{H}_3 = c_3(c_1 c_2 - c_3) > 0.$$

Assuming that the second of these conditions is met, the third obviously simplifies to $c_3 > 0$.

May (1973) gives the conditions for stability for values of k up to 5. Strobeck (1973) discusses the use of the Routh-Hurwitz criteria in judging the behavior of systems with k competing species and points out that for stable coexistence in such a system it is necessary that the elements of the community matrix satisfy $2(k-1)$ inequalities; [the number of conditions exceeds k in this case since Strobeck considers stability in the quadratic model (6.1) rather than the linear model (6.5)]. Volterra (1931) observed that if $\mathbf{A}$ is antisymmetric (that is, if $a_{ij} = -a_{ji}$ for all i, j) then the model (6.5) exhibits neutral stability: All the n_i values oscillate perpetually with constant amplitudes determined by the initial conditions.

Rescigno (1968) examined the way in which the populations behave close to their possible equilibrium points in systems of three competing species. He demonstrated that in certain circumstances the sizes of all three populations can oscillate even though there is no time lag in the response of each species' growth rate to changes in the system. This contrasts with two-species models in which oscillations do not occur unless delays are introduced.

2. Qualitative Stability in k-Species Systems

Using the Routh-Hurwitz criteria to decide whether a group of k interacting species can coexist stably requires, of course, that numerical values for all the k^2 elements of the community matrix be known. In ecological contexts this stringent requirement is rarely, if ever, met. Often, however, we do know (or are prepared to guess) which of the elements (the interaction coefficients) are positive, which are negative, and which are zero. Instead of a community matrix with numerical elements we then have a *"qualitative"* matrix in which each element is merely $+$, $-$, or 0.

Even with this very modest knowledge some assertions about the behavior of the system are possible. Conditions for stability have been derived (Quirk and Ruppert, 1965; Jeffries, 1974) and a system that meets the conditions is called *qualitatively stable*. Fulfilment of the conditions is, in fact, more than sufficient for stability since they ensure stability whatever the magnitudes of the a_{ij}; that is, community matrices that do not meet the conditions can sometimes be stable provided the values of the a_{ij} are suitably restricted.

Any of the following conditions is sufficient for qualitative *in*stability: (We give them here without proof; proofs are given by Quirk and Ruppert, 1965).

1. det **A** = 0.
2. $a_{ii} \geq 0$ for all i.
3. $a_{ij}a_{ji} > 0$ for at least one (i, j); $(i \neq j)$.
4. There exists a sequence of three or more different subscripts, say u, v, w, . . . , y, z, such that the cyclical product $a_{uv}a_{vw} \cdots a_{yz}a_{zu} \neq 0$.

Consider the implications of these conditions. Since we know nothing about the magnitudes of the a_{ij}, condition [1] implies that at least one row of the qualitative matrix consist entirely of zeroes. This amounts to asserting that at least one species is neither self-regulating nor regulated by another species. Obviously a system containing such a species cannot be stable.

Condition [2] implies that none of the species is self-regulating.

Condition [3] implies that stability is impossible if for two species, say i and j, their interaction coefficients are of the same sign. If the species are symbiotic (that is, if both a_{ij} and a_{ji} exceed 0) this result is not surprising. It is, perhaps, surprising that competition between the two species (a_{ij} and a_{ji} both less than 0) is also incompatible with stability; but it should be recalled (see page 82) that this result applies only in the neighborhood of equilibrium and affects only the direction from which the trajectory can reach equilibrium.

Condition [4] is the most interesting. It is easily tested graphically by substituting a *signed digraph* for the qualitative matrix. A signed digraph is constructed as follows (see, Jeffries, 1974). Draw k dots and label them $1, 2, . . . , k$; (any arrangement suffices). Draw a line joining dot i and dot j, with an arrowhead pointing towards i, if $a_{ij} \neq 0$, to imply that species i's growth is affected by species j. Then, according as $a_{ij} > 0$ or < 0, write a + or − beside the arrowhead. Do this for all ordered pairs (i, j) for which $a_{ij} \neq 0$; when $a_{ij} = 0$ no link is drawn. Note that a separate directed and signed line, with an arrowhead pointing to j, is required if $a_{ji} \neq 0$. If $a_{ii} \neq 0$, a link circling from dot i directly back to itself is drawn. Figure 6.1 shows an example of a 5×5 qualitative matrix and its associated signed digraph. (Obviously it is easy, also, to construct the matrix given the digraph.)

Now define as a p-cycle any closed circuit connecting p distinct dots that may be followed in the digraph along links all directed the same way. Thus in Figure 6.1 there is a 4-cycle; it links dots 1–3–5–4–1; (it is convenient to list the dots in order travelling against the arrows as this automatically arranges the subscripts in the desired order when the cycle

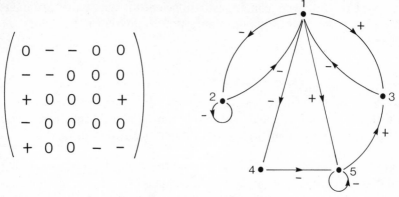

$$\begin{pmatrix} 0 & - & - & 0 & 0 \\ - & - & 0 & 0 & 0 \\ + & 0 & 0 & 0 & + \\ - & 0 & 0 & 0 & 0 \\ + & 0 & 0 & - & - \end{pmatrix}$$

Figure 6.1. A qualitative matrix and its signed digraph. (See text for discussion.)

is represented by symbols). But dots 1, 4, and 5, though all joined by links, do not form a 3-cycle since the links are not all in the same direction. Condition [4] for instability may now be stated, referring to the digraph, as follows. The existence in the digraph of any p-cycle with $p > 2$ implies instability of the system. Thus in Figure 6.1 the existence of the cycle 1–3–5–4–1 corresponds to the fact that

$$a_{13}a_{35}a_{54}a_{41} = (-) \times (+) \times (-) \times (-) \neq 0,$$

since none of the product's factors are 0. This accords with condition [4] for instability and we conclude that the system portrayed by both the qualitative matrix and the digraph in Figure 6.1 is unstable.

Notice that the existence of cycles is determined only by the directions, not the signs, of the links in the digraph; for testing condition [4], an unsigned digraph is therefore as useful as a signed one. However, attaching signs to the links allows the digraph to contain all the information in the qualitative matrix. In the example in Figure 6.1, for instance, the fact that $a_{13} < 0$ and $a_{31} > 0$ implies that $a_{13}a_{31} < 0$ (equivalently, that species 3 preys upon species 1) and condition [3] for instability is not met. It is met by the two competing species, 1 and 2, however; the signed digraph shows that $a_{12} < 0$ and $a_{21} < 0$ whence $a_{12}a_{21} > 0$ and this, by condition [3], implies instability.

For further discussions of qualitative stability, see Levins (1974) and DeAngelis, Goldstein, and O'Neill (1975).

Another method of investigating, graphically, the qualitative behavior of k-species systems has been suggested by Rapport and Turner (1975). It consists in examining the behavior of those species whose fate is of especial interest one at a time. For each such species, its *yield function* (representing its behavior if predators upon it were absent) and its *harvest function* (representing the mortality due to predation when predators are present) are modeled separately. Thus the co-occurring species are partitioned into two groups: those at or below the trophic level of the species of interest, and those at a higher trophic level. This may be a useful way of investigating the welfare of economically valuable species.

3.   A Critique of Ecological Model Building

It is now time to take a critical look at ecological models of the kinds discussed in Chapters 5 and 6. There are three chief motives for investigating ecological models: first, to explore the consequences of various sets of more-or-less plausible assumptions about the growth rates of co-occurring and interacting living populations. Second, to deduce what processes and interactions are, in theory, compatible with the observed behaviors of particular systems in nature; (and hence, also, what are incompatible). Third, to predict what will happen if a natural community is perturbed in various ways. We must now ask whether any of these goals can be achieved by the study of models.

All ecologists are uncomfortably aware of the numerous simplifying assumptions that underlie most models. It is worth drawing up a list (certainly not exhaustive) of the assumptions that are most frequently made, and that put the greatest strain on one's credulity. They are:

1. That the system under investigation occupies a spatially homogeneous environment, and also that conditions are temporally constant.

2. That the system is closed, so that the interacting populations are not reinforced by immigration or depleted by emigration.

3. That each population responds to changes in its own and the others' sizes instantly, without delay.

4. That variations in the age structures of the populations do not occur, or can be disregarded.

5. That (in community matrix models) the interaction coefficient between each pair of species is unaffected by changes in the species-composition of the remainder of the community; (see Neill, 1974).

6. That the genetic properties, and hence the competitive abilities, of a population are independent of its size; in other words, that a_{ij} is independent of N_i and N_j.

7. That stochastic effects can be neglected. Although the mean prediction of a model that allows for stochastic events is usually identical with the prediction of the corresponding deterministic model, the variance

attaching to this prediction is determined by the way stochasticity itself is modeled; (see Sykes, 1969b).

To repeat, the foregoing is a list of *some* of the simplifying assumptions that ecological models often make. Admittedly, many models do allow for at least a few of these complicating factors though the decision as to which simple assumptions are "too simple" and which are not is usually a matter of guesswork. Obviously, no model can allow for all conceivable complications; if it did, it would (by definition) cease to be a model. Moreover attempts to improve the realism of a model by relaxing those of its assumptions that are thought to be too simple involves the substitution for them of a much larger number of other assumptions; and any assumption in this context is merely a guess, perhaps a wild one. If the only alternative to an overly simple model is an elaborate framework of guesses, nothing much has been gained.

The investigation of mathematical models often seems to be motivated more by an interest in the mathematics of the model, and a striving for mathematical elegance, than by an interest in the model's ecological implications. A study of, say, the stability properties of a set of simultaneous differential equations, even if thinly disguised (and rendered picturesque) with some preliminary remarks about wolves and moose, or lynx and hares, is, notwithstanding, an item of mathematical research and should be judged as such, by mathematicians rather than biologists. Only if it subsequently turns out to have genuine ecological applicability should it be included in the subject matter of ecology.

Ecological models are easy to devise; even though the assumptions of which they are constructed may be hard to justify, the magic phrase "let us assume that..." overrides objections temporarily. One is then confronted with a much harder task: How is a model to be tested? The correspondence between a model's predictions and observed events is sometimes gratifyingly close but this cannot be taken to imply that the model's simplifying assumptions are reasonable in the sense that neglected complications are indeed negligible in their effects. Obviously, to test a model adequately requires that populations of plants and animals in their natural wild state be adequately sampled. This is often difficult and the attainable precision may be too low to permit discrimination between competing models (cf. Pielou, 1974c).

Discrimination between models is also made difficult by stochastic events in the populations themselves. For instance, two of the herbivore-predator models described in Chapter 5 (the Leslie-Gower model on page 91; and the Holling-Tanner model on page 92) are strikingly different as deterministic models. The first yields damped oscillations, with the sizes of the two populations tending towards steady equilibrium values;

the second yields a stable limit cycle. In nature, however, distinguishing between them might be impossible. In the first model, stochastic "jolts" would tend, every now and again, to shift the system in a way that increased its distance from equilibrium, and the damping process would then have to begin anew from this point; (a possible example has been given by Utida, 1957). In the second model, stochasticity would again spoil the perfect regularity of the deterministic model; in this case, stochastic jolts could on some occasions cause a temporary decrease, and on other occasions a temporary increase, in the amplitudes of the two species' oscillations, but the visible effect would be be very similar: unpredictable fluctuations in the magnitudes of the successive maxima and minima.

The tendency for theoreticians to examine a model's behavior in the neighborhood of an equilibrium state, or its responses to *small* perturbations around equilibrium, also raises doubts in the minds of experienced field ecologists. Large perturbations are probably common. And the tendency of environments to fluctuate, with fluctuations proceeding at several different rates simultaneously, ensures that theoretical equilibrium states are themselves nonstationary.

Some of the misconceptions that can arise from unwise "modeling" are suggested in the preceding paragraphs. They are dangerous, and the word is not meant only in the figurative sense (so common in scientific writing), when the only danger an ecologist faces is that of appearing ridiculous because an assertion he makes turns out to be spectacularly wrong. Now that the human population is so big, and its power to modify the environment (both deliberately and inadvertently) is so great that threats of ecological disaster are always present, it is urgently necessary that ecologists act responsibly. It is particularly important that they do not put unjustified faith in whatever are the newest, most fascinating, theoretical models in the stream of models being produced. Most theoretical models are *structurally stable*; that is, small changes in their parameter values cause only small changes in their outputs. If an apparent match between a structurally stable theoretical model and a real ecosystem led to the erroneous belief that the real system must be structurally stable also, action based on the belief could obviously be disastrous. Theoretical models are a part of theoretical ecology. The danger (I believe) is that they will be transferred uncritically to applied ecology.

Now let us consider the usefulness of models in theoretical ecology. In my opinion, their usefulness is great and it consists *not in answering questions but in raising them*. Models can be used to inspire field investigations and these are the only source of new knowledge as opposed to new speculation. The point is best illustrated by a few examples.

1. Investigations of qualitative stability have shown (see page 105) that a community matrix with too many nonzero elements is unstable. Perhaps, correspondingly, pairwise species interactions of appreciable magnitude are rarer in stable natural systems than has been thought.

2. The possibility that a many-species system can have a number of equilibrium points, or a number of possible limit cycles, has been shown theoretically. Natural examples are worth seeking; (See Holling, 1973; Sutherland, 1974).

3. It has also been shown theoretically (May, 1971) that in systems with several trophic levels, overall stability can be compatible with instability at one of the levels; and conversely, that one level can be stable even when the total system is not. But it is unknown how common in nature such mismatches are between the stabilities of a whole system and its parts.

4. The natural occurrence of systems isomorphic in trophic structure but with different species compositions appears to have been demonstrated by Heatwole and Levins (1972), and the implications deserve to be followed up.

This list of model-inspired speculations could be extended indefinitely. And, of course, the knowledge gained by model-inspired field studies leads to adjustments to the models and a further round of field studies. Meanwhile, applied ecologists must of necessity give the best advice they can, on a basis of incomplete information, to ensure that decisions on environmental action are as wise as is currently possible. Whenever such action is likely to produce effects on a wide scale, blind faith in insufficiently tested models must at all costs be avoided. Probably the safest course is to insist that actions whose effects are uncertain be executed in small steps, and that monitoring be continuous and vigilant.

II
Spatial Patterns in
One-Species Populations

7

Spatial Patterns and their Representation by Discrete Distributions

1. Introduction

Part II of this book must start with the acknowledgment that it brings an abrupt change in the styles of the investigations it describes and in the mathematical arguments used in them. This may seem unfortunate, and perhaps it is. In any case, it is unavoidable, in view of the way ecology has developed in the past half-century. Questions as to whether the history of the subject should have been, or could have been, different are not worth discussing. It is worthwhile, however, to bring together for comparison the tersest possible statements of the aims and methods of the two strongly contrasted halves of modern theoretical ecology; or rather, the contrast in their methods since their aims are presumably the same, namely the advancement of scientific knowledge of the biosphere, the living world treated at the highest organizational level. The two halves can be called "population dynamics" (which was the subject of Part I of this book) and "the spatial patterns and interrelationships of populations" (the subjects of Parts II, III, and IV).

In the study of population dynamics, as we have seen, it has been customary to envisage very concrete models that could account for observed temporal changes in the sizes of self-regulating and interacting populations and then to deduce the mathematical consequences of these models. In the formal exposition of any piece of research on the subject one finds, near the beginning, a statement which might be paraphrased thus: "We shall assume that [a list of assumptions here follows]." The argument then proceeds inexorably (provided it contains no errors) to a foregone, indeed tautological, conclusion; and if a reader does not like the assumptions, that is his own concern. Provided studies are limited to small, closed, genetically and phenotypically uniform populations of small, motile, short-lived organisms, all is well. But as soon as the populations studied do not have these properties, that is, as soon as we interest ourselves in the "messy" communities that must, because of their

113

ubiquity and practical importance, engage the attention of the majority of ecologists, the gap between pencil-and-paper processes and their counterparts in real life widens; the number of ecologists who regard a given model as relevant to their fields of interest quickly shrinks; and the connection between mathematical theory and ecology itself (which, to repeat, is the study of the biosphere) soon peters out altogether. Thus arguments of the kind discussed in Part I have, so far, only a very limited applicability to most of ecology. A breakthrough is needed; some hitherto unthought of way to proceed must be devised if the established, deductive methods of population dynamics are to take us much farther forward than they have already.

The topic of Parts II, III, and IV, or the study of the spatial patterns and spatial interrelations of living populations in their natural environments, has had a history very different from that of population dynamics. Customarily, work begins with the contemplation of, and collection of data from, some part of the biosphere itself; the ecologist then tries to argue back, inductively, from observed effects to hidden causes. The mathematics used is statistical. For example, the statistical study of the distances from the individual trees in a forest to neighboring trees might lead to the conclusion that the trees tended to occur in clumps. The argument might be faultless, and the conclusion correct and unsurprising, but it is dull. It appears not to lead anywhere. The clumping of the trees can certainly be regarded as the proximate cause of the observed statistical distribution of measured distances, but proximate causes by themselves are unsatisfying. What we want are ultimate causes. A breakthrough is needed; some hitherto unthought of way to proceed must be devised if the established, inductive methods of "spatial statistics" are to take us much farther backward (in the direction of ultimate causes) than they have already.

Before leaving generalities and proceeding with the subject matter of Part II, it is worth remarking on the fact that, for the most part, population dynamics has been concerned with temporal patterns and statistical ecology with spatial patterns. This is partly an accident of the history of ecology. Partly, also, it stems from the wide range of life spans and of motilities of the organisms studied by ecologists relative to those of the ecologists themselves. Life spans range from less than an hour (for some bacteria) to millennia (for some sequoias); motilities from the immotility of an adult barnacle to the world-girdling migrations of the arctic tern. It seems to me that the needed breakthroughs mentioned in the two preceding paragraphs are most likely to come from work that deliberately jumps the established rails. Beginnings have, of course, been made but it would be invidious to single them out for mention in a general discussion such as this.

We now proceed with the details of "statistical ecology."

2.   Spatial Pattern and the Poisson Distribution

In studying the spatial patterns of sessile and sedentary organisms throughout a specified tract of space, it is necessary first to distinguish three wholly different setups that depend on the nature of the space and the organisms.

1. Cases in which the organisms are confined to discrete habitable sites or "units"; for example, caterpillars of a pest species that attack the shoots of a tree. Each shoot constitutes a habitable site and is a natural sampling unit.The caterpillars will not be found elsewhere than on shoots, and therefore the space available to them is discontinuous. Migrations from one shot to another, even if possible, are assumed to be uncommon enough to be ignored. Thus, if we count the number of organisms per unit (in the example, caterpillars per shoot) for a large sample of the units, the observations clearly convey something about the spatial pattern of the species.

2. Cases in which the organisms have a continuum of space that they can occupy: for example, trees in a forest. There are now no natural sampling units such as the shoots of the preceding example. If we wish to count individuals per unit, the unit (a small plot of ground, or quadrat, for instance) has to be arbitrarily defined.

3. Cases in which, in addition to the absence of natural sampling units, there are no clearly delimited individuals that can be counted. This is a situation that often confronts plant ecologists. Not merely do plants occupy a continuum that cannot be subdivided naturally but also, owing to vegetative reproduction, a great many plant species do not occur as distinct individuals amenable to counting.

The problems raised by cases 2 and 3 are considered in later chapters. Chapters 7 and 8 deal with case 1, and we begin by supposing that we have available an observed frequency distribution of the number of individuals per unit for a sample of units selected at random from an extensive population. What will the observed frequency distribution be like and how can it be explained?

Let us first postulate a simple mechanism that might explain such an observed frequency distribution and deduce its consequences. (This is not to argue deductively from an "ultimate" model; arguments of the present kind are merely to uncover the most proximate of the causes for observed phenomena so that they may be used as a basis for induction.) If the individuals had been assigned independently and at random to the available units, we call their pattern (or dispersion) *random*, and expect to find that the number of individuals per unit is a Poisson variate. The probability that a unit will contain r individuals is then

$$p_r = \frac{\lambda^r e^{-\lambda}}{r!}, \qquad r = 0, 1, \ldots,$$

where λ is the mean number of individuals per unit.

To see this, suppose that every unit contains a large number, n, of locations, each of which can be occupied by a single individual. For every location, in every unit, the probability is the same, say p, that it will be occupied. Then the probability that exactly r locations in any one unit will be occupied is given by the binomial probability

$$\binom{n}{r} p^r (1-p)^{n-r}, \qquad r = 0, 1, \ldots, n.$$

Now suppose n is very large, p is very small, and the mean of the distribution np is of moderate magnitude. Write $np = \lambda$. Also assume that r is negligibly small compared with n. Then

$$p_r = \binom{n}{r} p^r (1-p)^{n-r} \sim \frac{(np)^r}{r!} (1-p)^n$$

$$= \frac{\lambda^r (1-\lambda/n)^n}{r!} \rightarrow \frac{\lambda^r e^{-\lambda}}{r!}$$

as n becomes indefinitely large. In other words, when the individual organisms are sparse in relation to what the whole collection of units could contain and every possible location within a unit has the same probability of being occupied, the number of individuals per unit will be a Poisson variate.

This argument assumes that the maximum number of individuals a unit could contain is the same for all units and is equal to n and also that the expected number per unit is the same for all units and is equal to $np = \lambda$. These are very restrictive assumptions and are seldom likely to hold. It is not surprising therefore that the Poisson distribution rarely fits observed frequency distributions of the number of individuals per unit.

The probability generating function (pgf) of the Poisson distribution is given by

$$g(z) = p_0 + p_1 z + p_2 z^2 + \cdots$$

$$= e^{-\lambda} + \lambda e^{-\lambda} z + \frac{\lambda^2 e^{-\lambda}}{2!} z^2 + \cdots$$

$$= e^{\lambda(z-1)}.$$

Therefore the mean is

$$\left. \frac{dg(z)}{dz} \right|_{z=1} = g'(1) = \lambda$$

and the variance is

$$g''(1) + g'(1)[1 - g'(1)] = \lambda,$$

where $g''(1)$ denotes the value of $d^2 g(z)/dz^2$ at $z = 1$. Thus for the Poisson distribution the mean and variance are equal. However, when we compare the sample mean and variance of observed frequency distributions of the number of organisms per unit, it is commonly found that the variance greatly exceeds the mean. When this is so the pattern is said to be "aggregated," "clumped," "clustered," or "patchy" and the frequency distribution itself is described as "contagious." We now search for more realistic hypotheses to account for the distributions actually found. These fall into two categories: generalized distributions and compound distributions.

Before describing them, however, some comment on terminology is desirable. Much confusion exists in the ecological literature because the word "distribution" is used in both its colloquial and statistical senses, even sometimes in a single sentence, and without any explanation of the meaning intended. Colloquially, "distribution" is synonymous with "arrangement" or "pattern." Statistically, it means the way in which variate values are apportioned, with different frequencies, in a number of possible classes. In this sense there is no implied reference to spatial arrangement; for instance, we may speak of the distribution of a variate such as tree-height without any thought of the location of the trees. Thus to talk of a population of insects, say, as having "a clumped distribution with a large variance" is nonsense; it is the insects themselves that are clumped or have a clumped pattern, and the large variance pertains to the distribution of the variate, the number of insects per sample unit. To avoid ambiguity in statistical ecology it is most desirable to use the word distribution in its statistical sense *only*. Then a variate has a distribution, whereas a collection of organisms has a pattern.

3. Generalized Distributions

A generalized distribution arises if we suppose that groups or clusters of individuals (rather than single individuals) constitute the entities having a specified pattern, and that the number of individuals per group is a random variate with its own probability distribution.

First let us derive the mean and variance of the generalized distribution, M and V say, in terms of the means and variances of the numbers of clusters per unit (m_1 and v_1) and of the number of individuals per cluster (m_2 and v_2). It will be shown that M and V are constant functions of m_1, m_2, v_1, and v_2 whatever the form of the two underlying distributions.

Let the pgf of the distribution of the number of clusters per unit be $G(z)$, and of the number of individuals per cluster be $g(z)$. Then the pgf, $H(z)$ say, of the generalized distribution, that is, of the distribution of the number of individuals per unit, is

$$H(z) = G(g(z)).$$

The mean and variance of the generalized distribution are given by

$$M = H'(1) \quad \text{and} \quad V = H''(1) + H'(1)[1 - H'(1)].$$

Now write P_i for the probability that a unit contains i clusters ($i = 0, 1, \ldots$) and π_j for the probability that a cluster contains j individuals ($j = 1, 2, \ldots$).

Then

$$G(z) = \sum_i P_i z^i, \qquad g(z) = \sum_j \pi_j z^j$$

and

$$H(z) = G(g(z)) = \sum_i P_i \left[\sum_j \pi_j z^j \right]^i,$$

We now obtain M and V. First,

$$H'(z) = \sum_i i P_i [g(z)]^{i-1} g'(z).$$

Since

$$g(1) = 1 \quad \text{and} \quad g'(1) = m_2,$$

it follows that

$$H'(1) = M = m_1 m_2. \tag{7.1}$$

Next,

$$H''(z) = \sum_i i P_i \{(i-1)[g(z)]^{i-2}[g'(z)]^2 + [g(z)]^{i-1} g''(z)\}.$$

Using the fact that $g''(1) = v_2 - m_2 + m_2^2$, it is seen that

$$H''(1) = V - M + M^2 = \sum_i i(i-1)P_i m_2^2 + \sum_i i P_i (v_2 - m_2 + m_2^2)$$

$$= (v_1 - m_1 + m_1^2)m_2^2 + m_1(v_2 - m_2 + m_2^2).$$

Therefore, substituting $m_1 m_2$ for M,

$$V = m_1 v_2 + m_2^2 v_1. \tag{7.2}$$

Now consider an example. Suppose female insects lay egg clusters that are dispersed at random among available units such as pine shoots. Denote the mean number of clusters per shoot by λ_1. Further, suppose that the number of larvae hatching from each cluster is itself a Poisson variate with mean λ_2. Then, using the fact that for a Poisson distribution the mean and variance are equal so that

$$m_1 = v_1 = \lambda_1 \quad \text{and} \quad m_2 = v_2 = \lambda_2,$$

the first two moments of the generalized distribution are

$$M = \lambda_1 \lambda_2 \quad \text{and} \quad V = \lambda_1 \lambda_2 (1 + \lambda_2).$$

This distribution is known as the Neyman Type A or Poisson-Poisson distribution. The probability, p_r, that a randomly chosen unit will contain r individuals is given by the coefficient of z^r in the expansion of $H(z)$.

As a second example of a generalized distribution we suppose, as before, that the number of clusters per unit is a Poisson variate and hence has pgf

$$G(z) = e^{\lambda(z-1)}.$$

Now, instead of assuming that the number of larvae per cluster has a Poisson distribution, we let it have the logarithmic distribution with parameter α. This is equivalent to saying that $P(x)$, the probability that a cluster will contain x larvae, is proportional to α^x / x with $0 < \alpha < 1$. Here x takes the values $1, 2, \ldots$, and no "empty" clusters, devoid of larvae, are postulated, as they were when x was assumed to be a Poisson variate. Since

$$\sum_{x=1}^{\infty} P(x) = 1$$

and

$$\sum_{x=1}^{\infty} P(x) \propto \alpha + \frac{\alpha^2}{2} + \frac{\alpha^3}{3} + \cdots = -\ln(1-\alpha),$$

we see that

$$P(x) = \frac{-1}{\ln(1-\alpha)} \frac{\alpha^x}{x}, \qquad x = 1, 2, \ldots.$$

The pgf of the logarithmic distribution is then

$$g(z) = \frac{-1}{\ln(1-\alpha)} \left[\alpha z + \frac{(\alpha z)^2}{2} + \frac{(\alpha z)^3}{3} + \cdots \right]$$

$$= \frac{\ln(1 - \alpha z)}{\ln(1 - \alpha)}.$$

For the generalized distribution, therefore, we obtain the pgf $H(z)$ by writing

$$H(z) = G(g(z)) = \exp\left\{\lambda\left[\frac{\ln(1-\alpha z)}{\ln(1-\alpha)} - 1\right]\right\}.$$

This is most easily simplified by redefining the two parameters λ and α as follows: Put $\lambda = k \ln Q$ and $\alpha = P/Q$, where $Q = 1 + P$. Now, in place of λ and α we have two other parameters, k and Q (or $P = Q - 1$). Then, since

$$\ln(1-\alpha) = \ln\frac{Q-P}{Q} = \ln\frac{1}{Q},$$

$$H(z) = \exp\left[-k\ln\left(1-\frac{Pz}{Q}\right)\right] \cdot \exp[-k \ln Q]$$

$$= \left(1-\frac{Pz}{Q}\right)^{-k} Q^{-k} = (Q-Pz)^{-k}.$$

This is the pgf of the negative binomial distribution. It may be contrasted with the pgf of the ordinary positive binomial for which q, $p < 1$, $p + q = 1$, and k is positive. For the negative binomial, on the other hand, the index is negative and $Q - P = 1$

In this case the mean and variance are most easily found directly from the pgf of the generalized distribution. As before, we find the mean, variance, and general term p_r of the distribution. The mean is $H'(1) = kP$; the variance is $H''(1) + H'(1)[1 - H'(1)] = kP(1 + P)$, or kPQ. Instead of taking P and k as the parameters of the distribution, it is often preferable to take, as one parameter, the mean $m = kP$. The variance is then $V = m + m^2/k$. Clearly the smaller the value of k, the greater the variance; but, if $k \to \infty$, $V \to m$, and we see that, as for the Poisson distribution, the mean and variance are equal. Indeed, as $k \to \infty$, $p_r \to m^r e^{-m}/r!$, as proved below.

The probability p_r of finding exactly r individuals in a unit is the coefficient of z^r in the expansion of $H(z)$. We shall now derive it.

Recall that

$$H(z) = Q^{-k}\left(1-\frac{Pz}{Q}\right)^{-k}$$

$$= Q^{-k}\left[1 + \frac{kP}{Q}z + \frac{k(k+1)}{2!}\cdot\left(\frac{P}{Q}\right)^2 z^2 + \cdots\right].$$

Therefore the coefficient of z^r is

$$p_r = Q^{-k}\frac{k(k+1)\cdots(k+r-1)}{r!}\frac{P^r}{Q^r} = \frac{\Gamma(k+r)}{r!\Gamma(k)}\cdot\frac{P^r}{Q^{k+r}}$$

Since the mean is $m = kP$, we may write instead

$$p_r = \frac{\Gamma(k+r)}{r!\Gamma(k)} \cdot \left(\frac{m}{k}\right)^r \left(\frac{k}{k+m}\right)^{k+r}$$

$$= \frac{\Gamma(k+r)}{\Gamma(k)k^r} \cdot \frac{m^r}{r!}\left(1+\frac{m}{k}\right)^{-(k+r)}, \qquad r = 0, 1, \ldots.$$

Then, as k becomes indefinitely large and r becomes negligible in comparison with k, $p_r \to m^r e^{-m}/r!$, which is the general term of the Poisson series.

We have now considered two examples of generalized distributions: the Neyman type A or Poisson-Poisson and the negative binomial or Poisson-logarithmic. In these double names for generalized distributions the first denotes the distribution of the number of clusters per unit and the second, the distribution of the number of individuals per cluster.

As an example of how these two distributions fit field observations, consider the data given by Bliss and Fisher (1953) in which both theoretical distributions were fitted to the observed distribution of the number per quadrat of *Salicornia stricta* plants growing in a salt marsh. Although in his case the sampling units are not discrete natural entities but arbitrary

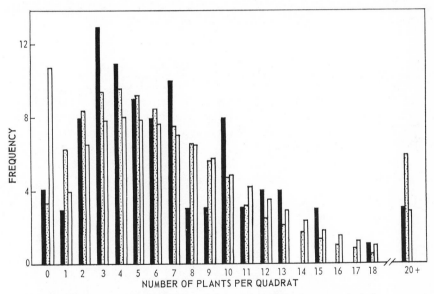

Figure 7.1. The distribution of the number of plants per quadrat of *Salicornia stricta*. Solid bars: observed. Stippled bars: fitted negative binomial distribution for which $P(\chi^2) = 0.48$. Open bars: fitted Neyman Type A distribution for which $P(\chi^2) = 0.17$. (Data from Bliss and Fisher, 1953.)

quadrats, the data serve for purposes of illustration. The way in which the parameters of the two theoretical distributions were estimated from the observations is described in the original paper by Bliss and Fisher and is not dealt with here, but, as can be seen from Figure 7.1, both theoretical distributions fit the data fairly well. By judging the goodness of fit with χ^2 tests it was found that for the negative binomial $P(\chi^2) = 0.48$ and for the Neyman type A, $P(\chi^2) = 0.17$. Thus, although the negative binomial appears to give a better fit, either of the hypotheses might be accepted. In other words, *if* we are willing to accept that the individuals occur as randomly dispersed clusters or clumps, the number of individuals per cluster may equally well be a Poisson variate or a logarithmic variate. A definite conclusion cannot be reached. Quite possibly, however, neither of these explanations accounts for the observed pattern. We now show how an entirely different hypothesis can lead to the negative binomial distribution.

4. Compound Distributions

Suppose that the organisms are independent of one another (or not clustered) and that if all the units available to them were identical their pattern would be random. If the mean density were λ, the probability that any unit would contain r individuals would therefore be the Poisson term $\lambda^r e^{-\lambda}/r!$.

Now suppose that the units are dissimilar. Some provide more favorable environments than others so that the parameter λ, the expected number of individuals in a unit, varies from unit to unit; that is λ is itself a random variable. Let us assume it has a Pearson type III distribution. The reason for choosing this standard curve to represent the distribution of λ is that whatever λ's true distribution, it is likely that some type III curve can be found to approximate it closely. A true type III variate may have any nonnegative value and the curve may be unimodal or J-shaped. Therefore we assume that the probability density function of λ is

$$f(\lambda) = \frac{1}{\Gamma(k)} \left(\frac{1}{P}\right)^k \lambda^{k-1} e^{-\lambda(1/P)}, \qquad (\lambda \geq 0).$$

Then

$$p_r = \frac{1}{r!\Gamma(k)} \left(\frac{1}{P}\right)^k \int_0^\infty \lambda^{r+k-1} e^{-\lambda(1+1/P)} \, d\lambda$$

$$= \frac{P^{-k}}{r!\Gamma(k)} \cdot \frac{\Gamma(r+k)}{\left(\frac{1+P}{P}\right)^{r+k}} = \frac{\Gamma(r+k)}{r!\Gamma(k)} \frac{P^r}{Q^{k+r}}, \qquad r = 0, 1, \ldots,$$

where $Q = 1 + P$. Notice that this is the distribution that we considered before—the negative binomial.

It is indeed true, in general, that every compound Poisson distribution corresponds to a generalized Poisson distribution and vice versa. A proof of this appears in Feller (1968). We have given only one well-known example here. Others have been described by Skellam (1952). Therefore it is futile to try to reach conclusions about the mechanism underlying a particular observed pattern simply by examining the observed distribution of the number of individuals per unit. Even when only one of the theoretical contagious distributions fits the observations there are two explanatory mechanisms to choose from. Also, since many of the theoretical series resemble one another closely, it is often found that two or more provide an adequate fit to a single set of observations.

Since every generalized, and compound, distribution is based on two (at least) assumptions, it is scarcely surprising that a single set of observations is inadequate to confirm both assumptions. To try to reach a decision on the acceptability of two independent assumptions by examining a single observed frequency distribution is to attempt too much; for example, the mechanism, described above, which led to the derivation of the Neyman type A distribution contained two assumptions: (a) that the number of clusters per unit was a Poisson variate and (b) that the number of individuals per cluster was also a Poisson variate. Only if one of these assumptions were believed, from *independent* evidence, to be true could the acceptability of the other be judged by fitting a Neyman type A series and seeing how good a fit was obtained.

It must be concluded that the fitting of theoretical frequency distributions to observational data can never by itself suffice to "explain" the pattern of a natural population.

8

The Measurement of Aggregation

1. Introduction

It was shown in Chapter 7 that even when one of the theoretical contagious distributions fits an observed frequency distribution closely it is still not possible to draw any conclusions regarding the mechanism that gave rise to it. However, we may still wish to measure the degree of aggregation (clumping, clustering, or contagion) of a population's spatial pattern without attempting to explain it. It would then be possible to compare the aggregation exhibited by a single species at different times or in different places or to compare the aggregation found at a single place and time in populations of two different species. Such observations are of obvious ecological interest and various methods of measuring aggregation have been devised. It should be emphasized that we are still considering organisms that occur only in discrete habitable units (Case 1 on page 115).

Usually, two populations that are to be compared will differ in mean density as well as in degree of aggregation. Attempts have been made to define some measurable property of a population's spatial pattern that can be thought of as equivalent in some way to aggregation but independent of mean density. In fact, of course, we cannot really contemplate aggregation without at the same time thinking of the things that are aggregated, hence of their numbers. Indeed, the phrase "degree of aggregation" describes a vague, undefined notion that is open to several interpretations. If aggregation is to be measured, we must first choose from a number of possibilities some measurable property of a spatial pattern that is to be called its aggregation, and the method of measurement is then implicit in the chosen definition. Thus the several existing ways of measuring aggregation are not different methods of measuring the same thing: they measure different things. All the measures make use of the observed frequency distribution of the number of individuals per unit.

2. The Variance/Mean Ratio

The defining property of the contagious discrete distributions is that the variance exceeds the mean for all of them, whereas for the Poisson

124

distribution the variance and mean are equal. This immediately suggests use of the variance:mean ratio V/m as a measure of aggregation. The sample value of this ratio is $(1/n\bar{x})\sum_{j=1}^{n}(x_j - \bar{x})^2$, where x_j is the number of individuals in the jth of n units sampled and $\bar{x} = (\sum x_j)/n$. For large n the expected value of the ratio when the individuals are dispersed at random is $E(V/m) \sim 1$. If a population yields a value of V/m that exceeds 1 only slightly, we may be tempted to enquire whether the ratio exceeds 1 significantly or whether the value obtained would quite likely have been yielded by a randomly dispersed population. This is easily tested in virtue of the fact that $\sum(x_j - \bar{x})^2/\bar{x}$ (known as the index of dispersion) is the sum of n terms of the form $(O - E)^2/E$, where O and E are the observed and expected frequencies of the individuals in each unit. The sum is therefore approximately distributed as a χ^2 variate with $n - 1$ degrees of freedom. The number of degrees of freedom is one less than the number of variate values observed, since the values are subject to the constraint $\sum x_j = n\bar{x}$ but are otherwise independent. Taking as null hypothesis that the pattern is random, the probability of obtaining any value of the index of dispersion may then be found by consulting a table of percentage points of the χ^2-distribution. Unless the observed value of the index were improbably high, we should accept the null hypothesis.

However, it is often *un*reasonable to postulate that a pattern is random; or, if not positively unreasonable, there may at least be no particular grounds for favoring the hypothesis of randomness over any other imaginable pattern. The V/m ratio should then be regarded not as a test criterion but merely as a sample statistic descriptive of a population's pattern. It is important to distinguish between the use of V/m (or, equivalently, of the index of dispersion) as a test criterion and its use as a measure of aggregation when no hypothesis is being entertained. In the latter case V/m is simply an estimate of a population parameter on a par with an estimate of, say, mean density. If V/m is found to be close to 1, this should not prompt the conclusion that the pattern is truly random in the sense that the individuals are independent and the expected number per unit is the same for all units. The latter conclusion is justified only if there are a priori reasons for supposing it to be true and if application of the test gives no reason for rejecting it.

David and Moore (1954) have suggested, as a measure of aggregation, $I = (V/m) - 1$. They call I the "index of clumping" and describe a method for comparing two values of I, I_1 and I_2, say, from two different populations. The comparison may be made regardless of whether the means differ. Suppose samples of the same size n have been collected from both populations. Let m_1 and m_2 be the means of the two sets of observations and let V_1 and V_2 be their variances; then $I_j = (V_j/m_j) - 1$

with $j = 1, 2$. Now evaluate

$$w = -\tfrac{1}{2} \ln\left(\frac{V_1/m_1}{V_2/m_2}\right).$$

David and Moore state that if w lies outside the range of $-2.5/\sqrt{(n-1)}$ and $+2.5/\sqrt{(n-1)}$ then I_1 and I_2 differ significantly at the 5% level. Therefore, if we choose to regard I as a measure of aggregation, a method of comparing the degrees of aggregation of two populations is provided; but, as already explained, we may prefer to define aggregation in some other manner. So it would be truer to say that David and Moore's test is to judge the significance of the difference between two values of a certain function, I, of the means and variances. If and only if we have chosen this particular function as our preferred measure of aggregation can we interpret a difference between I_1 and I_2 as being equivalent to a difference between the two populations in their aggregation.

In examining the suitability of I as a measure of aggregation, it is interesting to visualize what would happen if, in a given population, a proportion $(1 - \theta)$ of the individuals were selected at random and killed or removed. Now consider the pattern of the survivors. Should this pattern be thought of as having the same or a lesser degree of aggregation when compared with that of the original population? Either answer is reasonable.

On the grounds that the greatest number of deaths will occur in what were originally the most densely populated units, with the result that the clumps are less dense than they were before, we could argue that the deaths had reduced aggregation. On the other hand, the facts that the survivors are still at their original locations and that the only change in the population has been the removal of randomly chosen individuals lead to the argument that a measure of aggregation should be used that is *not* affected by random deaths. Random deaths could be said to alter only the mean density of the population while leaving other aspects of its pattern unchanged. This shows that we are free to choose how aggregation is to be defined and that the properties of any particular measure depend on the definition.

We describe below a measure that remains unaltered when deaths take place at random among population members. First we shall show that David and Moore's I decreases linearly with decreasing population density (provided the deaths are at random) and that this is true whatever the initial frequency distribution.

Suppose the pgf of the initial distribution is $G_0(z)$ and of the final distribution, after the deaths, $G_1(z)$. Since θ is the probability that an individual will survive and is the same for all individuals, we have

$$G_1(z) = G_0(\theta z + 1 - \theta).$$

We now wish to find the first and second moments of $G_0(z)$ and $G_1(z)$ and determine how they are related. This is most easily done by first deriving the factorial moments. The factorial moment generating function (fmgf), denoted by $\Phi(u)$, is easily seen to be the same as $G(1+u)$; that is,

$$\Phi(u) = \sum_{i=0}^{\infty} \frac{u^i}{i!} \mu'_{(i)} = G(1+u),$$

where $\mu'_{(i)}$ is the ith factorial moment about the origin.

So, writing $\Phi_0(u)$ and $\Phi_1(u)$ for the fmgf of the initial and final distributions respectively, we have

$$\Phi_0(u) = G_0(1+u)$$

and

$$\Phi_1(u) = G_1(1+u) = G_0[\theta(1+u) + 1 - \theta] = G_0(1 + \theta u) = \Phi_0(\theta u).$$

Now put $\mu'_{(i),0}$ and $\mu'_{(i),1}$ for the ith factorial moments of the initial and final distributions. Then, since $\Phi_1(u) = \Phi_0(\theta u)$,

$$\sum_0^{\infty} \mu'_{(i),1} \frac{u^i}{i!} = \sum_0^{\infty} \mu'_{(i),0} \frac{(\theta u)^i}{i!}.$$

Equating coefficients of $u^i/i!$ for $i = 1, 2$, we see that

$$\mu'_{(1),1} = \theta \mu'_{(1),0} \quad \text{and} \quad \mu'_{(2),1} = \theta^2 \mu'_{(2),0}.$$

Now, the first factorial moment of a distribution is identical with the mean m; and the variance V is given by

$$V = \mu'_{(2)} + \mu'_{(1)} - \mu'^2_{(1)} \quad \text{or} \quad \mu'_{(2)} + m - m^2.$$

So

$$I = \left(\frac{V}{m}\right) - 1 = \left(\frac{\mu'_{(2)}}{m}\right) - m,$$

and for the initial and final populations the indices of clumping are given by

$$I_0 = \left(\frac{\mu'_{(2),0}}{m_0}\right) - m_0 \quad \text{and} \quad I_1 = \left(\frac{\theta^2 \mu'_{(2),0}}{\theta m_0}\right) - \theta m_0 = \theta I_0.$$

Thus it follows that as the density of a population decreases owing to random deaths the index of clumping also decreases and is θ times its original value when a proportion θ of the original population remains.

This provides a method for determining whether the deaths in a cohort of sedentary organisms are density dependent, provided we can be certain there is no migration from unit to unit. Suppose the fate of an even-aged population of organisms (i.e., a cohort) were being followed; for instance, they might be pest caterpillars on tree shoots. The size of the population will gradually dwindle as time advances owing to the deaths of some of its members. If, at a succession of times, a sample is taken from the population of habitable units, we may determine how I varies with m. If there is no density dependence, the relation will be a straight line passing through the origin. If individuals belonging to dense aggregates are more likely to die than those in less crowded units, then I will decrease more rapidly; conversely, if there is inverse density dependence, that is, if living in dense clumps favors survival, I will decrease less rapidly. Iwao (1970) has used this method to study the effects of different causes of mortality on the fate of colonies of the western tent caterpilar (*Malacosoma californicum*). He found that deaths from the attack of a parasitic wasp tended to be density-independent whereas those from disease were density-dependent.

3. The Negative Binomial Parameter, k

When the number of organisms per unit has a negative binomial distribution, we may use the parameter k of the series as a measure of aggregation (Waters, 1959). Since, for the negative binomial, $V = m + m^2/k$ (see page 120), in terms of David and Moore's index I, $k = m/I$; that is, low values of k indicate pronounced clumping and high values, slight clumping. To obtain an index of aggregation that increases with increasing clumping some authors use a function of k such as its reciprocal.

An interesting property of k is that it remains unaltered when a population decreases in size owing to random deaths. This is the property we suggested earlier might be desirable in measures of aggregation: the measure may be thought of as representing some intrinsic property of a spatial pattern whatever the density.

To see this, consider a negative binomial series with pgf $(Q - Pz)^{-k}$ having mean $m = kP$ and variance $V = m(1 + m/k)$. Then

$$p_r = \binom{k+r-1}{r} \frac{P^r}{(1+P)^{k+r}},$$

where, for convenience, we have written $\binom{k+r-1}{r}$ in place of $[\Gamma(k+r)]/r!\Gamma(k)$. Now let a proportion $1 - \theta$ of the individuals be selected at random and destroyed so that a proportion θ survives in the final population.

Writing p'_r for the probability that a unit will contain r individuals in the final population, we tabulate initial and final probabilities (see Table 8.1). Clearly

$$p'_r = \theta^r \sum_{i=r}^{\infty} \binom{i}{r} (1-\theta)^{i-r} p_i.$$

TABLE 8.1

INDI-VIDUALS PER UNIT i	INITIAL PROB-ABILITY p_i	FINAL PROBABILITY p'_i
0	p_0	$p'_0 = p_0 + (1-\theta)p_1 + (1-\theta)^2 p_2 + \cdots$
1	p_1	$p'_1 = \theta p_1 + \binom{2}{1}\theta(1-\theta)p_2 + \binom{3}{1}\theta(1-\theta)^2 p_3 + \cdots$
2	p_2	$p'_2 = \theta^2 p_2 + \binom{3}{2}\theta^2(1-\theta)p_3 + \binom{4}{2}\theta^2(1-\theta)^2 p_4 + \cdots$
...	...	...
r	p_r	$p'_r = \theta^r p_r + \binom{r+1}{r}\theta^r(1-\theta)p_{r+1}$ $\qquad + \binom{r+2}{r}\theta^r(1-\theta)^2 p_{r+2} + \cdots$

Substituting the negative binomial term for p_i

$$p'_r = \theta^r \sum_{i=r}^{\infty} \binom{i}{r}(1-\theta)^{i-r} \binom{k+i-1}{i} \frac{P^i}{(1+P)^{k+i}}$$

$$= \theta^r \left\{ \binom{k+r-1}{r} \frac{P^r}{(1+P)^{k+r}} + \binom{r+1}{r}(1-\theta)\binom{k+r}{r+1}\frac{P^{r+1}}{(1+P)^{k+r+1}} + \cdots \right\}$$

$$= \frac{(\theta P)^r}{(1+P)^{k+r}} \binom{k+r-1}{r} \left\{ 1 + \binom{k+r}{1}\left(\frac{P(1-\theta)}{1+P}\right) \right.$$

$$\left. + \binom{k+r+1}{2}\left(\frac{P(1-\theta)}{1+P}\right)^2 + \cdots \right\}$$

$$= \binom{k+r-1}{r}(\theta P)^r \frac{1}{(1+P)^{k+r}}\left[1 - \frac{P(1-\theta)}{1+P}\right]^{-(k+r)}$$

$$= \binom{k+r-1}{r}\frac{(\theta P)^r}{(1+\theta P)^{k+r}},$$

and this is the rth term of a negative binomial series with parameters θP and k. The original mean was $m_0 = kP$, the new mean is $m_1 = \theta kP$, and the value of the exponent k is unchanged.

This is the reason for choosing k, or some function of it such as k^{-1}, as a measure of aggregation in patterns that yield a negative binomial distribution when sampled. However, this argument holds only for negative binomial populations. It is true that if we were to use as an (inverse) measure of aggregation the function k' defined as $k' = m^2/(V - m) = m/I$, then k' would remain unchanged if deaths occurred at random in a population. This is true whatever the parent distribution and follows from the fact that $k'_1 = m_1/I_1 = \theta m_0/\theta I_0 = k'_0$, where, as before, the subscripts 0 and 1 denote initial and final values. Further, if the parent distribution were negative binomial, k' would be the moment estimator of the parameter k. However, when the parent distribution is not negative binomial, it is meaningless to speak of it as having the parameter k; and if we were to assume that a distribution was negative binomial when it was not and estimate "k" by maximum likelihood, this estimate $\hat{k}$ would *not* remain unchanged when random deaths occurred.

A numerical example illustrates this point. The initial frequencies shown in Table 8.2 have been chosen arbitrarily and do not represent any

TABLE 8.2

i	INITIAL FREQUENCY np_i	FINAL FREQUENCY np'_i
0	40	63.59
1	30	44.06
2	10	40.78
3	20	31.88
4	30	15.16
5	40	4.06
6	30	0.47
	$n = 200$	200.00

theoretical contagious distribution. The final frequencies were obtained by assuming that half the population had been removed at random. In Table 8.2:

$$m_0 = 3.05, \qquad m_1 = 1.525,$$
$$V_0 = 4.7475, \qquad V_1 = 1.9494,$$

$$I_0 = \frac{V_0}{m_0} - 1 = 0.557, \qquad I_1 = \frac{V_1}{m_1} - 1 = 0.278 \ (= \tfrac{1}{2}I_0),$$

$$k_0' = k_1' = 5.480,$$

$$\hat{k}_0 < 3.0, \qquad\qquad \hat{k}_1 > 4.0.$$

The maximum likelihood estimates $\hat{k}_0$ and $\hat{k}_1$ were calculated by the method described by Bliss and Fisher (1953), and it is seen that $\hat{k}_1 > \hat{k}_0$. Thus, if we had assumed, mistakenly, that the parent distribution was negative binomial, we should be led to conclude that aggregation had decreased as a result of the deaths; this, in turn, would suggest that the deaths had been density dependent.

4. Lloyd's Indices of Mean Crowding and Patchiness

On page 126 we advanced arguments in favor of using a measure of aggregation that (a) does and (b) does not change when some of the population members are removed at random. Obviously, two different things are being envisaged and each should be measured separately. Lloyd (1967) has proposed an "index of mean crowding" and an "index of patchiness" that seem to meet the requirements. He defines mean crowding as the mean number per individual of other individuals in a unit; these other individuals may be thought of as co-occupants of the unit with the first individual. The "crowding" $\overset{*}{m}$ is an average over individuals instead of over units. It is calculated by counting, for each individual in a total population of N individuals, say, the number of co-occupants X_i that share the unit with it; $(i = 1, 2, \ldots, N)$.

Then the mean crowding is

$$\overset{*}{m} = \frac{1}{N} \sum_{i=1}^{N} X_i.$$

If there is a total of n units and x_j $(j = 1, 2, \ldots, n)$ denotes the number of individuals in the jth unit, then

$$\sum_{i=1}^{N} X_i = \sum_{j=1}^{n} x_j(x_j - 1); \qquad \text{also} \qquad \sum_{j=1}^{n} x_j = N.$$

Then

$$\overset{*}{m} = \frac{\sum x_j^2}{\sum x_j} - 1.$$

In terms of the mean and variance of the x's, m and V, say,

$$\overset{*}{m} = m + \left(\frac{V}{m} - 1 \right) \qquad \text{or} \qquad m + I,$$

since

$$\frac{\sum x_j^2}{\sum x_j} = \frac{V + m^2}{m}.$$

Thus the crowding is numerically equal to the sum of the mean density and David and Moore's index of clumping I. As with I itself, $\overset{*}{m}$ must remain proportional to density as a population is depleted by random deaths. Using, as before, the subscripts 0 and 1 for the initial and final values in a dwindling population, we have

$$\overset{*}{m}_0 = m_0 + I_0, \qquad \overset{*}{m}_1 = m_1 + I_1.$$

When only a proportion θ of the initial population survives, we know that

$$m_1 = \theta m_0 \qquad \text{and} \qquad I_1 = \theta I_0, \qquad \text{whence } \overset{*}{m}_1 = \theta \overset{*}{m}_0.$$

The patchiness is defined as $\overset{*}{m}/m$, or the ratio of mean crowding to mean density. Random deaths leave patchiness unaltered, since $\overset{*}{m}_0/m_0 = \overset{*}{m}_1/m_1$; this remains true whatever the form of the parent distribution. Notice also that

$$\frac{\overset{*}{m}}{m} = 1 + \frac{I}{m} = 1 + \frac{1}{k'}.$$

Thus crowding is something that is experienced by each individual and depends on the total number present. Patchiness, on the other hand, is a property of a spatial pattern considered by itself without regard to density, and two populations can exhibit the same degree of patchiness even though their densities differ.

In the numerical example already given (page 130) it can be seen that the crowding decreases from $\overset{*}{m}_0 = 3.607$ to $\overset{*}{m}_1 = 1.803$, when half the population is destroyed. The patchiness remains constant and equal to 1.182.

Iwao and Kuni (1971) showed that knowledge of the values of Lloyd's indices for several independent populations of a single species enables one to draw conclusions as to the mechanism underlying the typical spatial pattern of the species. Their argument is as follows.

Recall that, as shown on page 118, if individuals tend to occur in clusters or groups, then whatever the distributions of, respectively, the number of clusters per unit (with mean and variance m_1 and v_1) and of individuals per cluster (with mean and variance m_2 and v_2), the number of individuals per unit has mean $M = m_1 m_2$ and variance $V = m_1 v_2 + m_2^2 v_1$. Then $\overset{*}{M}$, the index of mean crowding of the individuals, is

$$\overset{*}{M} = \frac{V}{M} + M - 1 = \overset{*}{m}_2 + \overset{*}{m}_1 m_2$$

$$= \overset{*}{m}_2 + \left(\frac{\overset{*}{m}_1}{m_1}\right) M.$$

It follows that, if it is found from observations on several populations of a species that $\overset{*}{M}$ is a linear function of M, then the value of $\overset{*}{M}$ at $M = 0$ is an estimator of $\overset{*}{m}_2$, the index of crowding of the individuals within the clusters; and the slope of the line is an estimator of $\overset{*}{m}_1/m_1$, the index of patchiness of the pattern of clusters.

As an example, Figure 8.1 shows the relationship between $\overset{*}{M}$ and M for a species of oribatid mite (*Oppia ornata*) occurring in forest soil [data

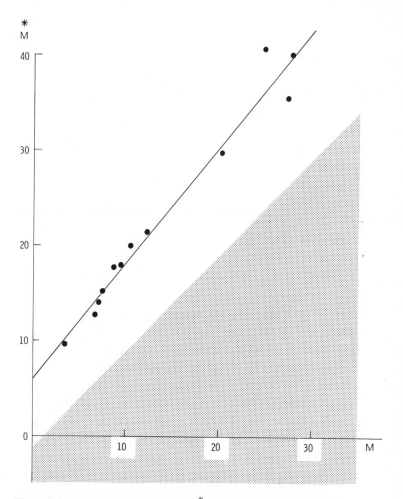

Figure 8.1. The relationship between $\overset{*}{M}$ and M for 12 populations of the mite *Oppia ornata*. The fitted line is $\overset{*}{M} = 5.96 + 1.22M$. Points cannot fall in the region (stippled) where $\overset{*}{M} < M - 1$. (Redrawn from Iwao and Kuni, 1971.)

from Berthét and Gérard (1965)]. The fitted line cuts the ordinate at $\overset{*}{M} = 6$ approximately, and has slope 1.22; it therefore seems reasonable to infer that the mites occur as clusters and that the clusters themselves are somewhat aggregated.

It should be noticed that since

$$\overset{*}{M} = \frac{V}{M} + M - 1$$

we must have $\overset{*}{M} \geq M - 1$. Therefore the points in a graph of $\overset{*}{M}$ versus M are constrained to lie above the line $\overset{*}{M} = M - 1$ and cannot fall in the region shown stippled in Figure 8.1.

9

The Pattern of Individuals in a Continuum

1. Introduction

So far we have considered only the spatial patterns of organisms that occupy small isolated units; the space available to the organisms was discrete. Now we turn to the case in which an extended continuum, either an area or a volume, is available to the organisms, and they may be found anywhere throughout it; this is Case 2 on page 115. Examples are individual plants scattered over an area of ground or microarthropods dispersed through a volume of soil. There are now no natural sampling units such as the discrete habitable units afforded and sampling units have to be arbitrarily defined. To fix ideas we suppose that the population to be studied consists of all the plants growing on an apparently homogeneous tract of level ground. It is assumed, further, that the plants are small in relation to the space available to them, that they are roughly equal in size, and that they reproduce entirely by seed and never vegetatively so that there is no difficulty in recognizing true individuals.

The commonest method of investigating the pattern of such a population is to sample it with randomly placed quadrats, which are small sample areas that are usually, but not necessarily, square. We then count the number of individual plants in each quadrat, compile a frequency table to show the observed numbers of quadrats that contained 0, 1, 2, ..., individuals, and examine the observed distribution. The method has two great drawbacks. In the first place, the results are greatly affected by the size of quadrat used; and, in the second, when the observations from all the quadrats are pooled to compile a frequency table, no record is kept of the locations of the quadrats. Even though *spatial* pattern is ostensibly being studied, the spatial relationship of the sparsely occupied quadrats and densely occupied quadrats is often ignored. The records normally kept do not show, for instance, whether sparse and dense quadrats were randomly mingled with one another or whether, instead, they tended to occur as contiguous groups on the ground.

We now consider how the size of the quadrats affects the results of quadrat sampling.

2. The Effect of Quadrat Size

To say that plants on the ground have a random pattern or are randomly dispersed is equivalent to saying that every point on the ground (within the area of study) is as likely as every other to be the site of an individual plant. When such a pattern is sampled with randomly thrown quadrats, the expected distribution of the number of plants per quadrat is Poisson with parameter λ, where λ is the mean number of plants per quadrat. Changing the quadrat size simply alters the magnitude of λ, which is proportional to quadrat size, and the distribution remains Poisson for all quadrat sizes. Now suppose that the plants are clumped: on certain patches of ground their density is high and on others, low. When such a pattern is sampled, the results will, in general, be influenced by quadrat size. A large quadrat will often contain the whole of a densely occupied patch or even several of these patches whereas a small quadrat may contain only part of a dense patch. As a result, measures of aggregation based on quadrat data will not as a rule be unique; different values will be obtained with different quadrat sizes.

There are, however, two rather special types of pattern for which this does not hold; that is, if the measure of aggregation to be used is suitably chosen, it will not vary with quadrat size.

1. Consider first a population that occurs in the form of clumps so compact and widely spaced that they are hardly ever cut through by the edges of a quadrat. If a measure of aggregation is used that depends only on the parameter (or parameters) of the distribution of the number of individuals per clump, the measure will be unaffected by changes in quadrat size; for example, suppose that the number of clumps per quadrat is a Poisson variate with parameter λ_1, say, and the number of individual plants per clump is a Poisson variate with parameter λ_2. Then the distribution of the number of plants per quadrat will be Poisson-Poisson with parameters λ_1 and λ_2 (see page 119). The mean and variance of this distribution are $\lambda_1\lambda_2$ and $\lambda_1\lambda_2(1+\lambda_2)$, respectively, and the V/m ratio is thus $1+\lambda_2$. Equivalently, David and Moore's index of clumping, I, has the value λ_2. In these circumstances, then, I is identical to the mean number of plants per clump. Obviously, changes in quadrat size will affect only λ_1 and leave λ_2 unaltered, and we see that by using I (or the V/m ratio) we shall have a measure of aggregation that does not depend on quadrat size. This argument applies, of course, only if we assume that the

frequency with which clumps are cut through by quadrat edges is negligibly small. However, if the clumps were compact enough for this assumption to be justified, they would presumably be sufficiently distinct to be easily recognizable. We could then investigate the spatial pattern of the clumps and, separately, the distribution of the number of plants per clump. It would be futile to treat the number of plants per quadrat as a variate to be investigated and needless to attempt indirect inferences based on such observations.

2. Next, consider a pattern consisting of a mosaic, or patchwork, of several phases. Assume that the pattern is very coarse; that is, the patches are large in relation to the size of the quadrats. Within any one phase, which may be represented by a number of distinct patches, the pattern is random, but different phases have different densities or Poisson parameters. Now let the area be sampled with quadrats small enough for us to assume that nearly all of them lie wholly within one or another of the patches and not across patch boundaries. The resultant distribution of plants per quadrat will obviously consist of a mixture of Poisson distributions. Suppose that there are k different phases and that in the jth phase, which occupies a proportion π_j of the total area, the Poisson parameter is $\lambda_j (j = 1, 2, \ldots, k)$. The probability that a quadrat will contain n plants is then

$$P_n = \frac{\pi_1 \lambda_1^n e^{-\lambda_1}}{n!} + \frac{\pi_2 \lambda_2^n e^{-\lambda_2}}{n!} + \cdots + \frac{\pi_k \lambda_k^n e^{-\lambda_k}}{n!}$$

or

$$P_n = \pi_1 p_{1n} + \pi_2 p_{2n} + \cdots + \pi_k p_{kn}, \tag{9.1}$$

where we have put $\lambda_j^n e^{-\lambda_j}/n! = p_{jn}$ for brevity. Evidently, if the quadrat size is changed (but still kept so small that there is a negligible chance that any quadrat will lie across patch boundaries), the only change in the resultant distribution will be that stemming from the fact that each of the Poisson parameters will be multiplied by a constant factor. Therefore it should be possible to find some index of aggregation that is unaffected by this change. Lloyd's (1967) index of patchiness meets this requirement, as we now show.

Consider the distribution whose general term is given by (9.1). Its mean and variance may be found as follows: the mean is

$$m = \sum_{j=1}^{k} \left(\pi_j \sum_{n=1}^{\infty} n p_{jn} \right) = \sum_{j=1}^{k} \pi_j \lambda_j.$$

Similarly, the second moment about the origin is

$$\sum_{j=1}^{k} \left(\pi_j \sum_{n=1}^{\infty} n^2 p_{jn} \right) = \sum_{j=1}^{k} \pi_j (\lambda_j + \lambda_j^2),$$

whence the variance V is

$$V = \sum_{j=1}^{k} \pi_j(\lambda_j + \lambda_j^2) - \left(\sum_{j=1}^{k} \pi_j\lambda_j\right)^2.$$

Recalling that the patchiness (see page 132), which we here denote by C, is

$$C = 1 + \frac{V - m}{m^2},$$

we now have

$$C = 1 + \frac{\sum \pi_j\lambda_j^2 - (\sum \pi_j\lambda_j)^2}{(\sum \pi_j\lambda_j)^2} = \frac{\sum \pi_j\lambda_j^2}{(\sum \pi_j\lambda_j)^2}.$$

If the population is sampled again with quadrats r times the size of those originally used, the Poisson parameter in the jth phase will become $r\lambda_j$ for all j. Denoting by C_r the patchiness that is observed when quadrats of r units of area are used, we see that

$$C_r = \frac{\sum \pi_j(r\lambda_j)^2}{(\sum \pi_j r\lambda_j)^2} = C,$$

and thus the patchiness is the same as before.

Had we used the index of clumping I instead of the patchiness as a measure of aggregation, it would have been multiplied by the factor r. Writing I and I_r for the index of clumping obtained with quadrats of 1 and r units of area, respectively, it is seen that

$$I = \frac{V - m}{m} = \frac{\sum \pi_j\lambda_j^2 - (\sum \pi_j\lambda_j)^2}{\sum \pi_j\lambda_j}.$$

If we multiply every λ_j by the factor r, the index becomes

$$I_r = \frac{\sum \pi_j(r\lambda_j)^2 - (\sum \pi_j r\lambda_j)^2}{\sum \pi_j r\lambda_j} = rI.$$

Similarly, Lloyd's index of crowding, $\overset{*}{m}$, is multiplied by the factor r. With quadrats of unit area the crowding is $\overset{*}{m} = m + I$, and with quadrats of r units of area it is $\overset{*}{m}_r = r(m + I) = r\overset{*}{m}$.

It is interesting to note that Lloyd's "patchiness" is almost identical with Morisita's (1959) "index of dispersion," I_δ. Beginning from the same premise that the population under investigation consists of a mosaic of large (relative to the quadrats) patches within which the pattern is random, Morisita also sought for an index that would be unaffected by changes in quadrat size. He derived his index I_δ from entirely different considerations, however. To explain his derivation it is necessary first to mention the notion of "diversity" and Simpson's (1949) method of measuring it.

Suppose we have a collection of N objects of s different kinds, of which n_1 are of the first kind, n_2 of the second kind, ... and n_s of the sth kind, with $\sum n_i = N$. If two objects are picked at random, and without replacement, from the whole collection, the probability that both will be of the same kind is clearly

$$\frac{\sum_{i=1}^{s} n_i(n_i - 1)}{N(N-1)}.$$

It is reasonable to call the diversity of the collection great if this probability is low and slight if it is high. Now suppose we sample a population with s quadrats and attach a label to each of the N individuals encountered to show in which of the quadrats it was found. It is assumed that there is no overlapping of the quadrats. Let x_i of the individuals be from the ith quadrat ($i = 1, \ldots, s, \sum_i x_i = N$); then these x_i individuals are classified as being of the ith kind, since they belong to the ith quadrat. The probability that any two individuals chosen at random (from the total of N individuals) belong to the same quadrat is therefore

$$\delta = \frac{\sum_i x_i(x_i - 1)}{N(N-1)}.$$

If the individuals are crowded into comparatively few of the quadrats, that is, if they are aggregated, δ will be high; conversely, if the individuals are fairly uniformly spaced so that they are more or less evenly apportioned among the s quadrats, δ will be low.

Consider now the expected value of δ when a population of random pattern is sampled. In this case the probability that a randomly selected individual will come from a given quadrat is the same for all quadrats and is therefore $1/s$. Since the individuals are independent of one another, the same is true for a second individual. Therefore the probability that two randomly picked individuals will both come from the same given quadrat is $1/s^2$. Summing over all quadrats, we see that the expected value of δ, given a random pattern, is $\delta_{\mathrm{ran}} = \sum(1/s^2) = 1/s$. Morisita's I_σ is defined as $\delta/\delta_{\mathrm{ran}} = s\delta$ and so has the value 1 in a random pattern. In an aggregated pattern, in which a high proportion of the individuals is concentrated into only a few of the quadrats, $I_\delta > 1$.

It may be seen that

$$I_\delta = \frac{s}{N-1} \frac{\sum_{i=1}^{s} n_i(n_i - 1)}{N} = \frac{s}{N-1} \overset{*}{m} = \frac{N}{N-1} \cdot C,$$

since $C = \overset{*}{m}/m$ and $sm = N$.

It follows that if we know that a pattern consists of a mosaic of patches with different densities, within each of which the individuals are randomly dispersed, either C or I_δ may be used as a measure of aggregation and that, provided the quadrats are sufficiently small, the values will be independent of quadrat size. To assume without evidence, however, that a pattern is of mosaic form is usually unjustified, but we can test the validity of the assumption by sampling the population repeatedly, using quadrats of several sizes and judging whether C (or I_δ) remains constant for several of the smallest quadrats.

Much can indeed be learned about a pattern by examining the way in which some measure of aggregation varies with quadrat size. Besides showing that for mosaic patterns with large patches the I_δ versus quadrat size curve is horizontal when the quadrats are small, Morisita (1959) also discusses the form this curve will take, given other types of mosaic pattern. Its shape depends on the sizes of the patches and on the pattern of the individuals within the patches.

We shall not explore the matter further here but shall turn to a consideration of Greig-Smith's method of pattern analysis.

3. Grids of Contiguous Quadrats

Greig-Smith's (1952, 1964) method of pattern analysis is also based on the fact that the way in which a measure of aggregation varies with quadrat size provides information on pattern.

To sample an area over and over again with quadrats of successively larger sizes is exceedingly time-consuming, not to mention the fact that the vegetation becomes increasingly trampled as work proceeds. Greig-Smith therefore proposes the use of a grid, or lattice, of small square unit cells that completely cover the area to be studied. The grid units are, in fact, contiguous quadrats. The number of individuals in each grid unit is first counted. Adjacent pairs of units are then combined to give oblong two-unit blocks which are twice as big and half as numerous as the original units. Adjacent two-unit blocks are next combined to give square four-unit blocks; now these are combined to give oblong eight-unit blocks, and so on. In this way from a single examination of the area a sequence of "quadrat" sizes is obtained in which the quadrat area is doubled at each step. The size of the units is chosen so that the total number in the area shall be a power of 2.

The total sum of squares about the mean for the single grid cells (which are one-unit blocks) may now be apportioned as in an analysis of variance. We obtain sums of squares that derive from the difference between each pair of single units within the two-unit blocks, from the difference between each pair of two-unit blocks within the four-unit blocks, ..., from the difference between each pair of 2^j-unit blocks within the 2^{j+1}-unit blocks, and so on. Thus the sum of squares for blocks of $r = 2^j$ units within blocks of $2r = 2^{j+1}$ units is

$$(SS)_r = \frac{1}{r} \sum_{i=1}^{n/r} x_i^2(r) - \frac{1}{2r} \sum_{i=1}^{n/2r} x_i^2(2r)$$

where $x_i(r)$ is the number of individuals in the ith of the r-unit blocks and n is the total number of units in the grid. These within-block sums of squares may now be divided by their degrees of freedom to give mean squares that are equivalent to V/m ratios. The mean square for r-unit blocks is $(MS)_r = 2r(SS)_r/n$.

A graph of mean square versus block size, that is, of $(MS)_r$ versus r, would then be expected to show which block sizes yielded the strongest evidence of aggregation. While the blocks are small relative to the mosaic patches of the pattern, adjacent half-blocks within each block are likely to be within a single patch and the mean square will be low. Thereafter, as block size increases, so will the mean square until a block size is reached of area close to that of the mean area of the patches. If block size is increased still further, the mean square will remain at this high level if the patches themselves are at random or aggregated; but if the patches are regularly arranged, the mean square will fall off again. Hierarchical clumping will produce a succession of peaks in the mean square versus block size graph.

Although this method of studying pattern has been favored by a number of plant ecologists [e.g., Phillips (1953), Kershaw (1960), Cooper (1961)], it has several drawbacks and it is worthwhile to list them:

(a) The method can be used only on areas small enough to be examined in their entirety. The whole of the area to be investigated must be included in the grid.

(b) Since the two half-blocks within each block are always contiguous, the center-to center distance between them is doubled at each second doubling of their size. This fact alone would be expected to result in an increased V/m ratio if the population were nonuniform (Goodall, 1963) and it is impossible to judge the relative importance of this effect and that of patch size on the value of the mean square.

(c) The graph sometimes has a sawtoothed shape because oblong blocks consistently give mean squares less than those of the square blocks on either side of them in the sequence of sizes.

(d) Block size is doubled at each step; therefore there is no means of knowing what the form of the graph would have been had blocks of intermediate sizes been used as well. A peak for blocks of 16 units, say, can be interpreted as meaning only that the mean patch area lies somewhere between 8 and 32 units. Also, it is not clear what sort of graph should be expected if the patches vary greatly in size.

(e) Consider a population formed of a number of patches (or clumps) of individuals dispersed over otherwise empty ground and visualize also the same pattern in reverse: what were clumps in the first population are now lacunae and what was empty ground is now occupied by randomly dispersed individuals with the same density as those that formerly occupied the patches (see Figure 9.1). What might be called the "grain" of these two patterns is the same, and, as a result, they have very similar mean-square-versus-block-size graphs when analyzed by Greig-Smith's method. The maps of these artificial populations shown in the figure were analyzed and the result is also shown; it can be seen that for blocks of less than 128 units the graphs are alike, although in map B clumps in the ordinary sense do not exist. The divergent results at the largest block sizes occur simply because the difference in numbers of individuals between the left and right halves of the areas is greater in map B than in map A.

(f) The number of degrees of freedom on which each variance estimate is based is halved at each step, being reduced to one for the last pair of blocks (the largest); there is therefore a concomitant unavoidable decrease in precision as the blocks become larger.

(g) Since the successive mean squares are not independent one cannot use variance ratio tests to judge whether different values differ significantly. However, Mead (1974) has devised a test to decide whether, as each pair of blocks is combined to form a larger block, the evidence of aggregation is indeed significantly increased. Zahl (1974) has proposed using overlapping blocks to estimate the sizes of mosaic patches.

Another way of analyzing the data from a grid of contiguous quadrats has been proposed by Goodall (1974). It consists in picking from the grid random pairs of quadrats spaced at each of several chosen distances apart. From each pair one calculates an estimate of the variance of the number of individuals per quadrat; then these estimates are averaged over any one spacing distance. The result is a graph of variance versus spacing. The method has obvious merits. In particular, it overcomes objections (b), (c), (d), (f), and (g) to the old method listed above. Thus, since quadrats are not combined into blocks, there is no confounding of the effects of center-to-center distance (i.e., spacing) and block size, and no distortion of block shape. One is free to examine a variety of different quadrat spacings instead of being forced to allow spacing to double at every second step. Also, even with the larger spacings, variance estimates

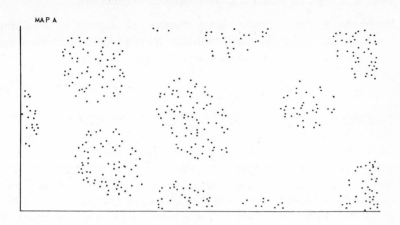

MAP A

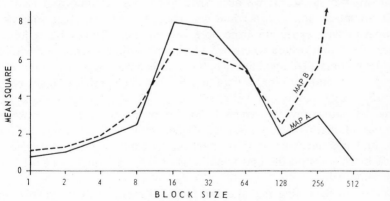

MAP B

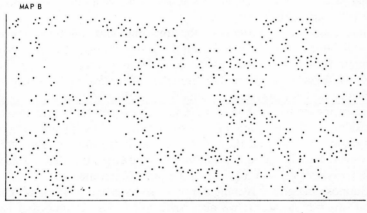

Figure 9.1. Two patterns (artificial) that give very similar graphs of mean square versus block size.

based on a comparatively large number of degrees of freedom are possible; and since they are mutually independent, standard statistical methods can be used to test the significance of differences.

4. Random Mingling of Sparse and Dense Grid Cells

Consider again a grid map of a population's pattern. As before, the map shows the number of individuals in each cell of the grid. Let us now subdivide the cells into two classes, say, sparse and dense, and color them white and black, respectively. If many cells are empty, the subdivision may be into empty versus occupied cells; or if nearly all the cells are occupied, the subdivision may be into cells containing fewer than x individuals (sparse cells) and cells containing x or more individuals (dense cells), with x being assigned a value that ensures a fairly equal division of the cells into whites and blacks.

It is clear that, regardless of the frequency distribution of the number of individuals per cell, indeed even if this distribution were well fitted by a Poisson series, we cannot regard the pattern as random unless the black and white cells are randomly mingled. Therefore it is worthwhile to consider how we may test for random mingling. A way of doing this has been proposed by Krishna Iyer (1949). To exemplify the use of his test he considered the mingling of diseased and healthy plants in a regular plantation in which all the plants were at lattice points. The method is, however, equally good for determining whether the dense (black) and sparse (white) cells in a grid map are randomly mingled.

Suppose the grid is rectangular, with m rows and n columns. There are then, $mn = b$ say, cells; we also write $m + n = a$. Assume that r_1 of the cells are black and r_2 are white so that $r_1 + r_2 = b$. The test consists in comparing with expectation the observed number of black-black joins: a black-black join occurs when two black cells adjoin each other within a row or column or diagonally. Thus in the example shown in Figure 9.2 $m = 5$, $n = 6$ and therefore $a = 11$ and $b = 30$. The number of black cells is $r_1 = 15$ and the number of white cells is $r_2 = 15$. There are 19 black-black joins, as shown by the arrows. To determine whether the observed number of black-black joins exceeds expectation significantly we must first find the mean and variance of the distribution of the number of black-black joins on the null hypothesis of random mingling of the black and white cells.

We begin by noting that the kth factorial moment of the distribution is $k!$ times the sum of the probabilities of the different ways of obtaining k black-black joins in the grid. [For a proof of this see Krishna Iyer (1952).]

The probability that any two given cells will be black is obviously

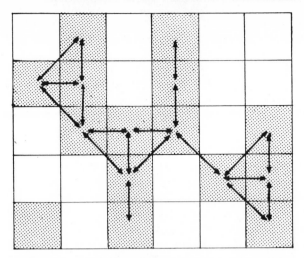

Figure 9.2. A grid map of an area in which the grid cells have been classified as dense (stippled) and sparse (white). The arrows show the joins between adjoining dense cells.

$r_1(r_1-1)/[b(b-1)]$. Thus by putting A for the number of ways of choosing two cells so that they are adjoining, the first factorial moment of the number of black-black joins is

$$\mu'_{(1)} = \frac{1! A r_1^{(2)}}{b^{(2)}},$$

where $r_1^{(x)} = r_1(r_1-1)\cdots(r_1-x+1)$ and similarly for $b^{(x)}$. This follows since for each of the A possible ways of choosing a pair of adjoining cells the probability is the same, namely $r_1^{(2)}/b^{(2)}$, that both members of the pair will be black.

Likewise, two black-black joins will occur (a) whenever three adjoining cells (not necessarily in a straight line) are all black and (b) whenever there are four black cells in the form of two adjoining pairs. (The fact that in these cases there may be more than two joins is irrelevant.) For case (a) the probability that any three given cells will be black is $r_1^{(3)}/b^{(3)}$ and we denote by B the number of ways in which these three cells may be chosen so that they will be connected by at least two joins. For case (b) the probability that any four given cells will be black is $r_1^{(4)}/b^{(4)}$ and we denote by C the number of ways of choosing two pairs of adjoining cells. Then the second factorial moment about the origin of the distribution is

$$\mu'_{(2)} = 2!\left[B\frac{r_1^{(3)}}{b^{(3)}} + C\frac{r_1^{(4)}}{b^{(4)}} \right].$$

It is now necessary to determine A, B, and C. Consider again the grid of $mn = b$ cells. It has four corner cells, $2(m + n - 4)$ cells that are on the edges of the grid but not at the corners, and $(m - 2)(n - 2)$ interior cells.

To determine A note that if any one corner cell is black there are three ways of choosing an adjoining cell that could also be black. If any one edge cell is black, there are five adjoining cells that could also be black, and if any interior cell is black there are eight adjoining cells that could also be black. Therefore $2A = 4 \times 3 + 2(m + n - 4) \times 5 + (m - 2)(n - 2) \times 8$; (the factor 2 on the left-hand side occurs because all joins have been counted twice). Then

$$A = 2 - 3(m + n) + 4mn$$

or

$$A = 2 - 3a + 4b.$$

Next we find B. If a corner cell is to be connected by joins to two other cells, there are $\binom{3}{2}$ ways of choosing the two others; if an edge cell is to be joined to two others, there are $\binom{5}{2}$ ways of choosing the two adjoining cells; and for an interior cell there are $\binom{8}{2}$ ways of choosing the two adjoining cells.

Then

$$B = 4\binom{3}{2} + 2(m + n - 4)\binom{5}{2} + (m - 2)(n - 2)\binom{8}{2}$$

$$= 44 - 36a + 28b.$$

To find C we note that there are A ways of obtaining a single adjoining pair of cells, hence $\binom{A}{2}$ ways of obtaining two such pairs; but these include the trios already discussed, of which there are B. Thus there are $\binom{A}{2} - B$ ways of obtaining a couple of distinct pairs (i.e., four cells altogether).

Therefore

$$C = \binom{A}{2} - b.$$

It now follows that the mean and variance of the number of black-black joins, on the null hypothesis that the black and white cells are randomly mingled, are

$$m = \mu'_{(1)} = \frac{A r_1^{(2)}}{b^{(2)}}$$

and

$$V = 2B \frac{r_1^{(3)}}{b^{(3)}} + 2C \frac{r_1^{(4)}}{b^{(4)}} + m - m^2,$$

since

$$V = \mu'_{(2)} + m - m^2.$$

The distribution is asymptotically normal. Although it seems to be unknown how large m and n must be for an assumption of normality to give a good approximation, Krishna Iyer uses the test for a lattice with $m = 15$ and $n = 20$. If, then, we have a grid with a sufficient number of cells, it is easy to test whether the black and white cells (or dense and sparse quadrats) are randomly mingled.

If the cells were very small and the pattern aggregated, we should expect an excessive number of black-black joins, since any two adjoining cells are likely to be both within or both outside of the densely populated patches. With large cells, on the other hand, even if the pattern were nonrandom the grid might be so coarse that the number of individuals in any one cell would be independent of the number in the cells adjoining it. The sparse and dense cells would then be randomly mingled.

It is not permissible (owing to dependence among the results) to sample the same area repeatedly with coarser and coarser grids in search of a cell size at which evidence of aggregation would vanish, but, if we wished to test a previously held hypothesis about the grain of a tract of vegetation, the test could be useful. It has not been used in this way so far by ecologists.

10

Studying Pattern by Distance Sampling

1. Introduction

Quadrat sampling is only one of the methods that may be used to study the spatial pattern of a population of organisms dispersed over a continuous surface. As we have remarked, it suffers from the grave disadvantage that quadrats are not natural sampling units but are necessarily arbitrary. There is, however, a wholly different method of investigating the pattern of points in a plane. This is by so-called "plotless sampling." What is examined is the spacing of the individuals, and there are two ways to proceed. We may locate sampling points at random throughout the area and measure the distance from each point to the individual nearest it, or, alternatively, select individuals at random from the whole population and from each of these measure the distance to its nearest neighboring individual. In either case the data consist of an empirical frequency distribution of a continuous variate, distance. Moreover, if the pattern of the population is random, the results are unaffected by whether random points or random individuals are used as the origins of measurement.

2. The Distribution of Distance-to-Neighbor in a Randomly Dispersed Population

Consider the distance from a random point (or a randomly chosen individual) to its nearest neighbor. We wish to find the probability distribution of this distance, given that the pattern of the population is random.

The required distribution function (cumulative distribution function) is, by definition,

$$F(r) = \Pr(\text{distance to nearest neighbor} \leq r).$$

This is the probability that a circle of radius r centered on the point will contain at least one individual or, equivalently, that this circle is not

148

empty. Therefore

$$F(r) = 1 - e^{-\lambda r^2},$$

where the Poisson parameter λ denotes the mean number of individuals per circle of unit radius. The term $e^{-\lambda r^2}$ is, of course, the probability that a Poisson variate will take the value 0.

The probability density function (or frequency function) of the distribution and its mean and variance follow immediately. Thus the probability density function (pdf) is

$$f(r) = F'(r) = 2\lambda r e^{-\lambda r^2}.$$

The mean is

$$E(r) = \int_0^\infty 2\lambda r^2 e^{-\lambda r^2}\, dr.$$

Using the substitution $r^2 = x$, hence $dr = dx/2\sqrt{x}$ gives

$$E(r) = \lambda \int_0^\infty \sqrt{x}\, e^{-\lambda x}\, dx$$

$$= \frac{1}{2}\left(\frac{\pi}{\lambda}\right)^{1/2} \tag{10.1}$$

The second moment about the origin is

$$E(r^2) = \int_0^\infty 2\lambda r^3 e^{-\lambda r^2}\, dr = \frac{1}{\lambda},$$

whence the variance is

$$\mathrm{var}(r) = E(r^2) - [E(r)]^2 = \frac{4 - \pi}{4\lambda}. \tag{10.2}$$

It is simpler and more convenient to use the square of the distance rather than the distance itself as variate. Putting $\omega = r^2$, the distribution function of the squared distance is then

$$F(\omega) = 1 - e^{-\lambda\omega}$$

and the pdf is

$$f(\omega) = \lambda e^{-\lambda\omega}.$$

The mean and variance of ω are easily found to be

$$E(\omega) = \frac{1}{\lambda} \quad \text{and} \quad \mathrm{var}(\omega) = \frac{1}{\lambda^2}.$$

We next inquire into the sampling distribution of $\bar{\omega}$, the mean of a random sample of n values of ω. Since the distribution function of ω is

$$F(\omega) = 1 - e^{-\lambda\omega},$$

if we put $y = 2\lambda\omega$, the distribution function and the pdf of y are, respectively,

$$G(y) = 1 - e^{-y/2} \quad \text{and} \quad g(y) = \tfrac{1}{2}e^{-y/2}.$$

We now obtain the moment generating function (mgf) of y, namely,

$$E(e^{ty}) = \tfrac{1}{2}\int_0^\infty e^{-y/2} e^{ty}\, dy$$

$$= (1 - 2t)^{-1}.$$

It will be seen that this is the mgf of the χ^2-distribution with two degrees of freedom (e.g., Hoel, 1954).

We can now obtain the mgf of $n\bar{y}$, the sum of n independent values of y, since it is given by

$$[E(e^{ty})]^n = (1 - 2t)^{-n},$$

and this is the mgf of the χ^2-distribution with $2n$ degrees of freedom. This, then, is the distribution possessed by $n\bar{y} = 2n\lambda\bar{\omega}$; that is, $2n\lambda\bar{\omega} = z$, say, has pdf

$$h(z) = \frac{1}{2^n \Gamma(n)} z^{n-1} e^{-z/2},$$

which is the pdf of the χ^2-distribution with $2n$ degrees of freedom. Then $\bar{\omega} = z/2n\lambda$ has pdf

$$f(\bar{\omega}) = 2n\lambda \frac{1}{2^n \Gamma(n)} (2n\lambda\bar{\omega})^{n-1} e^{-n\lambda\bar{\omega}}$$

$$= \frac{(n\lambda)^n \bar{\omega}^{n-1} e^{-n\lambda\bar{\omega}}}{\Gamma(n)}, \tag{10.3}$$

which is the distribution we set out to determine.

Since $2n\lambda\bar{\omega}$ has a χ^2-distribution with $2n$ degrees of freedom and since, also, for the χ^2-distribution the expectation is equal to the number of degrees of freedom, it follows that

$$E(2n\lambda\bar{\omega}) = 2n \quad \text{or} \quad E(\bar{\omega}) = \frac{1}{\lambda}. \tag{10.4}$$

This suggests that we might use $1/\bar{\omega}$ as an estimator of λ, the mean number of individuals per circle of unit radius. However, it is a biased estimator, as Moore (1954) has shown.

Thus

$$
\begin{aligned}
E\left(\frac{1}{\bar{\omega}}\right) &= \int_0^\infty \frac{1}{\bar{\omega}} \frac{(n\lambda)^n \bar{\omega}^{n-1} e^{-n\lambda\omega}}{\Gamma(n)} \cdot d\bar{\omega} \\
&= \frac{(n\lambda)^n}{\Gamma(n)} \int_0^\infty \bar{\omega}^{n-2} e^{-n\lambda\bar{\omega}} d\bar{\omega} \\
&= \frac{n}{n-1} \cdot \lambda.
\end{aligned}
$$

So to obtain an unbiased estimator, $\tilde{\lambda}$, say, from the mean of n observed values of ω we should need to put

$$
\tilde{\lambda} = \frac{n-1}{n} \cdot \frac{1}{\bar{\omega}}. \tag{10.5}
$$

Unfortunately, it is not possible in practice to estimate the density of a natural population of organisms by substituting an observed value of $\bar{\omega}$ in (10.5). The formula $(n-1)/n\bar{\omega}$ is an estimator of the density only when the population has a random pattern, so it can be used only if we can safely assume, or know beforehand, that a pattern is indeed random. However, we are never justified in assuming randomness without evidence for it; the assumption must always be tested. It turns out that there is no method of performing such a test by using only an observed sample of ω values. Even if an empirical distribution of ω values appears to be well fitted by the theoretical distribution $f(\omega) = \lambda e^{-\lambda\omega}$ it cannot be concluded that the pattern is random, since empirical distributions from nonrandom patterns are often not distinguishably different from this negative exponential form. Randomness therefore cannot be assumed without a test based on additional observations, and these observations entail estimating, or determining, the density; that is, distance measurements alone are not enough, and we must also carry out quadrat sampling or a complete count of the population. Thus an estimate of density has to be obtained *before* a test for randomness can be made, and then there is no longer any need to use the distance measurements for density estimation.

A way around the difficulty has been suggested by Batcheler (1971). He examined a number of artificially constructed populations whose patterns ranged from regular to strongly aggregated and on each of them measured a sample of paired distances. These were r_p, the distance from a random point to its nearest individual, and r_i, the distance from *this* individual to its nearest neighbor. The statistic $\sum r_p / \sum r_i = B$, say, depends, of course, on the pattern of the population: a regular pattern yields a low B and an aggregated pattern a high B. Now let λ be the true density of the population; and let d be a biased estimate,* based solely on the r_p measurements, for which the bias would be zero only if the pattern were random. Batcheler found empirically that, over the range of populations he tested, $\log(d/\lambda)$ was a linear function of B, say $\log(d/\lambda) = c_1 B + c_2$, with $c_1 < 0$. Therefore, as an estimator of $\log \lambda$, one has $\log \lambda = \log d - c_1 B - c_2$, using empirical values for the constants. Unfortunately, the standard error of this estimator has not been derived.

3. Tests for Randomness Based on Distance Measurements

Combinations of distance measurements and density estimates have been variously used to construct tests for randomness and indices of aggregation. Three such procedures are described here and for conciseness we assume that the individuals are plants.

1. The test due to Hopkins and Skellam (1954) hinges on the fact that if, and only if, a pattern is random, the distribution of the distance from a random point to its nearest plant is identical with the distribution of the distance from a random plant to its nearest neighbor.

Denote by ω_1 the square of a point-to-plant distance and by ω_2 the square of a plant-to-neighbor distance and suppose a sample is obtained of n distances of each kind. The statistic $A = \sum \omega_1 / \sum \omega_2$ then has an expected value of 1 if the pattern is random and A may be used as a measure of nonrandomness. Clearly, if the plants are aggregated, we shall have $A > 1$; conversely, if they are more evenly spaced than in a randomly dispersed population, $A < 1$.

To determine whether A departs significantly from its expected value of 1 we determine the sampling distribution of

$$x = \frac{A}{1+A} = \frac{\sum \omega_1}{\sum \omega_1 + \sum \omega_2}.$$

To find the desired sampling distribution put

$$x = \frac{\sum \omega_1}{\sum \omega_1 + \sum \omega_2} = \frac{a}{a+b}, \text{ say.}$$

For convenience we may take $\lambda = 1$; this entails no loss of generality, since we are always at liberty to choose the unit of distance in such a way that the mean number of plants per circle of unit radius is 1. Then, using

---

*The estimate d is based on a subset of the measured r_p values, not on all of them. Details are given in Batcheler (1971).

(10.3), it is seen that $\sum \omega_1 = n\bar{\omega}_1 = a$ has pdf

$$f(a) = \frac{a^{n-1}e^{-a}}{\Gamma(n)} \, ;$$

likewise $\sum \omega_2 = n\bar{\omega}_2 = b$ has a pdf of the same form. Now consider $a = bx/(1-x)$. The probability that, for fixed b, a will lie in the interval da, hence x in the interval dx, is

$$f(a)\, da = \frac{a^{n-1}e^{-a}}{\Gamma(n)} \, da,$$

which is equivalent to

$$g(x)\, dx = \frac{1}{\Gamma(n)}\left(\frac{bx}{1-x}\right)^{n-1} \exp\left(\frac{-bx}{1-x}\right) \frac{b}{(1-x)^2} \, dx,$$

since

$$da = \frac{b}{(1-x)^2} \cdot dx.$$

Thus for a given value of b

$$g(x \mid b)\, dx = \frac{1}{\Gamma(n)} \frac{b^n x^{n-1}}{(1-x)^{n+1}} \exp\left(\frac{-bx}{1-x}\right) dx.$$

Now allow b to vary. The probability that it will lie in the interval db is

$$f(b)\, db = \frac{b^{n-1}e^{-b}}{\Gamma(n)}.$$

Therefore the probability that b will be in the interval db and x in the interval dx simultaneously is

$$f(b)\, db \cdot g(x)\, dx = \frac{1}{[\Gamma(n)]^2} \frac{x^{n-1}}{(1-x)^{n+1}} \left[\int_0^\infty b^{2n-1} \exp\left(\frac{-b}{1-x}\right) db\right] dx$$

$$= \frac{1}{[\Gamma(n)]^2} \frac{x^{n-1}}{(1-x)^{n+1}} \frac{\Gamma(2n)}{(1-x)^{-2n}} \, dx.$$

Then

$$g(x) = \frac{\Gamma(2n)}{[\Gamma(n)]^2} x^{n-1}(1-x)^{n-1}, \qquad \text{with} \qquad 0 \le x \le 1,$$

$$= \frac{x^{n-1}(1-x)^{n-1}}{B(n,\, n)}$$

and it is seen that x is a beta variate. Since

$$E(x) = \frac{B(n+1,\, n)}{B(n,\, n)} \qquad \text{and} \qquad E(x^2) = \frac{B(n+2,\, n)}{B(n,\, n)},$$

the mean and variance are

$$E(x) = \tfrac{1}{2} \quad \text{and} \quad \text{var}(x) = [4(2n+1)]^{-1}.$$

The distribution tends to normality rapidly with increasing n, and for $n > 50$ we may treat

$$\frac{x - E(x)}{\sqrt{\text{var}(x)}} = 2(x - \tfrac{1}{2})\sqrt{(2n+1)}$$

as a standardized normal variate.

It seems at first sight that this is a test that depends on distance measurements only and that no prior knowledge of the population's density is involved. This is not so, however. In order to choose a random individual from which to measure the distance to its nearest neighbor, the only satisfactory method is to put numbered tags on all the plants in the population and then to consult a random numbers table to decide which of the tagged plants are to be included in the sample. In doing this, we acquire willy-nilly a complete count of the population from which its density automatically follows. There is another method of picking random plants, but it, too, requires that the size of the total population be known. If a sample of size n, say, is wanted from a population of size N, the probability that any given plant in the population will belong to the sample is $p = n/N$. We must then take each population member in turn and decide by some random process whose probability of "success" is p whether that member is to be admitted to the sample. Even if we are willing to guess the magnitude of N intuitively and assign to p a value that will give a sample of approximately the desired size, it is still necessary to subject every population member to a "trial" in order to decide whether it should be included in the sample; as the successive trials are performed, a complete census of the population is automatically obtained.

It must be stressed that it is *not* permissible to take, as a "randomly chosen" member of the population, the plant nearest to a random point. This gives a biased sample, a fact to which we shall return.

2. The test proposed by Clark and Evans (1954) requires a knowledge of population density and a sample of n values of r, the distance from a random plant to its nearest neighbor. These distances are not squared.

Let ρ be the number of plants per unit area; that is $\rho = \lambda/\pi$, where λ is the measure of density we have used hitherto, the number of plants per circle of unit radius.

From (10.1) we see that in a randomly dispersed population $E(r) = (2\sqrt{\rho})^{-1}$ and from (10.2), that $\text{var}(r) = (4 - \pi)/4\pi\rho$.

Write $\bar{r}$ for the mean of the observed distances. Then, if n, the sample size, is large enough, we may assume that $\bar{r}$ is normally distributed with expectation $(2\sqrt{\rho})^{-1}$ and standard error $\sqrt{[(4-\pi)/4n\pi\rho]}$. This enables us to test for randomness, provided ρ is known exactly. No allowance is made for the sampling variance of an estimated value of ρ.

As an index of nonrandomness, we may use the ratio of the observed to the expected mean distance or

$$R = \frac{\bar{r}}{E(r)} = 2\bar{r}\sqrt{\rho}.$$

Then in a random population $E(R) = 1$; for aggregated populations $R < 1$, since $\bar{r}$ is the mean of *plant*-to-neighbor distances.

3. A third test was described by Pielou (1959) and Mountford (1961). Suppose that we have a sample of n distances measured from random *points* to their nearest plants and that we take the square of the distance, $\omega = r^2$, as variate. Let the population density (in terms of plants per unit area) be ρ. Taking as an index of nonrandomness $\alpha = \bar{\omega}\pi\rho$, we see from (10.4) that $E(\alpha) = 1$ in a population of random pattern. Now suppose that ρ, instead of being exactly known, is estimated. The estimate, denoted by $\hat{\rho}$, is obtained from a sample of m randomly placed quadrats of unit area. Thus $\hat{\rho}$ and $\bar{\omega}$ are both subject to sampling errors. Mountford has shown that

$$E(\alpha) = E(\pi\hat{\rho}\bar{\omega}) = 1$$

and that

$$\text{var}(\alpha) = \frac{1}{n}\left(1 + \frac{n+1}{m\rho}\right).$$

To obtain an estimate of $\text{var}(\alpha)$, we must substitute the estimator $\hat{\rho}$ in place of the population value ρ in the last formula.

4. Comparison of the Various Indices of Aggregation

Each of the indices of aggregation we have considered is only a single statistic and can therefore describe only a single aspect of pattern. Each should be thought of as providing only a measure of the extent to which a pattern departs from randomness and nothing more.

A spatial pattern in a continuum obviously has two quite distinct aspects: they may be called *intensity* and *grain*. By the intensity of a pattern we mean the extent to which density varies from place to place. In a pattern of high intensity the differences are pronounced and dense clumps alternate with very sparsely populated zones; when intensity is low, the density contrasts are comparatively slight. The grain of a pattern is independent of its intensity. If the clumps or patches in which the density is relatively high are large in area and widely spaced, we may say that the pattern is coarse-grained. Conversely, if the whole range of different densities is encompassed in a small space, the pattern is fine-grained.

The indices of aggregation calculated from data obtained by sampling with quadrats of one size are all measures of the intensity of a pattern. Lloyd's index of patchiness C and Morisita's index of dispersion I_δ depend on (a) the relative areas occupied by patches of different density (though not on the degree to which these areas are fragmented, which is a question of grain), and (b) on the ratios of the different densities to one another. The V/m ratio, David and Moore's index of clumping, I, and Lloyd's "mean crowding," $\overset{*}{m}$, depend on (a) and (b) and also on the absolute magnitude of the mean density which does not affect C and I_δ. It must be reemphasized that all of these indices measure only the intensity, and not the grain, of a pattern. To study grain by means of quadrat sampling it is necessary to use several sizes of quadrats as explained in Chapter 9.

Consider now the indices of aggregation based on distance measurements. Of these, Clark and Evans's index R clearly measures only the intensity of a pattern. Since the distances are measured from plant to plant, most of them will be within-clump distances: the denser the clumps the shorter the measured distances and the smaller the value of R. Thus this index is possibly the best if one particularly wishes to measure pattern intensity. However, it has two defects: it involves very troublesome field work in selecting individuals at random from which to measure distances, and the sampling properties of $\bar{r}$, when the population mean density ρ is estimated from quadrat data, have not yet been worked out.

The two other indices based on distances, Hopkins and Skellam's A and Pielou's and Mountford's α, use point to plant distances. They are therefore influenced by both intensity and grain. This is because some of the random points will fall within high density clumps and the point's nearest neighbor will be a clump member; others will fall where the density of the plants is low, and the nearest one may be in the same sparsely populated patch of ground or in the nearest dense clump.

Separate methods of studying intensity and grain by means of distance measurements have yet to be devised. One possibility, not yet explored, would be to consider not merely the means $\bar{r}$ and $\bar{\omega}$ of the measured distances and their squares but also their variances $\text{var}(r)$ and $\text{var}(\omega)$.

Another possibility has been explored by Morisita (1954) and Thompson (1956). They suggested that more detailed information on a population's pattern could be gained by measuring the distances from random

points (or plants) to the nearest, second nearest, third nearest, ...,
neighbors. The method seems to have been little used, probably because
the field work is exasperatingly difficult. In a forest, for instance, it is
usually quite easy to determine which trees are the nearest and second
nearest to a given point; even the third and fourth nearest may be readily
recognizable, but the nth neighbor for higher values of n is often difficult
to determine. So many trees are approximately equidistant from the point
that it becomes almost impossible to rank them or would require such
careful measurements that it would not be worth the effort.

It is intuitively apparent, however, that an index of aggregation based
on $\omega_n (= r_n^2$, where r_n is the distance of the nth nearest plant) would
decrease with increasing values of n. It would decrease most rapidly in
populations with clumps that were small in area, closely spaced, and of
few members. In any case, it remains to be discovered what can be
learned about a pattern from examination of a curve of index of aggregation versus rank of the neighbor used for its calculation.

The distribution of ω_n when a pattern is random is easily derived. As
before, let the mean number of individuals per circle of unit radius be λ.
By arguments analogous to those on page 148 we see that the distribution
function of ω_n is the probability that a circle of area $\pi \omega_n$ centered on a
random point contains at least n individuals; that is

$$F(\omega_n) = 1! - e^{-\lambda \omega_n} - \frac{\lambda \omega_n e^{-\lambda \omega_n}}{1!} - \cdots - \frac{(\lambda \omega_n)^{n-1} e^{-\lambda \omega_n}}{(n-1)!},$$

whence the pdf is

$$f(\omega_n) = F'(\omega_n) = \frac{\lambda^n \omega_n^{n-1} e^{-\lambda \omega_n}}{\Gamma(n)}.$$

The mean and variance of this distribution are

$$E(\omega_n) = \frac{n}{\lambda} \quad \text{and} \quad \text{var}(\omega_n) = \frac{n}{\lambda^2}.$$

This accords with what we should expect intuitively. Comparing this
result with (10.4), we see that the expectation of ω_n in a population of
density λ is the same as the expectation of ω_1 (the squared distance to the
first, or nearest, neighbor) in a population of density λ/n.

5. Sampling Isolated Individuals

It was remarked earlier (page 154) that we cannot obtain a random
sample of the individuals in a population by selecting those that are
nearest to random points. This procedure gives a biased sample in which

relatively isolated individuals are over-represented. To see this, consider a population of just three members, arranged as follows:

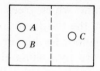

A random point placed in the area has equal probability of falling in the left or right halves, separated by the dashed line. Therefore the isolated individual C is twice as likely as either A or B to be the individual nearest to a random point.

This fact is useful if a sample biased in favor of isolated individuals is what we require. Suppose we want to compare isolated with crowded individuals in order to determine whether they differ in some attribute such as size, age, or susceptibility to disease. Classifying the attribute dichotomously (e.g., big-small, old-young, diseased-healthy) we may compare the proportion of individuals having the attribute in the whole population with the proportion having the attribute in a sample biased in favor of isolated individuals. The biased sample is obtained by taking the nearest individual to each of a number of random points. If the proportions differ significantly, we may conclude that possession of the attribute is related to the degree of isolation of an individual. The method is certainly rather crude, but it obviates the necessity of setting up some arbitrary definition of the "degree of isolation" of an individual and of devising and executing measurements to determine it for each of a large number of individuals (for an example of an application, see Pielou and Foster, 1962).

6. Reciprocity of the Nearest-Neighbor Relation

Whenever a population of individuals is dispersed over a plane, there will be some pairs of individuals such that each is the nearest neighbor of the other, that is, the two individuals are closer to each other than eith.er is to a third individual. For these pairs the nearest-neighbor relationship is reciprocal or reflexive (Clark and Evans, 1955). Given a random pattern, we may calculate the expected number of individuals that are members of reciprocal pairs.

In the diagram that follows, let I_1 and I_2 be two individuals a distance r apart. Then the probability that I_2 is I_1's nearest neighbor is the same as the probability that I_1's nearest neighbor will be at a distance in the range

$(r, r + dr)$. This probability is

$$2\pi\rho re^{-\pi\rho r^2}\, dr = p_1, \text{ say,}$$

where ρ is the density of the population in terms of numbers of individuals per unit area (not per circle of unit radius). From the diagram below we see that the probability that I_2 has I_1 as its nearest neighbor is the probability that the shaded crescent, of area $r^2(\sqrt{3}/2 + \pi/3)$, is empty, and this is

$$p_2 = \exp\left[-\rho r^2\left(\frac{\sqrt{3}}{2} + \frac{\pi}{3}\right)\right].$$

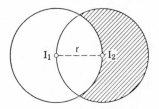

The probability that I_1 will form a reciprocal pair with I_2 (at a distance r apart) is then $p_1 p_2$. Integration over all possible values of r gives the probability P that any individual will be a member of a reciprocal pair. Thus

$$P = \int_0^\infty 2\pi r\rho \exp\left[-\rho r^2\left(\frac{\sqrt{3}}{2} + \frac{4\pi}{3}\right)\right]\, dr$$

$$= \frac{6\pi}{3\sqrt{3} + 8\pi} \quad \text{or} \quad 0.6215.$$

The expected proportion of individuals that belong to reciprocal pairs is thus 0.6215 in a population of random pattern. How we might interpret observed departures from expectation is not clear except for the obvious conclusion that an excessively high proportion of reciprocal pairs would support the hypothesis (if it were being tested) that individuals tended to occur as isolated couples. Situations in which such a hypothesis would be worth postulating and in need of testing are hard to think of.

7. The Detection of Regular Spacing

It has so far been assumed that nonrandom patterns always depart from randomness in the direction of aggregation. "Regular" patterns, those in which the individuals are more evenly spaced than they would be in a randomly dispersed population, are easy to visualize but surprisingly rare in nature. The pattern of the nests of colonially nesting birds is sometimes regular; an example (from Bartlett, 1974) is shown in Figure 10.1; (the variate in the figure is distance, r, not squared distance, ω).

Regular patterns in plant communities are very uncommon even though one might have expected that in a dense, even-aged forest, for instance, competition among the trees would cause those that survived beyond the seedling stage to be regularly spaced. Apparently this seldom,

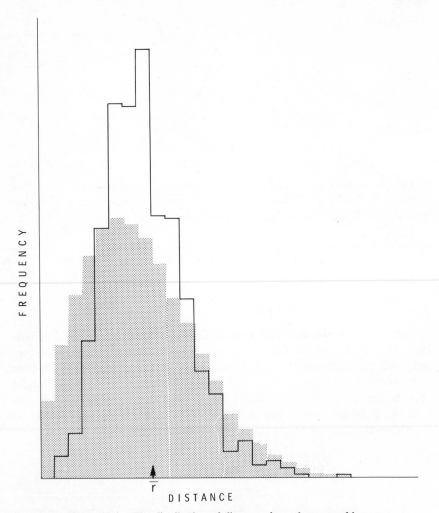

Figure 10.1. The distribution of distances from the nests of lesser black-backed gulls to the nearest neighboring nests; the expected histogram is stippled. (Redrawn from Bartlett, 1974.)

if ever happens. Evidence for regularity has been sought both by quadrat sampling and by using the tests based on distance measurements; the results of these searches have nearly always been negative.

The probable explanation is that, although in very dense parts of a forest short tree-to-neighbor distances may be rare (and distances shorter than a tree's diameter are obviously impossible, since we are assuming that distances are measured from center to center of the individuals), a dearth of short distances may go undetected if the forest as a whole has an aggregated or random pattern. In any comparison of observed and expected values of r or of ω by means of a χ^2-test subdivision of the observed distances into classes is always necessary. If the class containing the smallest variate values is defined to include not only the extremely low values that are impossible or rare but also the moderately low values that are numerous, the rarity of the former will be obscured. Figure 10.2a shows two theoretical probability distributions, A and B, both of which are representable by the same histogram shown in Figure 10.2b. From examination of the histogram we cannot tell whether the distribution of ω is like A or B. Curve B, which is exponential, is what we should expect if the trees' spacing were random and unaffected by competition. Curve A is what would result if competition prevented the occurrence of very short intertree distances. If an empirical distribution, portrayed as a histogram, were like that in Figure 10.2b, we should have no reason to suppose that the parent distribution is like A.

Failure to distinguish between the two kinds of curve will often occur if the data consist of observations on all tree-to-neighbor distances, the long as well as the short. Intertree competition is likely to affect spacing only in localized patches in which the tree density is especially high. To detect it, therefore, it is best to take a sample of short tree-to-neighbor distances only, choosing some arbitrary upper limit for distances to be admitted to the sample. We then have for comparison observed and theoretical distributions both of which are truncated and we wish to judge whether, within this sample, the relative frequency of very short distances fails significantly short of expectation.

It is now necessary to subdivide the observations into classes preparatory to doing a goodness of fit test. The subdivision must, of course, be done in a way that does not influence the outcome; class boundaries must be set objectively and this is difficult to do if one is aware, while setting them, how many observed variate values are going to fall in each class.

The best solution to the difficulty is to set class boundaries so as to give equal expected frequencies in all the classes. For example one might choose to have ten classes in each of which 10% of the observed variate values are expected to fall. Then if the observed frequency in the first of

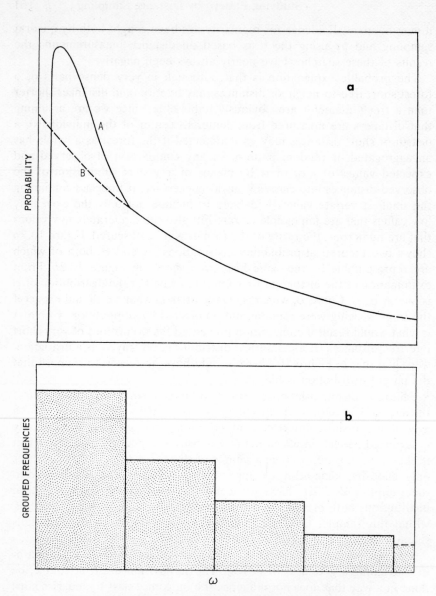

Figure 10.2a. Two possible distributions for ω, the square of the distance from a plant to its nearest neighbor. Curve B is exponential. For all but the shortest distances the curves are identical.

Figure 10.2b. Histogram corresponding to both curves in Figure 10.2a. The data are coarsely grouped for purposes of illustration.

these intervals (corresponding to the smallest ω values) were significantly less than expectation, the conclusion would follow that the trees were evenly spaced over very short distances.

The numerical values of the class boundaries are obtained as follows. Let the upper limit of the measured distances in the sample be $\sqrt{c}$ so that $0 \le \omega \le c$. Then under the null hypothesis that the untruncated population of ω's has the distribution function

$$F(\omega) = 1 - e^{-\lambda \omega}, \qquad 0 \le \omega,$$

the truncated form is

$$F(\omega \mid 0 \le \omega \le c) = \frac{1 - e^{-\lambda \omega}}{1 - e^{-\lambda c}}$$

with expectation

$$E(\omega) = \frac{1}{\lambda} - \frac{c e^{-\lambda c}}{1 - e^{-\lambda c}} \tag{10.6}$$

Now suppose we wish to subdivide the range of the variate into i class intervals in such a way that the probability that a randomly selected value of ω will fall into any one of them is the same for all and equal to $1/i$. Put ω_r for the upper bound of the rth interval $(r = 1, \ldots, i)$. Then it is seen that

$$\frac{F(\omega_r) - F(0)}{F(c) - F(0)} = \frac{1 - \exp[-\lambda \omega_r]}{1 - \exp[-\lambda c]} = \frac{r}{i}. \tag{10.7}$$

Solving for ω_r now gives the required class boundaries, namely

$$\omega_r = \begin{cases} \dfrac{-1}{\lambda} \ln\left[1 - \dfrac{r}{i}(1 - e^{-\lambda c}) \right] & \text{when} \qquad r = 1, 2, \ldots, (i-1) \\ c & \text{when} \qquad r = i. \end{cases} \tag{10.8}$$

In practice λ in (10.8) must be replaced by an estimate, say $\hat{\lambda}$, of the true population density. To obtain such an estimate from the data in hand one may substitute the observed mean, $\bar{\omega}$, for the theoretical expectation $E(\omega)$ in (10.6) and solve* the latter to give $\lambda = \hat{\lambda}$. [Use of an estimate of λ based on the sample, causes (10.8) to be approximate rather than exact, but for $n \ge 50$ the discrepancy is negligible; see David and Johnson (1948).]

The final step in testing for pattern regularity consists in comparing the observed frequency of ω values in the class $[0, \omega_1)$, say f_1, with its expectation n/i.

---

*To facilitate the solving of (10.6), a table of values of $(1/k) - [e^{-k}/(1 - e^{-k})]$ for $k = 0.1(0.1) \cdots 5.0$ is given in Pielou (1974).

Under the null hypothesis,

$$\frac{\left(\dfrac{n}{i}-f_1\right)}{\sqrt{\left\{n\dfrac{1}{i}\left(1-\dfrac{1}{i}\right)\right\}}}=\frac{n-if_1}{\sqrt{n(i-1)}}$$

is a standard normal variate. A one-tail test should be used since the alternative of interest is $f_1 < n/i$.

Figure 10.3 shows an example. Measurements of tree-to-nearest-neighbor distances in two-species conifer parkland in arid country had yielded an untruncated sample of 228 values of ω. Of these, 104 were less

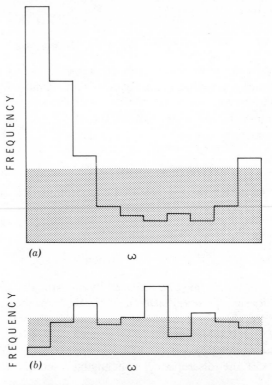

Figure 10.3. Distributions of ω values from open conifer woodland. The outline histograms show the observed frequencies and the stippled histograms show the expected frequencies. The abscissa scale is transformed to make the expected distribution rectangular. (a) Data untruncated; $n = 228$. (b) Data truncated at $c = 0.90$; $n = 104$. (Redrawn from Pielou, 1962a.)

than the chosen truncation level $c = 0.90$. The observed and expected histograms for the untruncated (above) and truncated (below) distributions of ω are shown in the figure. The boundaries of the class intervals were obtained from (10.8) with $i = 10$; thus the expected distributions are rectangular in each case, with equal expected frequencies in all 10 classes. As is obvious from the figure, comparison of the observed and expected distributions of the untruncated data gave no hint of regular spacing but, on the contrary, implied a strongly aggregated pattern: the observed frequency of the smallest ω values greatly exceeded expectation. Examination of the truncated data, on the other hand, yielded clear evidence that the frequency of very short tree-to-neighbor distances fell short of expectation; the discrepancy is significant at the 1% level.

This method of testing for regular spacing permits an objective test to be applied to "selected" (i.e., truncated) data. Thus it overcomes the commonly encountered difficulty that evidence for within clump regularity may be masked by that for overall aggregation in a clumped population.

Regular spacing in populations of sessile organisms, or in the nests of motile ones, is of obvious ecological interest. Regularity in plant population patterns may imply either that the individuals are competing with one another for a resource, such as water, in short supply; or that the plants are allelopathic, that is, that each individual secretes toxic exudates that prevent the establishment of other individuals close to it. In animal populations, territoriality obviously leads to regular spacing.

11

Patterns Resulting from Diffusion

1. Introduction

The discussion of spatial patterns in preceding chapters treated these patterns as static. No consideration has been given to the fact that any pattern must have a history. To reach the locations in which we observe them the organisms of a population must have moved, either actively or passively. The movements of animal populations are of two types: migration and diffusion. Well-known examples of migrations are those of many bird species, caribou herds, breeding seals, locust swarms, butterflies, and the spawning migrations of fishes such as eels and salmon; the list could be greatly extended. Although there is presumably some stochastic element in these mass movements, it seems reasonable to suppose it is overshadowed by the deterministic or "deliberate" elements.

In contrast to migration, diffusion* consists in the apparently aimless, undirected movements of animals that seem to be wholly random. For instance, suppose a group of organisms of one species is concentrated at some time $t = 0$ within a small space entirely surrounded by terrain suitable for the species; if we study how the group of organisms spreads out to occupy a larger area, we are studying diffusion. We may consider short-term diffusions, that is, what happens to the original animals within their lifetimes or an even shorter period, or long-term diffusions occupying many generations of the species.

Plant diffusion also invites investigation. Changes in habitat conditions are continually altering the suitability of areas of ground for different species of plants. During the sequence of successional stages at any place the area will become suitable to a succession of different colonizing species that invade from surrounding areas. More pronounced changes, for example, those wrought by fire, by the drying up of lakes, or by the

---

* An alternative word, favored by Odum (1959), for example, is "dispersal." Dispersal is to be contrasted with dispersion which, in ecological contexts, has the same meaning as pattern. Because of the similarity of the words *dispersal* and *dispersion*, it seems preferable to use *diffusion* when movement is meant.

melting of ice sheets, leave tracts of virgin ground ready to be colonized by plants from neighboring areas.

In short, the locations of animals and plants depend as much on how they got there in the first place as on how they establish themselves and survive once they have arrived. Considering, then, those population shifts that result entirely from random movements, we begin with a discussion of "random walks" and diffusion processes.

2. The Random Walk and Diffusion in One Dimension

Consider a particle that moves in discrete steps, at discrete times, along a line. Suppose at time $t = 0$ the particle is at the origin $x = 0$. To the left of the origin x takes the values $-1, -2, \ldots$, and to the right $+1, +2, \ldots$. At each of the times $t = 1, 2, \ldots$, the particle takes a step, to the right with probability p or to the left with probability $q = 1 - p$.

Let v_{rn} be the probability that after n steps the particle will be at $x = r$. To get there it must have done a total of j steps to the right and $n - j$ steps to the left, with $r = j - (n - j) = 2j - n$; then $j = (n + r)/2$ and $n - j = (n - r)/2$. The order of the steps is immaterial. Clearly, $n + r$ must be even, otherwise $v_{rn} = 0$.

We see that v_{rn} is a binomial probability; that is

$$v_{rn} = \binom{n}{j} p^j q^{n-j} = \binom{n}{\dfrac{n+r}{2}} p^{(n+r)/2} q^{(n-r)/2}.$$

Here $\binom{n}{j}$ is the number of ways in which the particle can reach $x = r$ in n steps.

Now let us pass to the limit, allowing both the step lengths and the time intervals to become infinitesimally small. First, let each step be of length Δx. We can now find the mean and variance of the displacement *per step*. The mean displacement will be positive if the net displacement is to the right of the origin and negative if to the left. We write the two possible events and their probabilities in the form of Table 11.1 and determine their mean and variance in the usual way.

The mean is $m = (p - q)\, \Delta x$ and the variance is

$$\sigma^2 = (p + q)(\Delta x)^2 - (p - q)^2 (\Delta x)^2$$
$$= 4pq(\Delta x)^2.$$

Suppose now that the time interval between steps is of length Δt. In a period of length t approximately $t/\Delta t$ steps will be taken (or exactly $t/\Delta t$ if

TABLE 11.1

DISPLACEMENT i	PROBABILITY π	πi	πi^2
$+\Delta x$	p	$p \, \Delta x$	$p(\Delta x)^2$
$-\Delta x$	q	$-q \, \Delta x$	$q(\Delta x)^2$

t is a multiple of Δt). Then the total displacement in time t has mean and variance given to a very close approximation by

$$m_t = \frac{t}{\Delta t}(p-q)\,\Delta x \qquad \text{and} \qquad \sigma_t^2 = \frac{t}{\Delta t} \cdot 4pq(\Delta x)^2.$$

Now let both $\Delta x \to 0$ and $\Delta t \to 0$. At the same time $\Delta x/\Delta t$ and also $(\Delta x)^2/\Delta t$ must be allowed to take suitably chosen values; otherwise nonsensical results will be obtained. Note that since p and q, the probabilities of steps to right and left, do not change with time and the steps occur at equal intervals the expected displacement must be proportional to the time elapsed. So we may put $m_t = 2ct$ where $2c$ is a constant of proportionality. Likewise, since the steps are independent, σ_t^2, the variance of the displacement at time t, is the sum of the variances pertaining to each step. Thus σ_t^2 is also proportional to t and we may write $\sigma_t^2 = 2Dt$ where $2Dt$ is another constant of proportionality. This is equivalent to putting

$$2c = (p-q)\frac{\Delta x}{\Delta t} \qquad \text{and} \qquad 2D = \frac{4pq(\Delta x)^2}{\Delta t}.$$

Thus we must allow Δx and Δt to tend to zero in such a way that c and D, hence the mean and variance of the displacement, remain finite. This is equivalent to requiring that the x- and t-scales be in appropriate ratio to each other. For the mean to remain finite it is also necessary that $p-q$ be small, of the same order of magnitude as Δx; that is, we must have $p-q = O(\Delta x)$. Then

$$4pq = 1 - (p-q)^2 = 1 - [O(\Delta x)]^2 \to 1 \qquad \text{as} \qquad \Delta x \to 0,$$

hence

$$\frac{(\Delta x)^2}{\Delta t} = 2D.$$

Also $p-q = c\Delta x/D$ and $p+q = 1$ and we have

$$p = \frac{1}{2} + \frac{c}{2D}\Delta x \qquad \text{and} \qquad q = \frac{1}{2} - \frac{c}{2D}\Delta x.$$

The two constants introduced, namely c and D, are known as the coefficients of *drift* and *diffusion*, respectively.

It will be recalled that v_{rn}, the probability that a particle will be at position r at time n, is a binomial probability. As the steps become short and numerous, this probability may therefore be approximated by the normal probability function with mean $2ct$ and variance $2Dt$; that is,

$$\phi(x, t)\, dx = \frac{1}{\sqrt{4\pi Dt}} \exp\left[-\frac{1}{4Dt}(x - 2ct)^2 \right] dx$$

is the probability that at time t, the particle's position will be in the interval $(x - \tfrac{1}{2}\, dx,\ x + \tfrac{1}{2}\, dx)$. If there is no drift, that is, if $c = 0$,

$$\phi(x, t) = \frac{1}{\sqrt{4\pi Dt}} \exp\left(\frac{-x^2}{4Dt}\right). \tag{11.1}$$

Starting from the same premises, we shall now derive the partial differential equation that describes diffusion with drift (the Fokker-Planck equation of physics), and we shall also show that (11.1) is a solution of it when the drift is zero.

Write $\phi(x, t)$ for the probability that at time t the particle will be at position x. We can immediately write down the difference equation

$$\phi(x, t + \Delta t) = p\phi(x - \Delta x, t) + q\phi(x + \Delta x, t),$$

since for the particle to be at x at time $t + \Delta t$ it must have been at either $x - \Delta x$ or $x + \Delta x$ at time t.

Expanding both sides by Taylor's theorem and writing ϕ for $\phi(x, t)$ gives

$$\phi + \frac{\partial \phi}{\partial t} \cdot \Delta t + \frac{1}{2!}\frac{\partial^2 \phi}{\partial t^2}(\Delta t)^2 + \cdots = \phi + (q - p)\frac{\partial \phi}{\partial x} \cdot \Delta x + \frac{1}{2!}\frac{\partial^2 \phi}{\partial x^2}(\Delta x)^2 + \cdots$$

or, on dividing through by Δt,

$$\frac{\partial \phi}{\partial t} + \frac{1}{2!}\frac{\partial^2 \phi}{\partial t^2} \cdot \Delta t + \cdots = (q - p)\frac{\partial \phi}{\partial x} \cdot \frac{\Delta x}{\Delta t} + \frac{1}{2!}\frac{\partial^2 \phi}{\partial x^2}\frac{(\Delta x)^2}{\Delta t} + \cdots .$$

As before let Δx and $\Delta t \to 0$ in such a way that $(p - q)\,\Delta x/\Delta t \to 2c$ and $(\Delta x)^2/\Delta t \to 2D$, whereas Δt and its powers (on the left-hand side) and $(\Delta x)^3/\Delta t$ and higher terms (on the right-hand side) tend to zero. Then

$$\frac{\partial \phi}{\partial t} = -2c\frac{\partial \phi}{\partial x} + D\frac{\partial^2 \phi}{\partial x^2}.$$

This is the Fokker-Planck equation for one-dimensional diffusion with drift.

If there is no drift so that $c = 0$,

$$\frac{\partial \phi}{\partial t} = D \frac{\partial^2 \phi}{\partial x^2}. \tag{11.2}$$

We now show that the normal probability function with mean zero, as given in (11.1), is a solution of (11.2). From (11.1)

$$\frac{\partial \phi}{\partial t} = \frac{e^{-x^2/4Dt}}{4\sqrt{\pi Dt^3}} \left(\frac{x^2}{2Dt} - 1 \right)$$

and

$$\frac{\partial^2 \phi}{\partial x^2} = \frac{e^{-x^2/4Dt}}{4\sqrt{\pi D^3 t^3}} \left(\frac{x^2}{2DT} - 1 \right),$$

so that $\partial \phi / \partial t = D(\partial^2 \phi / \partial x^2)$ as required.

So far we have considered only diffusion in one dimension; in the case of diffusion in the plane the diffusion equation is

$$\frac{\partial \phi}{\partial t} = D \nabla^2 \phi,$$

where

$$\nabla^2 = \frac{\partial^2}{\partial x^2} + \frac{\partial^2}{\partial y^2}.$$

The required solution is the joint normal distribution

$$\phi(x, y, t) = \frac{1}{4\pi Dt} \exp\left[\frac{-(x^2 + y^2)}{4Dt} \right], \tag{11.3}$$

with $\text{var}(x) = \text{var}(y) = 2Dt$ as before.

3. Alternative Derivation of the Two-Dimensional Diffusion Equation

In ecological contexts we are far more often concerned with two dimensions than with one. It is therefore worthwhile to derive the two-dimensional diffusion equation *de novo* in the way shown by Skellam (1951), for instance.

Imagine that a particle that can move in any direction in the plane is displaced through a distance ε at times t, $t + \Delta t$, $t + 2\Delta t$, ... Then at any moment $t + \Delta t$ it must lie somewhere on a circle of radius ε centered on the position that it occupied at time t. Therefore $\phi(x, y, t + \Delta t)$, the probability density at time $t + \Delta t$ at the point (x, y) is the mean of $\phi(\xi, \eta, t)$

over all the points (ξ, η) on a circle of radius ε with center (x, y). Thus

$$\phi(x, y, t+\Delta t) = \frac{1}{2\pi} \int_0^{2\pi} \phi(\xi, \eta, t)\, d\theta.$$

Putting $\xi = x + \varepsilon \cos \theta$ and $\eta = y + \varepsilon \sin \theta$, it is seen that

$$\phi(x, y, t+\Delta t) = \frac{1}{2\pi} \int_0^{2\pi} \phi(x + \varepsilon \cos \theta, y + \varepsilon \sin \theta, t)\, d\theta. \qquad (11.4)$$

Expanding the left-hand side in the form of a Taylor's series and writing ϕ for $\phi(x, y, t)$ gives

$$\phi(x, y, t+\Delta t) = \phi + \frac{\partial \phi}{\partial t} \cdot \Delta t + \frac{1}{2!} \frac{\partial^2 \phi}{\partial t^2} (\Delta t)^2 + \cdots.$$

Similarly, expanding the right-hand side gives

$$\frac{1}{2\pi} \int_0^{2\pi} \left[\phi + \varepsilon \left(\cos \theta \cdot \frac{\partial \phi}{\partial x} + \sin \theta \cdot \frac{\partial \phi}{\partial y} \right) \right.$$
$$\left. + \frac{\varepsilon^2}{2!} \left(\cos^2 \theta \frac{\partial^2 \phi}{\partial x^2} + 2 \cos \theta \sin \theta \frac{\partial^2 \phi}{\partial x\, \partial y} + \sin^2 \theta \frac{\partial^2 \phi}{\partial y^2} \right) + \cdots \right] d\theta,$$

but, since

$$\int_0^{2\pi} \cos \theta\, d\theta = \int_0^{2\pi} \sin \theta\, d\theta = \int_0^{2\pi} \cos \theta \sin \theta\, d\theta = 0$$

and

$$\int_0^{2\pi} \cos^2 \theta\, d\theta = \int_0^{2\pi} \sin^2 \theta\, d\theta = \pi,$$

this reduces to

$$\phi + \frac{\varepsilon^2}{2!} \cdot \frac{1}{2\pi} \left(\pi \frac{\partial^2 \phi}{\partial x^2} + \pi \frac{\partial^2 \phi}{\partial y^2} \right).$$

Equation 11.4 may now be written as

$$\frac{\partial \phi}{\partial t} + \frac{1}{2!} \frac{\partial^2 \phi}{\partial t^2} \Delta t + \cdots = \frac{\varepsilon^2}{4\Delta t} \left(\frac{\partial^2 \phi}{\partial x^2} + \frac{\partial^2 \phi}{\partial y^2} \right) + \text{negligible terms in } \varepsilon^3/\Delta t.$$

Therefore

$$\frac{\partial \phi}{\partial t} = D\, \nabla^2 \phi, \qquad (11.5)$$

where $D = \varepsilon^2/4\, \Delta t$.

Here ε is the distance in the *plane* covered by an infinitesimal step of the particle, so that $\varepsilon^2 = (\Delta x)^2 + (\Delta y)^2$. Thus, if, as before, we put

$(\Delta x)^2/\Delta t = 2D$, we have, analogously $\varepsilon^2/\Delta t = 4D$. Considerations of symmetry show that $\mathrm{var}(x) = \mathrm{var}(y) = 2Dt$. Thus the joint distribution of x and y, the coordinates of the particle at time t, has pdf

$$\phi(x, y, t) = \frac{1}{4\pi Dt} \exp\left[\frac{-(x^2 + y^2)}{4Dt}\right]$$

and this will be seen to be a solution of (11.5).

We now consider three applications of diffusion theory to ecological problems. In all three examples it is assumed that diffusion occurs without drift.

4. The Rate of Spread of a Population Over a Plane

We have hitherto considered the probability, given by $\phi(x, y, t)\, dx\, dy$, that a particular particle starting at the origin at time $t = 0$ will, at a later time t, be found in the element of area having coordinates in the ranges $x \pm \frac{1}{2} dx$ and $y \pm \frac{1}{2} dy$. If, instead of dealing with a single particle, we suppose that at $t = 0$ a whole population of particles has been concentrated at the origin, then $\phi(x, y, t)\, dx\, dy$ denotes the proportion of the population to be expected in this element of area at time t. The distribution is that of shots around a bulls-eye.

Since what interests us is the way in which density falls off with distance from the center of diffusion, it is convenient to convert to polar coordinates. Then

$$x = r\cos\theta; \qquad y = r\sin\theta; \qquad dx\, dy = r\, dr\, d\theta,$$

and

$$\phi(r, \theta, t)\, dr\, d\theta = \frac{r}{4\pi Dt} \exp\left(\frac{-r^2}{4Dt}\right) dr\, d\theta.$$

Following Skellam (1951), we now write $4D = a^2$. Thus a^2 is the mean-square displacement in a unit of time. Then

$$\phi(r, \theta, t)\, dr\, d\theta = \frac{r}{\pi a^2 t} \exp\left(\frac{-r^2}{a^2 t}\right) dr\, d\theta.$$

Integrating over θ gives $\phi(r, t)\, dr$, the expected proportion of the population of particles whose distance from the origin lies in the range $r \pm \frac{1}{2} dr$. Thus

$$\phi(r, t) = \int_0^{2\pi} \frac{r}{\pi a^2 t} \exp\left(\frac{-r^2}{a^2 t}\right) d\theta = \frac{2r}{a^2 t} \exp\left(\frac{-r^2}{a^2 t}\right). \tag{11.6}$$

We now wish to determine the rate at which a diffusing population spreads. Clearly, if there is no drift, the contours of equal density will move out in ever-expanding circles like ripples on a pond. To find the expected size of the boundary circle, which contains the whole population at time t, we evaluate p_t, the proportion of the population that, at time t, is expected to be farther than a distance R_t from the center of diffusion. Obviously,

$$p_t = \int_{R_t}^{\infty} \frac{2r}{a^2 t} \exp\left(\frac{-r^2}{a^2 t}\right) dr = \exp\left(\frac{-R_t^2}{a^2 t}\right).$$

Let the number of particles in the population be N and put $p_t = 1/N$. This amounts to choosing a value of p_t, hence of R_t, such that only one member of the population is expected to be farther than R_t from the origin or that all but one member of the whole population is contained in a circle of radius R_t. Then

$$\frac{1}{N} = \exp\left(\frac{-R_t^2}{a^2 t}\right)$$

or

$$R_t^2 = a^2 t \ln N.$$

The area of the circle containing the expanding population is thus proportional to the time elapsed since diffusion started. N is assumed to remain constant; therefore the density of the population must perforce get progressively less as time passes.

In the foregoing it was assumed that the diffusing organisms did not reproduce or die so that N did not vary with time. We next explore what happens if the particles are breeding organisms and observation is continued over many generations. Suppose the population, initially consisting of N_0 individuals, is increasing exponentially as the result of a simple birth and death process. Then at time t the size of the population is $N_t = N_0 e^{ct}$, where c is the intrinsic rate of natural increase (see Chapter 1).

Therefore $\ln N_t = ct + \text{constant}$, and for sufficiently small N_0 we may take $\ln N_t = ct$. The radius at time t of the circle containing the population, which is expanding both numerically and spatially, is then given by

$$R_t^2 = a^2 t \ln N_t = a^2 t^2 c.$$

We see that in this case the radius (rather than the area) of the circle containing the population is proportional to t.

Skellam (1951) examined the rate of spread of muskrats which were introduced into Central Europe in 1905. On five occasions in the succeeding 23 years the area occupied by the increasing population was mapped, thus enabling contours to be drawn to show the population's extent at five different times. As we should expect, the contours are not circular, but there is no evidence of drift, and it seems reasonable to treat the area within each contour as an estimate of πR_t^2. If we assume, further, that the muskrat population was increasing in numbers exponentially, as could well happen with an immigrant population spreading into areas not occupied by potential competitors, we should expect the square root of the area within each contour to be linearly related to time. Skellam found that this relationship did, in fact, exist; in a plot of $\sqrt{\text{area}}$ versus time the five points lie very close to a straight line passing through the origin.

Skellam also considered the rate of northward spread of oak trees in Great Britain after the melting of the last Pleistocene ice sheet. Assuming, very roughly, that the oaks advanced about 600 miles in 20,000 years, he relates this inferred rate of advance with (a) the presumed rate of numerical increase of the population of oaks and (b) the distance to which the acorns from a parent oak are disseminated. He concluded that (unless some oaks survived at sheltered unglaciated spots within the ice field) the speed of reinvasion of the oaks could have resulted only from acorns being transported by animals; the advance was far too rapid for one to assume that acorns fell to the ground close to their parent trees and germinated where they landed.

This last result is an excellent demonstration of the usefulness of mathematical models in ecology. Even though one would rarely expect such highly abstract arguments as those leading to the diffusion equations to be directly applicable to the complicated behavior of living organisms, they do permit one to reach convincing conclusions about the rates of spread that are possible in different circumstances. For many of the plant and animal species that invade hitherto unoccupied areas it would be interesting to know whether their spread has been purely passive or was accelerated by some extrinsic agency.

5. A Spatial Pattern Resulting from a Diffusion Process

Suppose an insect lays a compact cluster of eggs and that after hatching the larvae diffuse outward. Then, as already given in (11.6), the expected proportion of them in the range $r \pm \frac{1}{2} dr$ at time t is

$$\phi(r, t)\, dr = \frac{2r}{a^2 t} \exp\left(\frac{-r^2}{a^2 t}\right) dr.$$

Next assume that after moving (i.e., performing a random walk in the plane) for a time a larva stops and remains at the spot it has reached. For each larva the probability that it will stop in any short time interval Δt is

$\lambda \Delta t$, with λ a constant. Thus the travel times for the larvae are independent random variates with pdf $\pi(t) = \lambda e^{-\lambda t}$. (This is the pdf of the intervals between independent events occurring at a mean rate of λ per unit of time; the derivation is given on page 29.)

Then the distribution of the distances of the larvae from the center of diffusion after all movement has ceased is given by

$$g(r) \, dr = \left[\int_0^\infty \phi(r, t) \pi(t) \, dt \right] dr$$

$$= \left[\int_0^\infty \frac{2r}{a^2 t} \lambda \exp\left(-\lambda t - \frac{r^2}{a^2 t}\right) dt \right] dr.$$

This integral may be simplified. Put $\rho = 2r\sqrt{\lambda}/a$ so that $d\rho = 2\sqrt{\lambda} \, dr/a$. Then, since

$$g(r) \, dr = \left[\frac{1}{2} \int_0^\infty \frac{2r\sqrt{\lambda}}{a} \exp\left(-\lambda t - \frac{r^2}{a^2 t}\right) \frac{dt}{t} \right] \frac{2\sqrt{\lambda}}{a} \, dr,$$

we have

$$h(\rho) \, d\rho = \rho \left[\frac{1}{2} \int_0^\infty \exp\left(-\lambda t - \frac{\rho^2}{4\lambda t}\right) \frac{dt}{t} \right] d\rho,$$

where $h(\rho)$ is the pdf of ρ. Next put $\lambda t = \tau$. Then $\lambda \, dt = d\tau$ and $dt/t = d\tau/\tau$. Thus

$$h(\rho) \, d\rho = \rho \left[\frac{1}{2} \int_0^\infty \exp\left(-\tau - \frac{\rho^2}{4\tau}\right) \frac{d\tau}{\tau} \right] d\rho.$$

The integral in brackets is tabulated. It is $K_0(\rho)$, a modified Bessel function of the second kind, of which tables will be found in, for instance, Watson (1944). Therefore

$$h(\rho) = \rho K_0(\rho) = \frac{2r\sqrt{\lambda}}{a} K_0\left(\frac{2r\sqrt{\lambda}}{a}\right);$$

that is, the expected number of larvae in the annulus between circles of radius ρ and $\rho + d\rho$ is proportional to $\rho K_0(\rho) \, d\rho$. The area of this annulus is $2\pi\rho \, d\rho$; hence the expected number of larvae per unit area at distance ρ from the center of diffusion is proportional to $K_0(\rho)$. The pattern of the larvae is radially symmetrical, and the way in which density falls off with distance is shown in Figure 11.1. This is simply the curve of $K_0(\rho)$ versus ρ; [e.g., see Wylie (1951)]. Different curves of this form differ among themselves only in the relative scales of ordinate and abscissa or, equivalently, in the scale factor $2\sqrt{\lambda}/a$. Therefore the shape of the distribution depends only on the ratio $2\sqrt{\lambda}/a$ and not on the magnitudes of the

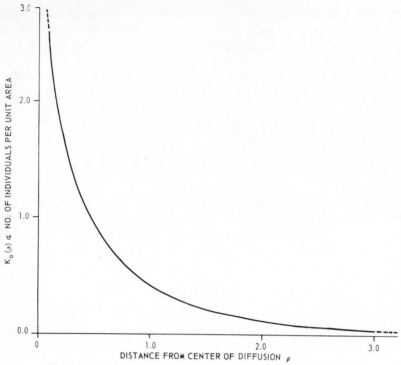

Figure 11.1. The relationship between density and distance from point of diffusion when organisms diffuse outward from a center and stop at random times.

separate components $\sqrt{\lambda}$ and a; λ is the reciprocal of the expected travel time of a larva and a^2 is the mean square displacement in unit time, given that no stopping occurs (see page 172). Thus we see that it would be possible for two species conforming to this model to have identical patterns if one of them diffused slowly and had a long mean travel time, whereas the other diffused rapidly and had a short mean travel time.

Broadbent and Kendall (1953) consider the applicability of this distribution to the pattern of the larvae of helminths parasitic on sheep and rabbits. The eggs are present in numbers in the excreta of the hosts and, after hatching, diffuse outward through the grass of a pasture. After an interval t [presumed to have distribution $\pi(t)$] each larva climbs a grass blade and stays there until ingested by a grazing animal, when the cycle starts again.

Williams (1961) discusses the appropriateness of the Bessel function distribution for describing the pattern of codling moth larvae in an apple orchard, where a large number of adult moths had been released at a central point. He gives a method of estimating the distribution's single parameter, the scale factor $2\sqrt{\lambda}/a$.

6.   The Probability of Reaching a Specified Destination and the Time Taken to Reach It

In this section we shall consider, for simplicity, only one-dimensional diffusion. Examples of ecological contexts in which the one-dimensional theory might apply are the movements of shore-dwelling animals or of aquatic animals in narrow watercourses (presumed to be stagnant so that the possibility of drift may be ignored).

Suppose a group of organisms is diffusing from a starting point at a distance x from a destination which we shall call "home." If they reach it, they stop. Two of the questions that may be asked (and, provided our assumptions hold, answered) are (a) what proportion of the starting population will have reached home by time t_0, assuming diffusion to have started at $t = 0$? And (b) what is the average speed of return of those that do reach home?

Consider first the random-walk model (see page 167) in which a particle is assumed to take discrete steps of constant length. For simplicity we also assume that there is no drift or, in other words, that steps to right and left are equiprobable. We now ask: what is the probability that the particle will reach "home" at a distance r steps to the right of the starting point, on the nth step *for the first time*? Let this probability be u_{rn}. It will now be proved that

$$u_{rn} = \frac{r}{n}\binom{n}{\frac{n+r}{2}}\frac{1}{2^n}, \tag{11.7}$$

with $u_{rn} = 0$ when $n + r$ is odd.

Figure 11.2 shows the set-up diagrammatically. The starting point is at the origin A, with coordinates $(0, 0)$. At every step the particle moves one unit upward, along the time axis, and one unit to right or left, toward or away from home. Any sequence of steps taken by the particle may therefore be plotted as a path on the graph like the one shown. If the particle reaches home on the nth step, the path must pass through B which has coordinates (r, n). For it to arrive there on the nth step *for the first time* the whole path must lie to the left of (must not touch or cross) the vertical line through B. Calling such a path an "admissible path," it follows that

$$u_{rn} = \frac{\text{number of admissible paths}}{\text{total number of possible paths with } n \text{ steps}}.$$

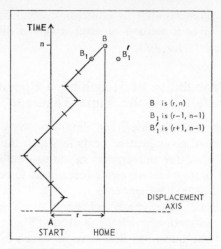

Figure 11.2. To illustrate the derivation of u_{rn}, the probability that a particle starting at 0 will reach "home" at a distance r for the first time on the nth step (see text).

Now, since at every step the particle has two possibilities (it may go right or left) the total number of possible paths is 2^n. It remains to determine the number of admissible paths, say N^*_{AB}. This may be done as follows:

Note first that all admissible paths must go through B_1 with coordinates $(r-1, n-1)$.

Put $N(AB_1)$ = total number of paths from A to B_1.

$N_L(AB_1)$ = number of paths from A to B_1 that are wholly to the left of the vertical through B. (*Note.* This is B, not B_1.) These paths are "admissible."

$N_R(AB_1)$ = number of paths from A to B_1 that touch or cross the vertical through B. These paths are "inadmissible."

Then

$$N(AB_1) = N_L(AB_1) + N_R(AB_1)$$

and also

$$N^*_{AB} = N_L(AB_1).$$

Next, consider the point B'_1 at $(r+1, n-1)$, which is the reflection of B_1 in the vertical through B. Contemplation of the figure will show that

$$N_R(AB_1) = N(AB'_1),$$

that is, the number of *in*admissible paths from A to B_1 is equal to the total number of paths by any route whatever from A to B_1'. This follows, since inadmissible paths to B_1 must reach the vertical through B at some point (r, m), say, with $m \le n-2$. Thereafter, from symmetry, the number of paths to B_1 is the same as the number of paths to B_1'. Therefore

$$N^*_{AB} = N_L(AB_1) = N(AB_1) - N_R(AB_1)$$
$$= N(AB_1) - N(AB_1').$$

Recalling that the total number of paths from the origin to the point (r, n) is $\binom{n}{(n+r)/2}$ and consequently that

$$N(AB_1) = \binom{n-1}{\frac{n+r-2}{2}} \quad \text{and} \quad N(AB_1') = \binom{n-1}{\frac{n+r}{2}},$$

it finally follows that

$$N^*_{AB} = \binom{n-1}{\frac{n+r}{2}-1} - \binom{n-1}{\frac{n+r}{2}} = \frac{r}{n}\binom{n}{\frac{n+r}{2}},$$

whence

$$u_{rn} = \frac{N^*_{AB}}{2^n} = \frac{r}{n}\binom{n}{\frac{n+r}{2}}\frac{1}{2^n}$$

as in (11.7). Here

$$\binom{n}{\frac{n+r}{2}} = 0 \quad \text{if} \quad n+r \text{ is odd.}$$

An entirely different proof will be found in Bailey (1964). It is the solution of the famous Gambler's Ruin problem of probability theory: if a gambler with capital of \$$r$, playing against an infinitely rich adversary, bets \$1 on the outcome of each of a series of tosses with a fair coin, u_{rn} is the probability that he will be ruined at the nth toss.

We now pass to the limit. Denote the step lengths by Δx and the time intervals by Δt and allow $\Delta x, \Delta t \to 0$. Recalling that when $p = q = \frac{1}{2}$ the limiting form of

$$v_{rn} = \binom{n}{\frac{n+r}{2}}\frac{1}{2^n}$$

is

$$\phi(x, t) = \frac{1}{2\sqrt{\pi Dt}} e^{-x^2/4Dt} \quad \text{[see (11.1)],}$$

we see by analogy that the limiting form of $u_{rn} = (r/n)v_{rn}$ is

$$\psi(x, t) = \frac{x}{t} \frac{1}{2\sqrt{\pi Dt}} e^{-x^2/4Dt}.$$

[For a rigorous proof of this, see Feller (1968).] That is, the probability that the particle will, for the first time, reach "home" at a distance x from its starting point in the interval $t \pm \frac{1}{2} dt$ is $\psi(x, t) dt$.

We return now to the ecological questions posed at the beginning of this section. From the above arguments we may conclude that if a group of organisms is released at time $t = 0$ at a distance x from home and then diffuses in one dimension the expected proportion of the group that will reach home on or before time t_0 is

$$Q(t_0, x) = \int_0^{t_0} \psi(x, t) \, dt.$$

The average speed of those that do reach home can also be derived. Thus an organism that travels a distance x in time t has traveled at an average speed of x/t. Therefore the mean speed of all the organisms that reach home (i.e., excluding those that have not arrived by time t_0) is

$$\bar{s}(t_0, x) = \frac{\displaystyle\int_0^{t_0} (x/t)\psi(x, t) \, dt}{Q(t_0, x)}.$$

These results are due to Wilkinson (1952) who applied them to experimental studies on the homing ability of sea birds. The birds were taken along a coast to a distance x from home and then released to find their way back. Observational results included records of (a) the percentage of birds that was successful in returning home before a given time had elapsed and (b) the average speed of the birds that did reach home within this time. Wilkinson found that the observations accorded well with what we should expect on the hypothesis that the homing movements of the birds amounted to random diffusion in one dimension (i.e., along the coast). We could infer (though Wilkinson does not) that the homing birds had no navigational ability and that their movements were wholly random. What Wilkinson does conclude is this: that *if* the birds have navigational ability more detailed observations than those at his disposal will be needed to reveal it.

12

The Patterns of Ecological Maps: Two-Phase Mosaics

1. Introduction

In this chapter we return to a consideration of ecological patterns from the static point of view and discuss the third of the three types described on page 115. These "case 3" patterns are exhibited by vegetatively reproducing plants, which commonly occur as extensive clumps of shoots. Regardless of whether the individual shoots of a clump are densely packed or well separated, it is natural to treat the clumps rather than the shoots as the entities whose pattern is to be studied. A clump need not be a distinct, separate unit, nor need it be an "individual" in the genetic sense.

When the pattern of a clumped species can be mapped—and this can be done only if the clump boundaries are recognizable—the result is a two-phase mosaic with a patch-phase (where the plant occurs) and a gap-phase (where it is absent). The patches are of indefinite size and shape and lack definable centers. They do not necessarily constitute "islands" in the gap-phase; in some parts of an area the patch-phase may form a continuum and the gap-phase occur as lucunae. Both maps in Figure 9.1 show two-phase mosaics; in them the patch-phase is shown as occupied by separate but fairly closely spaced shoots. Many species of plants, for instance those that spread by long stolons or have rhizomes from which the above-ground shoots grow at wide intervals, do not present such easily distinguishable patch- and gap-phases. However, we defer a discussion of such plants to page 193 and confine our attention for the moment to species whose patterns can be easily mapped as mosaics. In this chapter only two-phase mosaics, obtained by mapping a single plant species, are discussed. Mosaics with more than two phases are considered in Chapter 16. It will be noticed that other kinds of mosaics, besides those showing the patches of the different species in tracts of vegetation, are often studied by ecologists. Examples are maps or aerial photos showing land versus water or forest versus grassland; also soil, physiographic, and geological maps. Thus the properties of mosaic maps are of interest in a wide range of contexts and merit detailed consideration.

It is clear that two-phase mosaic maps are entirely different from the "dot maps" that can be used satisfactorily to portray the patterns of species occurring as distinct, widely scattered individual plants, and it is worth inquiring what properties such a mosaic must have to be regarded as "random." There is no unique answer, as there is in the case of a dot map. A dot map is random if and only if the locations of the dots, which are treated as dimensionless, form a realization of a Poisson point process in the plane; that is, if every point in an area is as likely as every other to be the location of a dot and the dots are independent of one another. The number of dots in a small sample area or quadrat is then a Poisson variate. For two-phase mosaics, on the other hand, no such unambiguous definition of randomness is possible. There are various ways in which a random mosaic may be constructed, and we cannot define a mosaic as random without also specifying exactly what sort of randomness is meant. We now discuss two kinds of random mosaics which, for brevity, are here named L-mosaics and S-mosaics.

2. The Random-Lines Mosaic, or L-Mosaic

Suppose that in the map of a mosaic, the patch-phase is to be colored black and the gap-phase left white. One way of constructing a random mosaic is to draw a set of "random lines" that subdivide an area into a network of convex polygons or "cells." Each cell is then independently assigned its color with fixed probabilities, say b for a black cell and w for a white cell with $b + w = 1$. When contiguous cells receive the same color, they form a many-celled patch, if black, or a many-celled gap, if white. In the arguments that follow it is important to keep clear the distinction between a cell and a patch or gap. The cells, which are always convex, are the small areas formed when the random lines are drawn across the area and are the units of which the patches and gaps are composed. A patch consists of any number of contiguous cells that chance to be colored black; likewise, any number of contiguous white cells constitutes a gap.

A method of drawing random lines is as follows: suppose the area in which the lines are to be drawn is circumscribed by a circle of radius r. Take the center of this circle as the pole of a polar coordinate frame and draw an initial line through it. Now, from a random-numbers table, take pairs of random polar coordinates, (p, θ), say, with p in $(0, r)$ and θ in $[0, 2\pi)$. For each such coordinate pair, for example (p_1, θ_1), a line may be drawn which passes through the point with coordinates (p_1, θ_1) and is perpendicular to the line joining (p_1, θ_1) and the pole of the coordinates;

this is a random line. Most of these random lines will lie across the map area, though a few may fall entirely outside it. Those that cross the area will subdivide it into a network of convex polygonal cells. These are the cells that are to be colored black or white. Figure 12.1 shows a mosaic prepared in this way. It is convenient to call it a random-lines mosaic or an L-mosaic for short.

Now suppose that such a mosaic is sampled at equidistant points along a line transect and that the color at each point (B or W) is recorded. The observations will be a sequence such as $BBB\ WW\ BB\ WWW\ B\ WWW\ BB$ ..., for example. We shall now show that for a mosaic constructed in the manner just described, that is, an L-mosaic, this sequence constitutes a realization of a simple two-state Markov chain. To do this we must show that the probability that the ith point will be a B, say, depends only on the color at the $(i-1)$th point and is independent of the colors at earlier

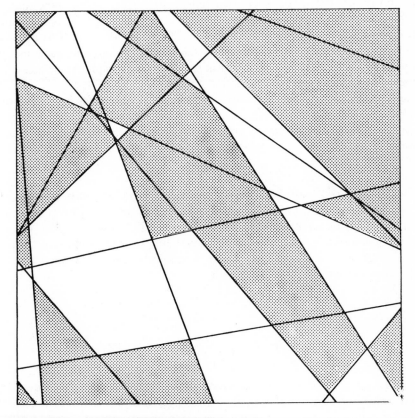

Figure 12.1.   A random-lines mosaic or "L-mosaic."

points, the $(i-2)$th, $(i-3)$th, and so on. This must hold for all i. A rigorous proof has been given by Switzer (1965) and here we are content with a heuristic approach.

Let $p(B_i, B_{i+1})$, for example, be the probability that both the ith point and the $(i+1)$th point on the transect will be B's. Denote by $\pi_{i,i+1}$ the probability that these two adjacent transect points will lie in the same *cell* of the random-lines network. Then

$$p(B_i B_{i+1}) = \pi_{i,i+1}b + (1 - \pi_{i,i+1})b^2; \qquad (12.1)$$

that is, $p(B_i, B_{i+1})$ is the sum of two terms. The first, $\pi_{i,i+1}b$, is the probability that both points will be in the same cell and that the cell is black. The second term, $(1 - \pi_{i,i+1})b^2$, is the probability that the two points will lie in different cells and that both cells are black; clearly, since the cells are colored independently and each has probability b of being black, the probability that any two cells will both be black is b^2. We therefore see that $p(B_i, B_{i+1})$ is a function of $\pi_{i,i+1}$ and b only, and it remains to show that $\pi_{i,i+1}$ is the same for all possible pairs of adjacent transect points. Now, the probability that the ith and $(i+1)$th points will lie in the same cell, namely $\pi_{i,i+1}$, is simply the probability that none of the random lines intersects the transect between them. And since the location of these network lines is random it is clear that $\pi_{i,i+1} = \pi$, a constant for all i; that is, every segment of the transect between two adjacent sampling points has the same probability of not being cut by a random line of the network. Thus $p(B_i, B_{i+1})$ does not depend on i and we may put

$$p(B_i, B_{i+1}) \equiv bp_{BB}, \qquad \text{say,}$$

where p_{BB} is the *conditional* probability that any point will be a B, given that the preceding point was a B. In other words, the probability that the second point of a pair will be black depends only on whether the first is black and *mutatis mutandis* for the other three ordered pairs of colors, BW, WB, and WW. The transition probabilities for the sequence may therefore be written as a Markov matrix:

$$\mathbf{P} = \begin{array}{c} \text{First point} \\ \text{of pair} \end{array} \left\{ \begin{array}{c} \text{Black} \\ \text{White} \end{array} \right. \overset{\overbrace{\text{Second point of pair}}}{\begin{pmatrix} \overset{\text{Black}}{p_{BB}} & \overset{\text{White}}{p_{BW}} \\ p_{WB} & p_{WW} \end{pmatrix}}. \qquad (12.2)$$

For a given mosaic the magnitude of these probabilities will depend only on the distance between the sampling points along the transect: as this distance is decreased, obviously p_{BB} and p_{WW} must increase. The sequence of states along a transect—black patches alternating with white gaps—represents, indeed, a continuous Markov process; instead of making observations of phase at separate equidistant points and treating the observed sequence as a discrete Markov chain, we could measure the lengths cut off on the transect by patches and gaps alternately. These form a realization of a continuous two-state Markov process.

The fact that L-mosaics have the Markov property makes them especially amenable to investigation. It is, conceptually, a straightforward matter to estimate the parameters of any particular mosaic of this form and to derive others of its properties. We pursue this investigation in Section 4, but before doing so we shall describe another kind of random mosaic and compare it with the L-mosaic.

3.   The Random-Sets Mosaic, or S-Mosaic

Matérn (1960) has described in detail the properties of what he calls a "random-sets" mosaic. For brevity we call it an S-mosaic. It is constructed as follows: a pattern of random dots is first drawn in the map

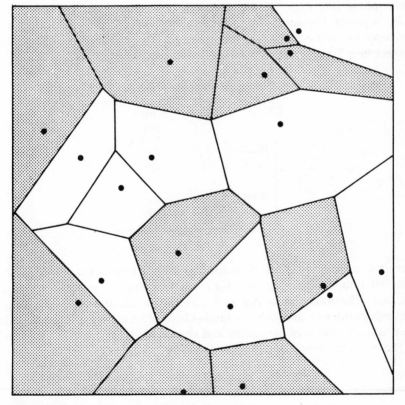

Figure 12.2.   A random-sets mosaic or "S-mosaic."

area. Then with each dot is associated a cell consisting of all those points in the area that are nearer to that particular dot than to any other. Each segment of a cell boundary is thus the locus of points that are equidistant from the two nearest dots. The cells are then independently colored, as with the L-mosaic, black with probability b and white with probability w, the probabilities being the same for all cells. Figure 12.2 is an example.

4. The Comparison of L-Mosaics and S-Mosaics

One way of comparing the two kinds of mosaic is to compare their theoretical autocorrelation functions. Unfortunately, for S-mosaics this function can be expressed in elementary form only in the one-dimensional case. Here, therefore, we derive and compare the autocorrelation functions only for one-dimensional forms of the mosaics.

We begin by describing results that are true for both kinds of mosaic of any dimensionality. Suppose we assign arbitrary scores to the two different phases, for example 1 to the black phase (the patches) and 0 to the white phase (the gaps). If the mosaic is now sampled at many ordered pairs of points, the interpoint distance being constant and equal to v, say, a sample of "score-pairs" will be obtained. The expected results can be set out in a 2×2 correlation table as follows:

		Second point		
		Black (1)	White (0)	
First point	Black (1)	a_{11} (1)	a_{12} (0)	a_1
	White (0)	a_{21} (0)	a_{22} (0)	a_2
		a_1	a_2	1

Here the a_{ij} $(i, j = 1, 2)$ are the expected proportions of the observations that fall into the four classes; they are functions of v, the interpoint distance. The numbers in brackets underneath are the product scores. Corresponding row and column totals obviously have equal expectations from considerations of symmetry and consequently $a_{21} = a_{12}$.

The expected autocorrelation is therefore

$$\rho(v) = \frac{a_{11} - a_1^2}{a_1 a_2},$$

and we must now evaluate the right-hand side of this equation as a function of v. Suppose, as before, that the black and white phases are present in proportions b and w; that is, $a_1 = b$ and $a_2 = w$. Denote by $\pi(v)$ the probability that two points a distance v apart will fall in the same cell of the network from which the mosaic was constructed. Then from (12.1), provided $\pi(v)$ is constant everywhere in the mosaic, it is seen that

$$a_{11} = b\pi(v) + b^2[1 - \pi(v)].$$

From similar arguments

$$a_{12} = a_{21} = bw[1 - \pi(v)]$$

and

$$a_{22} = w\pi(v) + w^2[1 - \pi(v)],$$

so that

$$\rho(v) = \pi(v).$$

The requirement that $\pi(v)$ be constant amounts to saying that every cell must have the same expected size, regardless of its location. (The size of a cell is its area if the mosaic is two-dimensional or its length if the mosaic is one-dimensional.) This is not the same as saying that the sizes of contiguous cells are independent of one another, an assertion that is true only for L-mosaics. Indeed, this assertion is what we proved on page 184 and it accounts for the fact that L-mosaics have the Markov property. For an S-mosaic it is easily seen that the sizes of contiguous cells are correlated. However, since the *unconditional* expectation of cell size is the same for all cells, $\pi(v)$ is constant throughout an S-mosaic and the above formulas hold.

The results given so far are quite general. They apply to both L- and S-mosaics, regardless of dimensionality. Thus we have now shown that both mosaics have autocorrelation function $\rho(v) = \pi(v)$; that is to say, the correlation between the phases at two points a given distance apart is equal to the probability that both points are in the same *cell* of the network. In other words, it depends only on the sizes of the network cells and not on b and $w = 1 - b$, the areal proportions of the black and white phases of the mosaic. If the cells are small, the correlation is low and the mosaic may be described as "fine-grained." Conversely, in a "coarse-grained" mosaic the cells are large and the correlation high. This amounts to saying that in a mosaic in which the cells are assigned their colors independently the "grain" of the mosaic depends only on the grain of the cell network. It now remains to determine $\pi(v)$ for one-dimensional mosaics of the two types.

Take L-mosaics first. A one-dimensional L-mosaic consists simply of a row of contiguous nonoverlapping linear cells (i.e., line segments) whose boundaries (which are points) are a realization of a one-dimensional Poisson point process. The lengths of the cells therefore have pdf $f(x) = \lambda e^{-\lambda x}$, where λ is the parameter of the process and is the expected number of points per unit length of line.

Now let a sampling segment of length v (hereafter called simply the segment) be laid at random on the line. The probability that the segment will fall entirely within one cell is $\pi(v)$ and may be determined as follows.

Clearly $\pi(v) = 0$ if the midpoint of the segment falls in a cell of length $x < v$. Suppose it falls in a cell of length $x \geq v$. Then the probability that the whole segment will lie within the cell is easily seen to be $(x - v)/x$. Thus, writing p_x for the probability that the segment's midpoint will fall in a cell of length x, we see that for $x \geq v$,

$$\pi(v) = \frac{x - v}{x} \cdot p_x.$$

It should now be noted that $p_x \neq f(x)$; that is, to select a point at random on the one-dimensional mosaic is *not* equivalent to selecting a cell at random from the population of all cells. Clearly, a point placed at random on the mosaic is more likely to fall in a long cell than a short one and thus the probability that any given cell will contain the segment's midpoint is proportional to the cell's length; that is $p_x = Cxf(x)$, where C is a constant of proportionality. Since we must have $\int_0^\infty p_x \, dx = 1$, it follows that

$$C = \frac{1}{\displaystyle\int_0^\infty xf(x) \, dx} = \frac{1}{E(x)} = \lambda.$$

Therefore

$$p_x = \frac{xf(x)}{E(x)}$$

$$= \lambda^2 x e^{-\lambda x}. \tag{12.3}$$

Then for fixed $x \geq v$

$$\pi(v \mid x) = \frac{x - v}{x} \lambda^2 x e^{-\lambda x}.$$

Integrating over all $x \geq v$ gives

$$\pi(v) = \int_v^\infty \lambda^2 (x - v) e^{-\lambda x} \, dx = e^{-\lambda v}.$$

We have now proved that for a one-dimensional L-mosaic the autocorrelation function is

$$\rho(v) = \pi(v) = e^{-\lambda v}.$$

For a two-dimensional L-mosaic $\rho(v)$ is of the same form. This follows because a transect across a two-dimensional L-mosaic is itself a one-dimensional L-mosaic. Both are realizations of a continuous Markov process.

Next we determine $\pi(v)$ [hence $\rho(v)$] for a one-dimensional S-mosaic. A diagram of a linear S-mosaic is shown below. The X's on the line are points generated by a Poisson point process of intensity λ; the dotted lines, which bisect the intervals between the X's, are the cell boundaries.

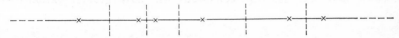

The lengths of the intervals between the X's have, as before, the pdf $f(x) = \lambda e^{-\lambda x}$. The length of a cell, say y, is therefore the mean of two independent values of x, so y has pdf

$$g(y) = (2\lambda)^2 y e^{-2\lambda y}$$

[see (10.3)] with

$$E(y) = \frac{1}{\lambda}.$$

We now obtain p_y, the probability that the midpoint of a randomly placed sampling segment will fall within a cell of length y. By the same argument that led to (12.3) it is seen that $p_y = yg(y)/E(y)$ so that in this case

$$p_y = 4\lambda^3 y^2 e^{-2\lambda y}.$$

Again, the probability that a sampling segment of length v will fall entirely within a cell of given length y is $(y-v)/y$ for $y > v$, and 0 otherwise. Therefore

$$\pi(v) = \int_v^\infty \frac{y-v}{y} \cdot p_y \, dy$$

$$= \int_v^\infty 4\lambda^3 (y^2 - vy) e^{-2\lambda y} \cdot dy$$

$$= (1 + \lambda v) e^{-2\lambda v},$$

and, since $\rho(v) = \pi(v)$, this is the autocorrelation function for a linear S-mosaic.

Writing the autocorrelation functions for L- and S-mosaics as $\rho_L(v)$ and $\rho_S(v)$, we see that $\rho_L(v) > \rho_S(v)$ for all v and λ. This follows since

$$\rho_L(v) - \rho_S(v) = e^{-\lambda v} - (1 + \lambda v) e^{-2\lambda v}$$

$$= e^{-2\lambda v}(e^{\lambda v} - 1 - \lambda v)$$

$$= e^{-2\lambda v} \sum_{j=2}^\infty \frac{(\lambda v)^j}{j!} > 0.$$

Therefore, for given λ, the correlation between the phases at a chosen fixed distance apart is greater in a linear L-mosaic than in a linear S-mosaic. Also, given a mosaic of each kind with the same mean cell length, the variance of the cell lengths is twice as great in the L-mosaic as in the S-mosaic; this results from the fact thay every cell in an S-mosaic is made up of two adjacent half-cells of an L-mosaic.

It is intuitively clear that in two-dimensional mosaics with equal mean cell areas the autocorrelation will be greater in L-mosaics than in S-mosaics, and, likewise, the cell areas will have greater variance in L-mosaics.

5. The L-Mosaic as a Standard of Randomness

Because of its greater mathematical tractability, the L-mosaic provides a better standard of randomness than the S-mosaic. Although it would be absurd to suppose that the boundaries of the patches in ecological maps resulted from the existence of random lines, actually occurring mosaics may often have patterns that are indistinguishable in many though not all of their properties, from two-phase two-dimensional L-mosaics. Admittedly, we should not expect the boundaries of the patches in a vegetation map, for example, to have sharp angles at the corners and straight edges between, as has a geometrically constructed L-mosaic (see Figure 12.1), but a mosaic in which the means and variances of the sizes of the patches and gaps are closely approximated by those of an L-mosaic is perfectly possible. If we choose to define such a mosaic as random, we have a useful definition of randomness. A precisely defined concept of randomness is as desirable for mosaic patterns as it is for dot patterns. It provides a standard with which natural mosaics may be compared. Also it should be possible, although this has not yet been done, to devise appropriate ways of measuring the departures from randomness that are exhibited by actual mosaics. Without such a standard, comparisons are difficult. This explains and justifies the preoccupation of ecologists with random patterns in general.

The contrast between dot maps and mosaic maps is this: genuine random dot patterns can certainly occur naturally, although they are uncommon; such a pattern will be found, for instance, whenever a sparse population of small annual plants grows in a homogeneous area from independently dispersed seeds. In contrast, true L-mosaics are undoubtedly nonexistent; it is hard to think of any natural process that would yield a patch and gap pattern with all the geometrical properties of an L-mosaic. This does not, however, detract from the value of an L-mosaic treated as a standard or basis for comparison.

Only one parameter is needed to specify a random dot pattern fully. This is the expected number of dots per unit area or, equivalently, the intensity of the Poisson point process in the plane that generated the pattern. Mosaics, on the other hand, need two parameters before they can be exactly specified. Thus suppose a unit of length has been chosen. In terms of this unit the pair of parameters necessary to specify a particular two-phase L-mosaic may be expressed in two ways:

1. If the mosaic is sampled along a row of equidistant points a unit distance apart, then, as already shown, the sequence of phases encountered (the B's and W's) are a realization of a two-state Markov chain with matrix **P**, as given in (12.2). The pattern of the mosaic is completely specified when any two independent transition probabilities, say p_{BW} and p_{WB}, are given.

2. Any transect across the mosaic will cut through black patches and white gaps alternately. Denote by $E(l_B)$ and $E(l_W)$, the mean lengths along the transect of these patch and gap intercepts. Then, if $E(l_B)$ and $E(l_W)$ are given, the pattern of the mosaic is fully specified.

We now consider how the transition probabilities of **P**, and the linear dimensions of the patches and gaps, are interrelated.

We begin by considering the distribution of the variate l_B, the length, as measured along a transect, of a black patch. Recall that each patch consists of one or more contiguous network cells that chance to have been colored black. The length of any patch is therefore the sum of the lengths of one or more contiguous black network cells. The lengths of these cells are independent. Writing x for the length of a *cell*, x has pdf $f(x) = \lambda e^{-\lambda x}$, where λ is the expected number of network lines cutting across the transect per unit length. We now wish to derive the pdf of l_B, the length of a *patch*.

If a patch consists of j contiguous cells (along the sampling transect), its length is the sum of j independent values of x; so the conditional pdf of l_B for given j is

$$g(l_B \mid j) = \frac{\lambda^j l_B^{j-1} e^{-\lambda l_B}}{(j-1)!}$$

(see page 150). Now, the probability of encountering an uninterrupted sequence, or run, of j black cells along a transect is $b^{j-1}w$. For suppose a black cell has been entered; a run of j black cells will occur if the succeeding $j-1$ cells are also black and then a white cell follows to terminate the run. The probability that this will happen is evidently $b^{j-1}w$.

The unconditional pdf of l_B, when j is allowed to vary, is therefore

$$g(l_B) = \sum_{j=1}^{\infty} g(l_B \mid j) b^{j-1} w$$

$$= \lambda w e^{-\lambda l_B} \sum_{j=1}^{\infty} \frac{(\lambda b l_B)^{j-1}}{(j-1)!}$$

$$= \lambda w \, e^{-\lambda w l_B},$$

since $b + w = 1$, or

$$g(l_B) = \lambda_B e^{-\lambda_B l_B}, \qquad \text{say,}$$

on putting $\lambda w = \lambda_B$. Thus the pdf of the length of a patch has the same form as that of the length of a single cell. The expected patch length is $E(l_B) = 1/\lambda_B$. Correspondingly, the pdf of l_W, the length of the gaps, is

$$g(l_W) = \lambda_W e^{-\lambda_W l_W}$$

where

$$\lambda_W = b\lambda \qquad \text{and} \qquad E(l_W) = \frac{1}{\lambda_W}.$$

Further,

$$\lambda_B + \lambda_W = \lambda(b + w) = \lambda,$$

the density of the network lines.

We have now shown that both the patch lengths and the gap lengths are exponentially distributed, with means $1/\lambda_B$ and $1/\lambda_W$, respectively. An alternative way of stating the same result is to say that the alternating patch lengths and gap lengths constitute a realization of a continuous two-state Markov process. The parameters λ_B and λ_W are known as the transition rates of the process, and the matrix $\mathbf{R}$, defined as

$$\mathbf{R} = \begin{pmatrix} -\lambda_B & \lambda_B \\ \lambda_W & -\lambda_W \end{pmatrix}$$

is the matrix of transition rates.

Now it may be shown (e.g., see Howard, 1960) that

$$\mathbf{P} = \mathbf{I} + \mathbf{R} + \frac{\mathbf{R}^2}{2!} + \frac{\mathbf{R}^3}{3!} + \cdots. \tag{12.3}$$

Thus the transition *probabilities* (the elements of $\mathbf{P}$) may be expressed in terms of the transition *rates*, and vice versa, by equating corresponding elements on both sides of (12.3). Writing, as before, $\lambda_B + \lambda_W = \lambda$, we see that $\mathbf{R}^2 = -\lambda \mathbf{R}$ and consequently that $\mathbf{R}^{j+1} = (-1)^j \lambda^j \mathbf{R}$.

Therefore (12.3) may be written

$$\mathbf{P} = \mathbf{I} + \mathbf{R} - \frac{\lambda}{2!}\mathbf{R} + \frac{\lambda^2}{3!}\mathbf{R} - \frac{\lambda^3}{4!}\mathbf{R} + \cdots$$

$$= \mathbf{I} - \frac{(e^{-\lambda} - 1)}{\lambda}\mathbf{R}.$$

It now follows that

$$p_{BW} = \frac{1 - e^{-\lambda}}{\lambda} \cdot \lambda_B \qquad \text{and} \qquad p_{WB} = \frac{1 - e^{-\lambda}}{\lambda} \cdot \lambda_W.$$

Also,

$$p_{BW} + p_{WB} = 1 - e^{-\lambda}, \qquad \text{hence} \qquad \lambda = -\ln(1 - p_{BW} - p_{WB}).$$

Therefore

$$\lambda_B = \frac{-p_{BW}}{p_{BW} + p_{WB}} \ln(1 - p_{BW} - p_{WB})$$

and

$$\lambda_W = \frac{-p_{WB}}{p_{BW} + p_{WB}} \ln(1 - p_{BW} - p_{WB}).$$

It will be recalled that $\lambda = \lambda_B + \lambda_W$ is the density of the network lines from which the mosaic is constructed. It is a measure of the grain of the mosaic (see page 187) and is independent of the areal proportions of the patch and gap (or black and white) phases. When λ is large, the sizes of the patches or gaps (or both) are small and the mosaic is fine-grained; when λ is small, the patches and/or gaps are large and the mosaic is coarse-grained.

6.   Practical Problems of Sampling Mosaics

To test whether a natural mosaic can be regarded as random we must either (a) obtain a sample of values of l_B and l_W and judge whether their distributions are fitted by the pdf's $g(l_B)$ and $g(l_W)$; or (b) sample the mosaic along a row of equidistant points and judge whether the observed sequence of phases forms a realization of a Markov chain. Although method (a) is conceptually more straightforward, method (b) is preferable in practice because natural mosaics rarely have clear-cut boundaries between the phases. When a vegetation mosaic is sampled in the field, it is usually difficult to decide exactly where a patch ends and the succeeding gap begins, so that measurements of the sample values of l_B and l_W are hard to make; even if a map of the mosaic has been drawn, making measurement easy, the results are probably unreliable and the seeming accuracy illusory. It therefore seems better always to use method (b). Methods of estimating the transition probabilities and of fitting the Markov hypothesis to the data are given in Pielou (1964).

When one is sampling natural vegetation on the ground, difficulties often arise with plant species that form very open clumps; the above-ground shoots of a vegetatively reproducing plant may be widely spaced, and it is unreasonable to treat the spaces among them as part of the gap phase. The best way to deal with this problem is to sample a mosaic along a row of very small circular quadrats instead of at true points; then, when any plant part falls within a circular quadrat, the center of the quadrat is treated as being within a patch. The choice of radius for the quadrats is necessarily arbitrary; their use will always lead to overestimation of the area of the patch phase, and the larger the quadrats, the greater the overestimation. To use circular quadrats instead of points is, in effect, to blur the pattern; it is as though one were examining the mosaic with an optical instrument of low resolving power. However, this is not necessarily a disadvantage. By deliberately using low resolution, we can prevent the minor details of a pattern from obscuring its major features.

7. Non-Random Mosaics and Anisotropic Mosaics

For a mosaic to be random in the sense we have defined it is necessary for both the patch lengths and the gap lengths to be exponentially distributed. Therefore a mosaic may be random in one of its phases and nonrandom in the other; for example, the gap lengths could have a variance either less than or greater than expected even though the patch lengths had an exponential distribution. The patches could then be said to be regularly spaced, or aggregated, in relation to the gaps. The reverse is also possible; the patch lengths may vary either to a lesser or a greater extent than expected, regardless of the distribution of the gap lengths. Any mosaic can, in fact, be thought of as having two patterns: that of the patches relative to the gaps and that of the gaps relative to the patches. One cannot say of a mosaic, as one can of a dot pattern merely that its pattern is regular, random or aggregated. Both the patches and the gaps have definable patterns and nine combinations can be envisaged.

The possibility that a mosaic may not be isotropic should also be borne in mind. If the mean lengths of the patches or gaps (or both) are longer in one direction than in another the mosaic is anisotropic. There are a number of possible causes of anisotropy in vegetation mosaics; for instance, elongated ridges and hollows, an anisotropic soil pattern, or the effects of the prevailing wind in an exposed situation may all produce vegetation patches that are not isodiametric. The data yielded by transect sampling will then vary with the direction of the transect.

8.   The Grain of a Mosaic

Regardless of whether a mosaic is random in any sense, we may wish to have some objective measure of its grain. For a mosaic of given area, any reasonable measure of grain should obviously depend upon the total length of interphase boundary in the area; in a fine-grained mosaic the total length of boundary is high and in a coarse-grained mosaic it is low. To estimate a parameter related to boundary length we may sample the mosaic (either on a map or on the ground) with a short sampling rod whose length is to serve as the unit of measurement. The rod is dropped at random on the mosaic several times, and the proportion of times it lies across a boundary is recorded. Call this proportion $\hat{P}$. The rod is assumed to be so short that, wherever it falls, the probability is negligible that it will cut the interphase boundary more than once. Then $\hat{d} = 2/\pi\hat{P}$ provides an intuitively clear measure of the mosaic's grain, as we now show.

The argument hinges on a famous problem of classical probability theory, Buffon's needle problem. Imagine a "mosaic" made by ruling parallel straight lines a distance d apart (with $d > 1$) on a plane surface and coloring the stripes between them alternately black and white. Now let a "needle" of unit length be dropped at random on the plane. The problem is to determine the probability, P, that the needle will cut a black-white boundary; obviously it cannot cut more than one since the width of the stripes exceeds the length of the needle.

Consider Figure 12.3a which shows part of the striped mosaic and a randomly dropped needle. We wish to find the probability that the needle will cut the only black-white boundary it could cut, the one nearest the center-point of the needle.

Let the perpendicular distance of the needle's center from the nearest boundary line be y. All values of y in the range $[0, d/2]$ are equiprobable. Let the angle between the needle and the parallel boundary lines be θ. All values of θ in the range $[0, \pi]$ are equiprobable; (it is assumed that the ends of the needle are indistinguishable). Thus specification of the observed bivariate (θ, y) completely describes the result of any throw of the needle and all possible values of (θ, y) are equiprobable. Clearly, the needle will cut the nearest boundary line only if $y < \frac{1}{2}\sin\theta$. Figure 12.3$b$ shows how the required probability may be derived. The rectangular area with sides of length $d/2$ and π, and hence of area $\pi d/2$, represents the set of all equiprobable outcomes of dropping the needle. The shaded area

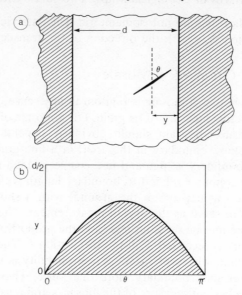

Figure 12.3. To illustrate Buffon's needle problem. (See text for details.)

under the curve $y = \frac{1}{2}\sin\theta$ is the set of all outcomes consistent with "success," that is, the needle's cutting a boundary line. This area is $\int_0^\pi \frac{1}{2}\sin\theta\, d\theta = 1$.

Hence

$$P = \frac{Shaded\ area}{Total\ area\ of\ rectangle} = \frac{2}{\pi d}$$

Now suppose d were unknown and were to be estimated by dropping a needle at random, n times, on the striped mosaic. An estimate of P would be given by $\hat{P}$, the observed proportion of successes, and we should take $\hat{d} = 2/\pi\hat{P}$ as the estimate of d. Hence if we use $2/\pi\hat{P} = \hat{d}$ as a measure of the grain of an irregular mosaic, sampled in the same way, we may visualize a striped mosaic with stripes of width $\hat{d}$ as being in an obvious sense "equivalent" to the irregular mosaic; for examples, see Pielou (1974). Further, since every throw of the needle is a Bernoulli trial with probability of success P, if n throws of the needle (with $n \geq 60$) gave $n\hat{P}$ successes, the approximate 95% confidence interval for P is $\hat{P} \pm 2\sqrt{(\hat{P}\hat{Q}/n)}$, where $\hat{Q} = 1 - \hat{P}$. Then, approximately, the corresponding confidence interval for d is $2\{\pi[\hat{P} \pm 2\sqrt{(\hat{P}\hat{Q}/n)}]\}^{-1}$.

9.   Estimated Maps

It is obviously conceptually possible to map any natural mosaic pattern. If the thing to be mapped is visible and shows up clearly in a vertical photograph, a map can easily be made by tracing the photograph, but if one wishes to map such invisible features as height of the water table, depth of litter, or soil texture, for example, the usual way to proceed is to

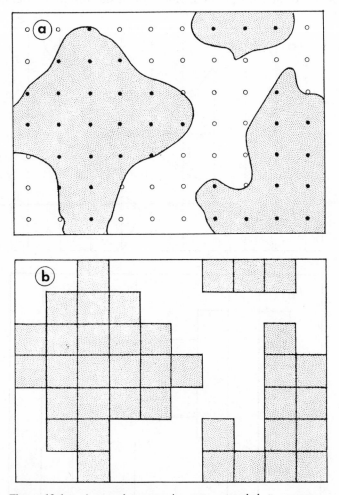

Figure 12.4a. A two-phase mosaic pattern sampled at a square lattice of points. These points are shown as black or white circles according as they occur in the black or white phase.

Figure 12.4b. The estimated map.

sample the ground at a number of points, mark the observed result at corresponding points on the map sheet, and attempt to reconstitute the pattern from the points. Switzer (1967) has investigated the accuracy of such estimated patterns and has explored the effect on accuracy of the different arrangements and spacings of the sample points.

We shall suppose that a two-colored map is to be made or, in other words, that two phases are to be distinguished. These phases may differ in the presence or absence of an attribute or, if the variate is continuous, in

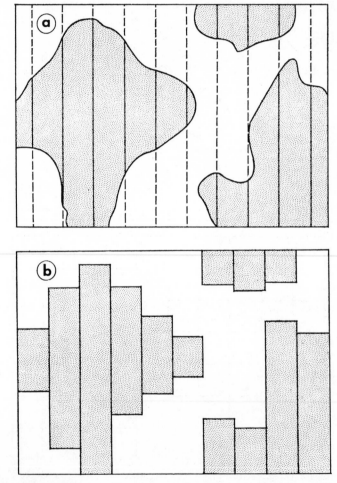

Figure 12.5a. A two-phase mosaic pattern sampled along the marked transect lines.

Figure 12.5b. The estimated map.

being high-valued or low-valued. The accuracy of a two-color estimated map may be measured by a loss function, L, defined as the proportion of the total area misclassified on the map. Misclassified areas consist of those parts shown black that should be white and vice versa. Switzer has derived the expected loss function EL for different kinds of mosaics mapped by various sampling methods, but here it will be possible to quote only two of his results without proof.

1. Suppose the area to be mapped is rectangular and can be subdivided into n squares each of area $1/n$. The area is sampled at the center of each square and the corresponding square on the map is colored black or white in accordance with the phase observed at the sampling point (see Figure 12.4). Then, if the pattern is an L-mosaic and if the proportions of the black and white phases on the ground are b and $w = 1 - b$,

$$\frac{EL}{2bw} = 1 - 8 \int_0^{1/2} \int_0^y \exp\left[-\lambda \left(\frac{x^2 + y^2}{n} \right)^{1/2} \right] dx\, dy.$$

2. Again assume that the pattern to be mapped is an L-mosaic but suppose now that it can be sampled continuously (rather than at a succession of points) along parallel transects a distance h apart. This is called line sampling with spacing h. The lines can be drawn on a map and marked to show the intercepts of the black and white phases as they occurred on the ground. Each line is then expanded to form a strip of width h with the sampling transect along its center (see Figure 12.5). For the estimated map that results it can be shown that

$$\frac{EL}{2bw} = 1 - \frac{2(1 - e^{-\lambda h/2})}{\lambda h}.$$

Further details and other results have been obtained by Switzer and will be found in his paper. Their potential usefulness to ecologists is obvious.

III

Spatial Relations of
Two or More Species

13

Association Between Pairs of Species
I: Individuals in Discrete
Habitable Units

1. Introduction

The spatial pattern exhibited by a single species within a limited area is often worth examining for its own sake. The factors controlling and determining pattern, however, are likely to affect many species rather than just one, and much may be learned by investigating the way in which species are associated with one another. If two co-occurring species are affected by the same environmental factors, or if they have some effect, either favorable or unfavorable, on each other, their patterns will not be independent; the species will be associated, either positively or negatively. Association or the lack of it among pairs and groups of species is therefore of obvious ecological interest. As in the study of pattern in one-species populations, it is desirable to consider separately those species that occupy discrete habitable units (e.g., pest insects in fruits, ectoparasites on rats) and those that may occur anywhere throughout an extended space or continuum (e.g., plankton organisms in a volume of water, plants in a meadow).

In this chapter we confine attention to organisms in discrete units and begin by discussing the association of a single pair of species. The far more difficult problem of investigating association in a group of more than two species is mentioned briefly in Sections 6 and 7.

2. Testing the Association Between Two Species

The standard method of testing for association between two species is as follows. Assume that we are examining a sample of N discrete units collected at random from a large population of possible units. Let the two species being studied be labeled species A and species B; for each unit note whether it contains species A alone, species B alone, both species, or neither species. The quantity of each species in each unit is disregarded; we record only presences and absences. The observed frequencies can then be set out in the form of a 2×2 table:

Species B

		Present	Absent	
Species A	Present	a	b	$m = a + b$
	Absent	c	d	$n = c + d$
		$r = a + c$	$s = b + d$	$N = m + n = r + s.$

The usual approach of ecologists to 2×2 tables is to carry out a χ^2-test and let it go at that. This is a slovenly approach, and it is important to realize that there are two entirely different questions we could ask on being confronted with a table such as this (see Pearson, 1947).

Question 1. Among the N units examined do species A and species B occur independently of each other?
Question 2. In the population as a whole are the two species independent of each other?

Assume first that Question 1 is being asked. We *know* that m of the N units contain species A and that r of them contain species B; the table's marginal totals are therefore fixed. The question thus becomes: for these given marginal totals what are the probabilities of the various possible sets of cell frequencies (or partitions of N)? Is the particular set of cell frequencies we have observed consistent with the hypothesis of independence? For given N, M, and r, the conditional probability that a of the units will contain both species is

$$\Pr(a \mid N, m, r) = \frac{m!n!r!s!}{a!b!c!d!N!};$$

that is, a has a hypergeometric distribution. To see this note that the number of ways of choosing m units out of N to contain species A is $\binom{N}{m}$; similarly, the number of ways of choosing r units to contain species B is $\binom{N}{r}$. Thus the number of arrangements that would give rise to the observed marginal totals is $\binom{N}{m}\binom{N}{r}$.

The number of different ways of partitioning N to produce the observed cell frequencies a, b, c, and d is $N!/(a!b!c!d!)$. Therefore

$$\Pr(a \mid N, m, r) = \frac{N!/(a!b!c!d!)}{\binom{N}{m}\binom{N}{r}} = \frac{m!n!r!s!}{a!b!c!d!N!}. \tag{13.1}$$

In this way we may calculate the probabilities for all the different sets of cell frequencies that give rise to the observed marginal totals.

We resume discussion of this case after considering how the second case (Question 2) differs from it. In asking Question 2, it is no longer assumed that the marginal totals are fixed. When a sample of N units is taken at random from a large population of units, not only are the cell frequencies free to vary but so also are their pairwise sums, the marginal totals. To determine the probability of obtaining any particular 2×2 table we must argue as follows.

Let $p(A)$ denote the probability that a unit will contain species A and $p(\bar{A}) = 1 - p(A)$, the probability that a unit will lack species A. The probabilities $p(B)$ and $p(\bar{B}) = 1 - p(B)$ are defined likewise for species B. Any unit must belong to one of four classes, AB, $A\bar{B}$, $\bar{A}B$, or $\overline{AB}$, and on the null hypothesis of independence of the species we must have

$$p(AB) = p(A) \cdot p(B), \qquad p(A\bar{B}) = p(A) \cdot p(\bar{B}),$$
$$p(\bar{A}B) = p(\bar{A}) \cdot p(B), \qquad p(\overline{AB}) = p(\bar{A}) \cdot p(\bar{B}).$$

The probability $\Pr(a, b, c, d)$ of obtaining the observed cell frequencies a, b, c, and d in a sample of N units is then a term from a multinomial distribution; it is given by the coefficient of $z_1^a z_2^b z_3^c z_4^d$ in the expansion of the probability generating function

$$[p(AB)z_1 + p(A\bar{B})z_2 + p(\bar{A}B)z_3 + p(\overline{AB})z_4]^N;$$

that is

$$\Pr(a, b, c, d) = \frac{N!}{a!b!c!d!} [p(AB)]^a [p(A\bar{B})]^b [p(\bar{A}B)]^c [p(\overline{AB})]^d$$

$$= \frac{N!}{a!b!c!d!} [p(A)]^{a+b} [p(B)]^{a+c} [p(\bar{A})]^{c+d} [p(\bar{B})]^{b+d}$$

$$= \frac{N!}{m!n!} [p(A)]^m [1 - p(A)]^n$$

$$\cdot \frac{N!}{r!s!} [p(B)]^r [1 - p(B)]^s \cdot \frac{m!n!r!s!}{a!b!c!d!N!}$$

$$\equiv b(m \mid p(A), N) \times b(r \mid p(B), N) \times \Pr(a \mid N, m, r).$$
$$(13.2)$$

Here the binomial term $b(m \mid p(A), N)$ denotes the probability that in N trials an event whose probability is $p(A)$ will occur m times; likewise $b(r \mid p(B), N)$ is the probability of r occurrences in N trials of an event with probability $p(B)$, and $\Pr(a \mid N, m, r)$ is the conditional probability we found in answering Question 1, namely, the probability of observing the partition of N into the parts a, b, c, and d, given that $a + b = m$ and $a + c = r$.

The probability of obtaining an observed 2×2 table thus depends on whether the table is assumed to have preassigned marginal totals, in which case it represents what Barnard (1947) has called a doubly restricted double dichotomy; this is the assumption made when Question 1 is asked. Or whether the marginal totals as well as the cell frequencies are treated as random variates, giving a table that is an unrestricted double dichotomy; this is the assumption made when Question 2 is asked.

3. Testing the Association in a Sample (Question 1)

If the empirical table is treated as a doubly restricted double dichotomy, we may calculate $\Pr(a \mid N, m, r)$ for all the possible values of a that could arise, subject to the restriction that the marginal totals are fixed. Denote the minimum and maximum possible values of a by $a(\min)$ and $a(\max)$.

Now suppose that the observed frequency a of the event AB (i.e., the observed number of joint occurrences of species A and B) is greater than its expectation $E(a)$; that is, $E(a) < a \leq a(\max)$. This leads us to believe that there may be significant positive association between the species. Then the probability of observing a deviation from expectation as great as or greater than $a - E(a)$, and in the same direction, is

$$P_{\text{upper}} = \sum_{i=a}^{a(\max)} \Pr(i \mid N, m, r).$$

Thus P_{upper} is the appropriate probability for a one-tail test for positive association. It is the probability, on the null hypothesis of independence, of obtaining evidence for positive association as strong as or stronger than that observed.

Likewise, if $a(\min) \leq a < E(a)$, so that the data suggest negative association, the probability required for a one-tail test is

$$P_{\text{lower}} = \sum_{i=a(\min)}^{a} \Pr(i \mid N, m, r).$$

To do a two-tail test we must sum the probabilities of obtaining a deviation as great as or greater than that observed in either direction. Thus, if the deviation of the observed a from expectation, $|E(a) - a|$, is x, the probability for the two-tail test is

$$\left\{ \sum_{i=a(\min)}^{E(a)-x} + \sum_{i=E(a)+x}^{a(\max)} \right\} \Pr(i \mid N, m, r).$$

This *exact* test is easily done with the help of tables such as those of Finney et al. (1963) and Bennett and Horst (1966), provided both column totals (or both row totals) are ≤ 50. Outside the range of these tables we may make use of the fact that the distribution of a tends to normality. As already remarked, a is a hypergeometric variate. Its mean and variance are

$$E(a) = \frac{rm}{N} \quad \text{and} \quad \text{var}(a) = \frac{mnrs}{N^2(N-1)}.$$

So we may treat

$$X = \frac{a - E(a)}{\sqrt{\text{var}(a)}}$$

as a standardized normal variate and refer to Normal tables to judge significance. Either a one-tail or a two-tail test may be done.

For large N it is permissible to substitute N for $N-1$ in the denominator of $\text{var}(a)$. Then X becomes

$$X = \frac{\sqrt{N}(ad - bc)}{\sqrt{mnrs}} \quad \text{and} \quad X^2 = \frac{N(ad - bc)^2}{mnrs}. \tag{13.3}$$

We see that X^2, being the square of a standardized normal variate, has the χ^2-distribution with one degree of freedom.

Since the continuous χ^2-distribution is being used to approximate a discrete distribution, it is desirable to make a continuity correction. In calculating X^2 this is done by subtracting $\frac{1}{2}$ from the two observed frequencies that exceed expectation and adding $\frac{1}{2}$ to the two frequencies that fall short of expectation. Then

$$X^2(\text{corrected}) = \frac{[|ad - bc| - N/2]^2 N}{mnrs}.$$

This ensures a closer approximation of the χ^2-integral to the sum of the tail terms of the true, discrete distribution of X^2.

It will be seen that the expression here denoted by X^2 is the one often described as χ^2. However, for a function of the observed cell frequencies (in other words, a sample statistic) it is preferable to use the noncommital symbol X^2; the symbol χ^2 should be reserved for the theoretical variate with the χ^2-distribution (see Cochran, 1954).

The foregoing arguments explain why we may use the χ^2-test as an approximation to the exact test appropriate to a doubly restricted double dichotomy. However, ecologists who use the test should never lose sight of the fact that a χ^2-test is automatically two-tailed. So if we sometimes use the χ^2-test and at other times (because of low observed frequencies) the exact test, the two-tail form of the exact test should be used. Otherwise the results are not comparable.

4. Testing the Association in a Population of Units (Question 2)

The test just described, in either its exact or approximate version, is strictly applicable only to a doubly restricted double dichotomy; it answers Question 1. Now suppose we require an answer to Question 2; that is, we wish to know whether the data yielded by a *sample* could have come from a *population* in which the two species are independent. The desired probability is a sum of terms of the form of $Pr(a, b, c, d)$ in (13.2), in which hypergeometric probabilities are weighted with binomial probabilities. The χ^2-test makes no allowance for the binomial terms and although it is the best test (for a proof of this, see Kendall and Stuart, 1967), the calculated tail probabilities are greater than their true values. This may lead to acceptance of the null hypothesis of independence when it should be rejected—a type II error. At the same time the risk of asserting that there is true association when there is not—a type I error—is reduced.

For sufficiently large samples the error introduced is usually negligible. However, there are two reasons for stressing the conceptual difference between the two situations. In the first place ecologists are often tempted to use the exact test when some of the cell frequencies yielded by a sample are very low. This is satisfactory when one is inquiring whether the species are independent within the particular sample of units being examined. But it is futile to use the exact test when an answer to Question 2 is sought; the exact answer to one question may be very inexact as an answer to a different question.

In the second place certain methods of classifying vegetation (to be described in Chapter 19) require repeated applications of the ordinary χ^2-test to 2×2 tables, often with quite low values of N. A risk of error that can reasonably be ignored when only one or a few tests are being done may then become appreciable.

5. Measurements of Association

Besides testing a 2×2 table to judge whether the null hypothesis of independence should be accepted or rejected, we may also wish to measure the strength of the association between two species. There are many ways of measuring association, some devised by statisticians for use with any 2×2 table; and others by ecologists in search of a coefficient particularly suited to measuring ecological association. Rather than catalog a great many of them, it will be more useful to consider a few in detail. For clarity, we shall speak only of positive association. The modifications to be made in the arguments when the species are negatively associated are obvious.

Two desirable properties of a coefficient to measure association are : (i) that it should be zero when the observed cell frequencies are equal to their expected values; and (ii) that it should range from -1 to $+1$, taking the value -1 when negative association is as great as possible and $+1$ when positive association is as great as possible. The second of these properties is ambiguous: what is meant by "as great as possible"?

The phrase can have two meanings. Suppose species B occurs in more of the units than does species A. Then positive association would be as great as possible if A was never found in the absence of B, though there would perforce be some units in which B was found without A. This degree of association can be called *complete* (see Kendall and Stuart, 1967). However, we might choose to assert that the association was "as great as possible" only when neither species ever occurred without the other. This is called *absolute* association. In terms of the cell frequencies in the 2×2 table complete positive association requires that either b or c (not necessarily both) be zero. For absolute association we must have both $b = 0$ and $c = 0$; then $m = r = a$ and $n = s = d$. Depending on whether we want the coefficient of association to be $+1$ when the association is complete or absolute, we can use the coefficients Q or V defined as follows (Yule, 1912):

$$Q = \frac{ad - bc}{ad + bc};$$

then $Q = 1$ when either $b = 0$ or $c = 0$. Or

$$V = \frac{ad - bc}{+(mnrs)^{1/2}}.$$

Then $V = \pm 1$ only if $mnrs - (ad - bc)^2 = 0$, but since

$$mnrs - (ad - bc)^2 = 4abcd + a^2(bc + bd + cd)$$
$$+ b^2(ac + ad + cd) + c^2(ab + ad + bd)$$
$$+ d^2(ab + ac + bc)$$

the expression on the right vanishes only if two of the cell frequencies are zero. We can exclude from consideration cases in which the two zeros occur in the same row or the same column, for there would be nothing to test. This leaves either $b = 0$ and $c = 0$, for which $V = +1$, or $a = 0$ and $d = 0$, for which $V = -1$.

Both coefficients are zero when the association is nil, that is, when observed and expected frequencies are equal, for then $a - E(a) = (ad - bc)/N = 0$.

The sampling variance of Q is

$$\text{var}(Q) = \frac{(1-Q^2)^2}{4}\left\{\frac{1}{a} + \frac{1}{b} + \frac{1}{c} + \frac{1}{d}\right\}.$$

The derivation is given in Kendall and Stuart (1967). We shall not consider Q further here. The fact that its use precludes any distinction between complete and absolute association makes it unsuitable as a measure of ecological association. To see this consider the two tables

		Species B						Species B		
		+	−					+	−	
Species	+	80	80	160		Species	+	80	0	80
A	−	0	15	15		A	−	0	15	15
		80	95	175,				80	15	95.

They are strikingly different. In the population tabulated on the left, although all the 80 units containing B also contain A, there are in addition 80 units with only A in them. By contrast in the population on the right neither species occurs without the other. For both tables $Q = 1$, whereas $V = 1$ only for the right-hand table; for the left-hand table $V = 0.281$. Any ecologist would assert that the association shown by the table on the right was by far the greater and V is therefore preferable to Q in ecological contexts.

Two other points to mention about V are the following: (a) $V^2 = X^2/N$, where X^2 is the test statistic defined in (13.3); V^2 is known as the mean-square contingency of the 2×2 table. (b) V is a correlation coefficient. Assign to each unit a pair of values (x, y) with

$$x = \begin{cases} 1 \text{ when species } A \text{ is present,} \\ 0 \text{ when species } A \text{ is absent,} \end{cases} \quad y = \begin{cases} 1 \text{ when species } B \text{ is present,} \\ 0 \text{ when species } B \text{ is absent.} \end{cases}$$

Then

$$\text{cov}(x, y) = \frac{a}{N} - \frac{mr}{N^2} = \frac{ad - bc}{N^2},$$

$$\text{var}(x) = \frac{mn}{N^2} \quad \text{and} \quad \text{var}(y) = \frac{rs}{N^2},$$

and therefore

$$V = \frac{ad - bc}{\sqrt{mnrs}} = \frac{\text{cov}(x, y)}{[\text{var}(x)\text{var}(y)]^{1/2}}.$$

In other words, V is the correlation coefficient between x and y. An estimate of its sampling variance, derived by Yule (1912), is given by

$$\text{var}(V) = V^2 \left\{ -\frac{4}{N} + \frac{ad(a+d) + bc(b+c)}{(ad-bc)^2} \right.$$

$$\left. -\frac{3}{4}\left[\frac{(m-n)^2}{Nmn} + \frac{(r-s)^2}{Nrs} \right] + \frac{(ad-bc)(m-n)(r-s)}{2Nmnrs} \right\}.$$

It should be noticed that although V is a correlation coefficient we cannot use the usual formula for the sampling variance of a correlation coefficient, r, namely, $\text{var}(r) = (1-r^2)^2/N$, because this formula is appropriate only when the parent distribution of the x's and y's is bivariate normal. In the present case each variate is discrete and can take only the values 0 and 1.

An example of a coefficient of association proposed especially for ecological use is one suggested by Cole (1949) and used by him to measure the association between two species of mites parasitic on rats trapped in food-processing establishments. He stipulated that a desirable coefficient for measuring the degree of association of a pair of species should have the following properties:

1. It should be zero when the observed frequencies are equal to those expected on the null hypothesis of independence; that is, when $a = E(a)$.
2. It should be $+1$ (or -1) when $a - E(a)$ has its maximum possible positive (or negative) value compatible with the observed marginal totals.
3. The coefficient should vary linearly with a.

Condition 3 amounts to requiring (see Figure 13.1) that the graph of the coefficient C plotted against a should be a straight line going through the points $[a(\min), -1]$, $[E(a), 0]$, and $[a(\max), +1]$, where $a(\min)$ and $a(\max)$ are the smallest and largest values that a could have. Obviously this requirement cannot be met unless the three points are colinear. This will be so only when $E(a) = \frac{1}{2}[a(\min) + a(\max)]$. Cole therefore proposed different formulas for C for use in different circumstances. We now require that positive values of C shall lie on the line joining $[E(a), 0]$ and $[a(\max), +1]$ and that negative values of C shall lie on the line joining $[a(\min), -1]$ and $[E(a), 0]$.

Suppose first that the association is positive or that $ad > bc$. Then C must fall on the line KL in the figure, and

$$C = \frac{a - E(a)}{a(\max) - E(a)}.$$

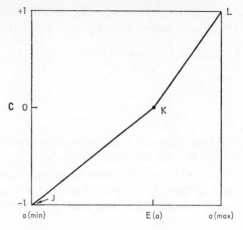

Figure 13.1. To illustrate the derivation of the different formulas for Cole's coefficient of interspecific association.

Let the species be so labeled that A is the less frequent species or $m \leq r$. Then

$$a(\max) = m \qquad \text{and} \qquad C = \frac{a - mr/N}{m - mr/N} = \frac{ad - bc}{ms}.$$

Next suppose that there is negative association or that $ad < bc$. Then C must fall on the line JK and

$$C = \frac{a - E(a)}{E(a) - a(\min)}.$$

The value of $a(\min)$ depends on whether $a \leq d$ or $a > d$. If $a \leq d$, $a(\min) = 0$. Writing C_1 for the coefficient in this case,

$$C_1 = \frac{a - mr/N}{mr/N} = \frac{ad - bc}{mr}.$$

If $a > d$, $a(\min) = a - d$. Writing C_2 for the coefficient,

$$C_2 = \frac{a - mr/N}{mr/N - (a - d)} = \frac{ad - bc}{ns}.$$

Cole also obtained the sampling variances of the three versions of his coefficient.

Property (2) of the coefficient (see page 211) requires that

$$C = \begin{cases} +1 & \text{when } a = a(\max), \\ -1 & \text{when } a = a(\min). \end{cases}$$

Therefore, like Yule's Q, it suffers from the defect that no distinction is made between complete and absolute association.

6.   Association Among k Species

Up to this point we have discussed association between two species. Now suppose we wish to explore the joint occurrences of several, say k, species. Suppose that for each of a sample of N units it has been noted whether it contains species i $(i = 1, 2, \ldots, k)$. Each unit then belongs to one of 2^k distinguishably different classes and there are 2^k comparisons to make between observed and expected frequencies. Grouping of cell frequencies is nearly always necessary before the goodness of fit can be tested and it is desirable that the grouping be done objectively, that is, according to a procedure specified before the observational data have been inspected. A natural method of grouping consists in combining the $\binom{k}{r}$ classes containing exactly r species (regardless of what those species are) for $r = 0, 1, \ldots, k$. This yields the distribution of the number of species per unit, say s, a variate of obvious interest.

To derive the distribution of s, consider first the case when $k = 1$; that is, there is only one species. Put p for the probability that a given unit contains the species and assume p to be constant over all units. Then examining any unit to see if it does contain the species is equivalent to performing a Bernoulli trial with probability p of success. The variate s, the number of species per unit, takes the value 0 or 1 according as the trial fails or succeeds, and the probability generating function (pgf) of s is $g(z) = q + pz$, where $q = 1 - p$.

Now suppose there are k independent species. Examining a unit is equivalent to performing k independent Bernoulli trials, with probabilities of success $p_1, \ldots, p_k$ respectively; the variate s takes the values $0, 1, \ldots, k$. Since the outcomes of the trials are independent, the pgf of s is

$$G(z) = \prod_{j=1}^{k} g_j(z) = \prod_{j=1}^{k} (q_j + p_j z).$$

That is, $G(z)$ is the product of the pgf's of k independent binomial distributions.

This distribution is intractable as it stands but Barton and David (1959) showed that, unless the p_j values vary over a very wide range, it can be approximated by a binomial distribution with the same mean and variance.

Let us write

$$H(z) = (Q + Pz)^K$$

for the pgf of the approximating binomial. The mean and variance of the approximating distribution are therefore KP and KPQ respectively.

The mean and variance of the exact distribution are easily seen to be

$$E(s) = \sum_{j=1}^{k} p_j = kE(p) \tag{13.4}$$

where $E(p)$ is the expectation of the p_j values; and

$$\text{var}(s) = \sum_{j=1}^{k} p_j q_j = kE(p)[1 - E(p)] - k\,\text{var}(p) \tag{13.5}$$

where

$$\text{var}(p) = \left(\frac{1}{k}\right) \sum_j [p_j - E(p)]^2$$

is the variance of the p_j values.

The p_j's for $j = 1, \ldots, k$ are estimable from the data. An estimate of p_j is given by the observed proportion of units in which the jth species was found. Substituting the observed mean and variance of these estimates, say $\bar{p}$ and $v(p)$, for their population values in (13.4) and (13.5) gives estimates of $E(s)$ and $\text{var}(s)$. Then, equating the latter estimates to the corresponding moments of the approximating (binomial) distribution of s, we have

$$k\bar{p} = KP \qquad \text{and} \qquad k\bar{p}(1 - \bar{p}) - kv(p) = KPQ.$$

Solving for P and K gives

$$P = \bar{p} + \frac{v(p)}{\bar{p}} \qquad \text{and} \qquad K = \frac{k}{1 + v(p)/\bar{p}^2}$$

as the desired parameters of the approximating binomial. One can now test the null hypothesis, that the species are mutually independent, by judging the fit of the approximating binomial distribution to the observed distribution of s. Examples will be found in Pielou (1971, 1974c).

The test is not satisfactory if there are numerous "rare" species, that is, if p_j is very low for many values of j; in such cases, sample estimates for the true values of the probabilities p_j are unreliable. We must therefore do a test that treats the observed frequencies of occurrence of the species as fixed, rather than as random variates. These frequencies are the marginal totals of a 2^k table. In other words, we test the significance of the association within the sample actually examined (as in Question 1 on page 204) without attempting to infer whether the conclusion holds for the parent population from which the sample was drawn. An exact test is required, and to avoid excessive computation when k is large it is necessary to carry grouping of classes to the limit and recognize only two classes: "empty" units with no species (for which $s = 0$) and units with at least one species (for which $s > 0$). One is then doing the k-dimensional analog of the exact test for a 2×2 table; in other words, one is treating a 2^k table as a k-tuply restricted polyotomy (cf. page 206), and then grouping the cells of the table into two classes: one class consists of the single cell corresponding to absence of all k species and the other class consists of all other cells combined. The test is laborious; Barton and David (1959) give derivations of the required probabilities, and Pielou and Pielou (1967) give an example of application of the test to field data.

When doing this test (as, indeed, when doing any test for ecological association) it is important to keep in mind that it leads to acceptance or rejection of what is merely a statistical hypothesis, not an ecological hypothesis. For example, if one were led to reject the null hypothesis (of independence among k species) because there were "too many" empty units in the sample, several possible conclusions would be suggested: that some of the units were inhospitable to all the species; that all k species were actively attracted to one another; or that some of the pairwise associations between species were positive and some negative but that positive associations predominated. The test by itself does not enable one to distinguish among these possibilities, but it should not be treated as worthless on that account. A single statistical test usually permits one to take no more than a single step, of the many that may be needed, from observed facts to a theoretical conclusion. It is a common mistake among ecologists to demand too much from one statistical test and to feel aggrieved when their demands are not met.

7. Segregative and Nonsegregative Association

Suppose two species have been found to be positively associated and it is suspected that the only cause is that some of the units in the sample were intrinsically unsuitable for either of the species. If this were so, the number of empty units (the frequency d in the 2×2 table) would necessarily be improbably large and would lead inevitably to positive association. Some ecologists would argue that such association was apparent rather than real and that the high value of d was, in a sense, artificial. This conclusion seems unwarranted (see page 221), but, granting its reasonableness for the moment, we may now enquire whether the species are independently distributed among the units than *can* contain them. When only two species are considered, it is impossible to make a judgment merely on the basis of statistical tests, for we need only assume that the d empty units consist of $d - bc/a$ that are "unoccupiable" and bc/a that, though "occupiable," are empty as the result of chance.

Substitution of bc/a for the observed d then immediately yields a 2×2 table in which observed and expected frequencies are identical. (Inconsequential adjustments are necessary, of course, if bc/a is not a whole number.)

When more than two species are observed, however, it does become legitimate to ask whether the species are independent of one another within the occupiable units only. This follows, since, given a 2^k table in which the observed number of empty units exceeds expectation, it may or may not be possible to find a number which, if substituted for the observed number of empties, will yield a table not significantly different from that expected on the null hypothesis of independence of the species.

When such a number can be found, there is no reason to reject the hypothesis that the species are independent, since the association could be ascribed simply to the presence of some unoccupiable units. This type of association may be called *nonsegregative*. Conversely, the association can be called *segregative* if no such number exists; that is, if the discrepancy between observed and expected frequencies cannot be attributed merely to the presence of an unknown number of unoccupiable units. The occurrence of segregative association implies either that the various species are responding differently to differences among the units or that the species are affecting one another in various ways. Some ecologists would regard only segregative association as "true" and nonsegregative as "spurious."

A test for judging whether the association among several species is segregative or nonsegregative has been described by Pielou and Pielou (1968). The test is designed for use with species of infrequent occurrence and entails the comparison of the observed and expected numbers of different classes of unit encountered when the units are classified according to the species they contain. The number of units in each class is disregarded.

14

Association Between Pairs of Species II: Individuals in a Continuum

1. Introduction

Studies of the association between species that occupy discrete units normally take no account of the spatial arrangement of the units. The mathematical methods for judging the association between, say, a pair of parasite species infesting a population of mammals are formally identical with those for testing the association between, for example, the eye colors of parents and children. The spatial arrangement of the sample units is disregarded except in so far as the population studied is defined in terms of the geographical area it occupies.

To sample individuals that are scattered through a continuum usually entails taking arbitrarily delimited bits of the continuum as sample units; some of the difficulties that arise were mentioned in Chapter 9, which dealt with the spatial patterns of single species.

Continuum sampling is needed in the study of communities of sessile or sedentary organisms such as plants (on land or in fresh water) and benthic organisms (in salt or fresh water). Quadrats are the usual sampling units and in testing for association between two species, it is customary to treat each quadrat as if it were a discrete sample unit and to use the same methods as those described in Chapter 13.

Most ecologists (e.g., Greig-Smith, 1964) are aware of the problems that may arise from treating an arbitrary quadrat as though it were a discrete natural entity, but many authors seem to confound two wholly different sources of difficulty which ought to be treated separately. These are the effects on the conclusions of (a) the spacing of the quadrats and (b) the sizes of the quadrats and it is these problems that this chapter considers. A method of testing for association that does not require the use of quadrats is described in Section 5 of Chapter 15.

2. The Spacing of the Quadrats

In this section we discuss only positive association. The same arguments, with obvious modifications, apply equally to negative associations.

When two species are found to be positively associated, the conclusion drawn is usually one or both of the following: (a) one of the species has a beneficial effect on the other, either directly or by modifying the environment in a way favorable to it; or (b) some independent environmental factors are variable over the area, and because the two species have identical or overlapping tolerance ranges for the factors, both are forced to occupy coincident or overlapping areas.

The fact that a statistical test gives evidence that two species are positively associated does not, of course, lead directly to the conclusion that one of these mechanisms must be operating. The test by itself suggests only that the null hypothesis should be rejected. The null hypothesis is this: the probability that a quadrat contains species A is independent of whether it does or does not contain species B, and vice versa. Rejection of the hypothesis, and the consequent acceptance of the alternative hypothesis, namely that the probabilities are *not* independent, does not automatically imply that it is the two species that are dependent. It may simply mean that the quadrats are dependent.

Thus an unexpectedly high number of joint occurrences of the two species could easily arise from the following cause. Suppose both species have patchy spatial patterns merely as a result of their modes of reproduction. Then, if, for both species, their patches are large relative to the study area, the extensive overlap of a large patch of species A with another of species B will result in the joint occupancy by both species of a large region of overlap, all this being purely a matter of chance. We need not assume that the species have any beneficial effect on one another, or that the environment is heterogeneous.

Imagine, for concreteness, that each pattern has the form of a random L- or S-mosaic (see Chapter 12). Then, if the two species are independent, their joint pattern will be that of two random mosaics haphazardly superimposed. The joint pattern will be a four-phase mosaic with phases that can be labeled (AB), $(A\bar{B})$, $(\bar{A}B)$, and $(\overline{AB})$. Now, if this mosaic were coarse-grained, that is, if its patches were large in relation to the whole area being sampled, then many of the quadrats placed in it would not be independent of one another. The fact that one quadrat proved to be an (AB), for example, would mean that nearby quadrats also had a high probability of being (AB). Only if the quadrats were so widely separated that their mutual dependence was negligible could we safely conclude that any observed association was not the result of chance overlap.

The association that may be observed when a coarse-grained mosaic is sampled with closely spaced quadrats is not in any sense "unreal," but it is due merely to chance and not to the operation of biological causes. In order to test for the existence of association having a true biological cause it is therefore necessary to space the sample quadrats far enough apart to ensure that there is only a negligible probability that any two of them will occupy the same patch of either species. Unless this is done, chance association is likely to be mistaken for biological association. It should be emphasized that association resulting from patch overlap should be ascribed to chance only if the patches themselves are the outcome of chance. If the patches owe their existence to especially favorable patches of ground, then association resulting from patch overlap has indeed a biological cause.

It is obvious that chance overlap might as easily produce an unexpectedly high proportion of $(A\bar{B})$, $(\bar{A}B)$, or $\overline{(AB)}$ quadrats. If the area sampled is large enough and there is no biological association between the species, the localized excesses of quadrats of one kind will tend to cancel one another out. It is these localized excesses that give rise to chance association within small areas. If, in a large area, we still find evidence of positive association, it follows that the two one-species mosaic patterns are not haphazardly superimposed. Then the association must have a biological cause and the overlap is not wholly due to chance.

We may, of course, choose to use closely spaced quadrats in studying a population of small area, but this should be done only when the objective is to study a spatial pattern formed of two superimposed one-species patterns. This is a legitimate endeavor, though it seems not to have been tried to date. Far more often tests of association are prompted by a desire to know if the two species are causally associated. What must then be avoided is the placing of several quadrats in any single mosaic patch.

Having stressed that quadrats must be widely spaced in a test for causal association, the question now arises: how can one tell whether a particular spacing is wide enough? Suppose sampling is done systematically, with quadrats at the corner points of a square lattice (see Figure 14.1). Now let a map of the quadrat layout be drawn and let the (AB) quadrats be marked and the rest left unmarked. If many of the (AB) quadrats are in a few large regions of patch overlap, they will tend to fall into groups; in other words, they will not be randomly dispersed among the available lattice points. Conversely, if the (AB) quadrats are in different regions of overlap, they will not form groups; instead they will be randomly and independently mingled with the other quadrats. Thus, if we take as a null hypothesis that the (AB) quadrats are independent of one another, we may easily test this hypothesis by means of Krishna Iyer's test (see page 144), applied now to a lattice of spaced quadrats instead of to a grid of contiguous quadrats. The power of the test is probably not great, but its use should at least permit the detection of pronounced dependence

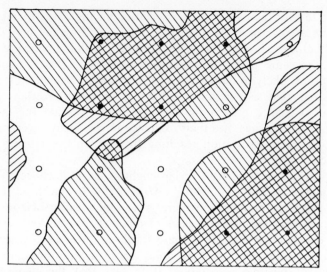

Figure 14.1a. Systematic sampling of a coarse-grained mosaic, with dependence among the quadrats. (*AB*) quadrats (shown ●) are in groups relative to the other quadrats (shown ○).

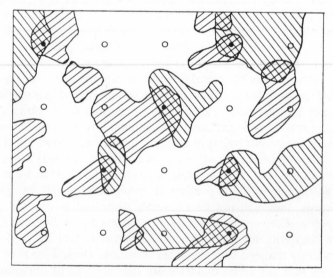

Figure 14.1b. Systematic sampling of a fine-grained mosaic; the quadrats are independent and (*AB*) quadrats are randomly mingled with the others.

220

among the (AB) quadrats. Judgment by intuition is likely to be unreliable in this context.

If we wish to examine a large sample of widely spaced quadrats, it is clear that the study area itself must be large. It is well known that judgments of association are strongly affected by the size of the area within which sampling is done. Thus suppose we define the population under study as a small area, throughout which both the species are common, and we find no evidence of association between them. If the population is now redefined to include large surrounding areas of ground from which both species are absent and the enlarged area is sampled, the two species will show strong positive association. This must obviously happen, since the proportion of $(\overline{AB})$ quadrats in the large area will greatly exceed the proportion in the small area (see page 215). It seems often to be believed that to include in a sample a great many quadrats that for unknown reasons may be incapable of containing either of the species invalidates an association test in some way. The notion seems to be that the evidence for positive association has been obtained by cheating. This is not so. If large areas of ground contain neither of the species, it is reasonable to conclude that the two species resemble each other in the conditions they will *not* tolerate. This in itself shows that they are associated in the sense that there is considerable overlap in their tolerance ranges. The search for association of this type should obviously not be restricted to small homogeneous areas. If we wish to know whether two species will react similarly to environmental conditions, it is absurd to insist that sampling be confined to a homogeneous area throughout which the conditions are constant. To do so would be to rule out the possibility of finding different responses to different conditions.

Coefficients of association, such as V, Q, and C, described in Chapter 13, are inevitably affected by the amount of ground unsuitable for both the species that happen to be included in a study area. We can only say that a measure of association must always be regarded as a property of a species pair and an area considered *jointly*. In an attempt to overcome this defect in the ordinary association coefficients some workers (e.g., Dice, 1945, and Bray, 1956) have proposed what amounts to a coefficient of overlap, though the different authors use various names and definitions for their coefficients. Denoting the observed frequencies of (AB), $(A\overline{B})$, $(\overline{A}B)$, and $(\overline{AB})$ quadrats by a, b, c, and d, as in a 2×2 table (see page 204), all of these coefficients are functions of a, b, and c only; d is disregarded. Thus, for instance, Bray's (1956) "coefficient of amplitudinal correspondence" is $2a/(2a + b + c)$; that is, it is the ratio of the number of quadrat occurrences of A's and B's, when they occur jointly, to the total number of occurrences of both species.

The use of a coefficient of overlap is, however, no solution to the difficulty just mentioned—that a coefficient of association is strongly influenced by d, the number of empty quadrats. A coefficient of overlap contains no new information: one cannot judge whether it departs significantly from expectation, on the null hypothesis of independence of the species, without taking d into account.

3. The Effect of Quadrat Size

We turn now to a consideration of the effects of quadrat size on indications of association, assuming quadrat spacing to be wide enough for no problems to arise in that connection.

Clearly, only a limited range of sizes is permissible. The quadrats must not be so small that they are incapable of containing at least two individuals of the larger species. Nor must they be so large that one of the two species will occur in every quadrat; this would cause one of the marginal totals of the 2×2 table to be zero and make a test impossible. For practical reasons the feasible range of quadrat size will often lie well within the theoretically permissible range.

Suppose that the two species whose relationship is being studied are associated; that is, they are not independent. Suppose also that this association is caused by the fact that each is responding to the same controlling factor in the environment. For concreteness imagine this factor to be soil moisture. We could, at least in theory, mark the tolerance range of each species on a linear scale of moisture values. Various relationships are possible. The range of one species could be wholly within that of the other; the ranges of the two species might be coincident, they might overlap to a greater or lesser degree, or they might be disjunct but contiguous. Lastly, they might be disjunct and separated by a gap representing soil too dry for one species and too wet for the other.

When two species are controlled in this way by the same factor, they will certainly show association at some quadrat sizes, but the magnitude and even the sign of the observed association must depend on quadrat size. If the tolerance ranges are coincident or overlap markedly, even small quadrats will show positive association, but if the tolerance ranges are disjunct or only slightly overlapped small quadrats may give evidence of negative association and larger quadrats evidence of positive association. The way in which association varies with quadrat size therefore depends on the interrelation between the species' tolerance ranges for the factor and also on the gradient of this factor on the ground; that is, on whether its spatial variation is abrupt or gradual. Thus, if two species showed negative association at a given quadrat size in a region in which the factor varied only gradually, the same two species might be found to be positively associated, at the *same* quadrat size, in a region in which the controlling factor varied abruptly.

It is clear that any observed relation between association and quadrat size is capable of a variety of interpretations. This does not mean that association tests are useless; for example, suppose we were interested in three species, *A*, *B*, and *C*, and found that with a given size of quadrat *A* and *B* and also *B* and *C* showed marked positive association, whereas for *A* and *C* the association was only weakly positive or perhaps negative. We could then conclude that *B*'s tolerance range for some factor lay between those of *A* and *C*.

In any case, whenever the association between two species is tested or measured, the size of quadrat used for sampling must always be stated. This is merely a matter of defining the sample units, a thing one would always do whether the units were discrete natural objects or arbitrary pieces of a continuum.

It must again be emphasized that if the tolerance ranges of two species overlap there will be regions on the ground in which the actual overlap of patches of plants may result only from chance. This is because, within a region suitable for it, the pattern of a species is likely to be patchy because of its mode of reproduction. What is observed is not patches of suitable ground but familial patches *within* the patches of suitable ground. (Because of this even total coincidence of two species' environmental requirements may give coefficients of association significantly less than one.) Therefore only if the quadrats are spaced widely enough to encompass unsuitable as well as suitable ground and an extensive area of each can observed association be ascribed to biological causes.

To study association by means of grids of contiguous quadrats, as suggested by Grieg-Smith (1964) and Kershaw (1960), can only lead to confounding of two effects, those of quadrat spacing and quadrat size. For investigating association, as well as for investigating pattern, these authors use grids of contiguous quadrats and then combine the quadrats in pairs to form successively larger blocks (see page 140). Thus interquadrat distance and $\sqrt{\text{(quadrat area)}}$ are directly proportional and the results become impossible to interpret.

15

Segregation Between Two Species

1. Introduction

It was shown in Chapter 14 that when we examine the association between two species of plants the results will be strongly influenced by both the spacing of the quadrats and their sizes. This is because what is being investigated is not so much interspecies relationships per se but rather joint, two-species patterns. Other factors, besides the relationship between the species, affect these joint patterns. This suggests that it would be worthwhile to attempt to study the pattern of each species in *relation to the other* without regard to the pattern of either in relation to the ground.

We assume that the plants occur as discrete, genetically distinct individuals, reproducing by seed, and that therefore we shall not be misled by the presence of clumps of vegetative shoots which are, or may have been in the past, organically connected. What we now enquire is: do the two species form "relative clumps"? A relative clump of species A, for instance, occurs in a group of plants in which the proportion of A's is greater than their proportion in the whole population. Likewise for species B. A relative clump may or may not be a spatial clump also. Thus consider Figure 15.1. In Figure 15.1a, although the plants as a whole have a random spatial pattern, there is clear evidence of relative clumping. Conversely, in Figure 15.1b, although the population as a whole is strongly clumped, the two species are not clumped in relation to one another, since within each clump the A's and B's are present in the same proportions and are randomly mingled.

The objective now is to study the relative patterns of two species independently of their spatial patterns. The first question that arises is: are the two species randomly mingled or are they relatively clumped? If they are randomly mingled, they may be described as unsegregated; if not, they are to some extent segregated from each other.

One way of answering this question would be to sample the population with quadrats and adjust the size of each quadrat so that it contained exactly n individuals. Then, if the two species were randomly mingled and

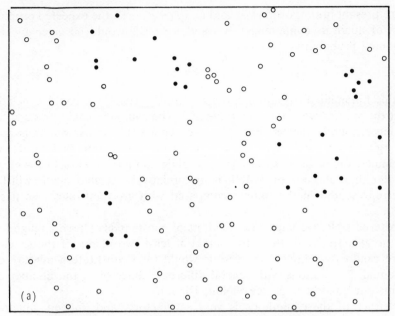

Figure 15.1a. A two-species population in which all plants together have a random pattern, although the species are segregated.

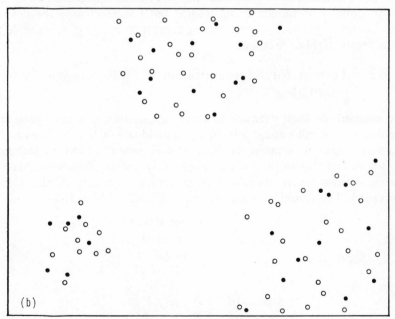

Figure 15.1b. A two-species population in which the plants have clumped patterns but the species are unsegregated.

were present in proportions p_A and p_B ($p_A + p_B = 1$), the expected proportion of quadrats containing r A's and $n - r$ B's would be given by the binomial probability

$$\binom{n}{r} p_A^r p_B^{n-r}.$$

If this binomial distribution fitted the observations, we could conclude that the species were randomly mingled. The sampling could be done by taking randomly located points as the centers of circular quadrats and enlarging each circle until exactly n plants were contained in it. As explained on page 157, this is very difficult to do in practice with n greater than about 2 or 3. If n is low, relatively isolated plants will be overrepresented in the sample compared with crowded ones (see page 158).

Alternatively, we might use quadrats of constant size, large enough for the great majority of them to contain at least two plants. If the species were randomly mingled, the results would then constitute a mixture of binomial distributions with several different values of n, the number of plants per quadrat (e.g., see Pielou, 1963a).

Neither of these methods is very satisfactory, and, indeed, quadrat sampling is an inappropriate method for studying relative patterns. A quadrat is, after all, a sample piece of space; it is an appropriate sampling unit if spatial patterns are being studied but not necessarily the best in other contexts. For studying relative patterns it seems more reasonable to use nearest-neighbor methods.

2. Testing for Segregation in a Fully Known Population

The method we shall examine in detail is one that is useful when the population is small enough for all the individuals of both species to be taken into account. Suppose we examine each individual in turn and note its species and that of its nearest neighbor. In judging which neighbor is nearest, distances are measured from center to center of the plants concerned. The results may be set out in a 2×2 table as follows:

		Species of nearest neighbor		
		A	B	
Species of base plant	A	a	b	m
	B	c	d	n
		r	s	N

A χ^2-test may be used to judge whether the observed cell frequencies depart significantly from expectation. Thus, if $a > mr/N$ significantly, A's have other A's as nearest neighbors unexpectedly often, and the species are partly segregated.

Notice that m and n are the population frequencies of species A and B; they are not estimates, since every plant in the whole two-species population has been examined. The column totals r and s are the numbers of times individuals of species A and of species B have served as the nearest neighbor of another individual. There is no reason to expect r and s to be identical with m and n. It is clear that in any plant population some individuals will be so placed that they are not the nearest neighbor of any other individual; others will serve as nearest neighbors to 1, 2, ..., 5 individuals. (Since distances are measured from center to center, it is impossible for any individual to serve as nearest neighbor to more than five others.) Thus the "population of nearest neighbors" is not the same as the "population of base plants" (i.e., the population in the ordinary sense). The population of nearest neighbors does not contain those base plants that are not the nearest neighbor of any other plant, whereas the remaining base plants contribute 1, 2, ..., 5 members to the nearest neighbor population according as they are the neighbors of 1, 2, ..., 5 base plants. For this reason we do not expect to find $r = m$ and $s = n$ exactly. If both species have random spatial patterns, the expected proportions of A's and B's in the nearest-neighbor population will be the same as their proportions in the base population; in other words, the column totals will be homogeneous with the row totals. In general, however, even this will not be true.

To see this, suppose the individuals of species A tend to be larger or to require more space than those of species B; that is, the A's tend to be more isolated. The A's will then serve as nearest neighbors relatively less often than the B's and the column totals will not be homogeneous with the row totals. Such a population may be called unsymmetrical.

Random mingling of the species may occur in an unsymmetrical population as well as in a symmetrical one. In testing for random mingling, each plant is treated as having two attributes; its own species and the species of its nearest neighbor. The two species are unsegregated if the two attributes are independent.

There are many ways in which we might define a coefficient of segregation. The one originally suggested (Pielou, 1961) is S, defined as

$$S = 1 - \frac{\text{observed number of mixed pairs}}{\text{expected number of mixed pairs}},$$

where a "mixed pair" denotes an individual of one species having an individual of the other as nearest neighbor.

Then

$$S = 1 - \frac{N(b+c)}{ms+nr}.$$

In an unsegregated population $E(S) = 0$. In a fully segregated population, in which no A's have B's as nearest neighbors and vice versa, or $b = c = 0$, $S = +1$.

Negative segregation is also theoretically possible, though in plant populations it is very unlikely. Imagine a population of G widely spaced groups of plants, each consisting of a single central individual of species A with a "halo" of g B's around it. We suppose that each B has the A at the center of its group as nearest neighbor, so necessarily $g \leq 5$. The 2×2 segregation table is

		Species of neighbor		
		A	B	
Species of base plant	A	0	G	G
	B	Gg	0	Gg
		Gg	G	N

Then

$$S = 1 - \frac{(G+Gg)N}{G^2 + G^2 g^2}$$

$$= 1 - \frac{(1+g)^2}{1+g^2} < 0,$$

since $g > 0$. Differentiating with respect to g, we find that S takes its minimum value when $g = 1$, that is, when the whole population consists of isolated A-B pairs; then $S = -1$. Thus S ranges from -1 for the maximum possible negative segregation, through 0 for no segregation, to $+1$ for the maximum possible positive segregation.

If S is determined in the way described above, that is, by inspecting every member of a two-species population, it is a population parameter and has no sampling variance. If the observations are made on a random sample of the population, the calculated S will be only an estimate of the true population value and will be subject to sampling variation. The sampling distribution of S in an unsegregated population has not so far been derived.

The following is an example of an ecological problem on which segregation studies threw light (Pielou, 1966a). Consider a dense stand of seedling trees of two or more species growing up on burned-over land. Initially the trees are likely to be relatively clumped (or partly segregated) for two reasons: (a) the seeds from which they grew may have had clumped patterns and (b) the habitat may be heterogeneous so that one species is favored in one small area and another in another. As the trees grow, their number is bound to decrease due to natural thinning that results from suppression of the less successful trees. If the deaths are preponderantly in crowded one-species clumps that have grown from clumps of seeds, segregation will decrease as the population ages. Conversely, if the trees that die out are those that chanced to germinate in comparatively unsuitable sites, the deaths will tend to sort the species into more clearly defined one-species groups; then segregation will increase. In the work mentioned above the degree of segregation was determined (by a different method from that described here) in five dense stands of young trees at the beginning and end of a period of several years. In all cases segregation was found to decrease, which suggests that natural thinning was the result of intraspecific competition within dense one-species clumps.

3. Plant Sequences in Transects

To identify only the nearest neighbor of a plant in a two-species population will obviously give less information about the relative patterns of the species than would the identification of the 1st, 2nd, ..., nth nearest neighbors. As already explained (page 157), the larger the value of n, the more difficult it is to make the field observations. A way around the difficulty consists in studying the sequence of plants (of the two species concerned) along a belt transect; the transect should be just narrow enough for there never to be any doubt of the order of the plants within it. The resulting observations then consist of a sequence of occurrences such as

$$AAA\ BB\ AAA\ BBBBB\ A\ BBB....$$

Assuming that we have such an observed sequence, we now wish to test whether the A's and B's are randomly mingled. Suppose there are a A's and b B's. For a short sequence (a, $b \leq 20$) an exact test may be done as follows (Feller, 1968):

Let every uninterrupted sequence of A's or of B's, be called a *run*. If the observed number of runs is small, it is reasonable to suspect that the species are segregated. Therefore, if the observed sequence contains R runs of both species, we wish to determine the probability of there being R runs or fewer on the null hypothesis of random mingling of the species. If this probability is α, we may assert that the species are significantly segregated at the $100\alpha\%$ level.

We now need to determine the probability $P(k)$ that there will be exactly k runs. Note first that the total number of visibly different arrangements of a A's and b B's is

$$\frac{(a+b)!}{a!b!} = \binom{a+b}{a}.$$

Each of these distinguishable arrangements represents the same number, namely $(a!b!)$, of different *in*distinguishable permutations of the A's and B's. Therefore, since all permutations have equal probability, so also have all the distinguishable arrangements.

Next we must find the number of arrangements of the A's and B's that will give k runs.

Suppose, first, that k is even and put $k = 2m$. Then there are m runs of A's and m runs of B's. Now the number of ways of placing a objects into m cells so that every cell contains at least one object is $\binom{a-1}{m-1}$. (For a proof of this see Feller, 1968.) This is the number of ways in which the A's can be partitioned to give m runs. Likewise, the B's can be partitioned into m runs in $\binom{b-1}{m-1}$ ways. Therefore the number of arrangements that will give rise to m runs of each species when the species of the first run is specified is

$$\binom{a-1}{m-1}\binom{b-1}{m-1}.$$

Then, since a run of either species is equally likely to start the sequence, the number of ways of obtaining m runs of both A's and B's, regardless of which species starts the sequence, is

$$2\binom{a-1}{m-1}\binom{b-1}{m-1}.$$

To determine $P(k)$ we note that

$$P(k) = \frac{\text{number of arrangements of } a \text{ } A\text{'s and } b \text{ } B\text{'s that give } k \text{ runs}}{\text{number of distinguishable arrangements of } a \text{ } A\text{'s and } b \text{ } B\text{'s}},$$

so that, when $k = 2m$,

$$P(2m) = \frac{2\binom{a-1}{m-1}\binom{b-1}{m-1}}{\binom{a+b}{a}}. \tag{15.1}$$

Next, suppose k is odd and put $k = 2m+1$. Then either we must have $m+1$ runs of A's and m runs of B's, which can happen in $\binom{a-1}{m}\binom{b-1}{m-1}$ ways or there must be m runs of A's and $m+1$ runs of B's which can happen in $\binom{a-1}{m-1}\binom{b-1}{m}$ ways. Then

$$P(2m+1) = \frac{\binom{a-1}{m}\binom{b-1}{m-1} + \binom{a-1}{m-1}\binom{b-1}{m}}{\binom{a+b}{a}}. \tag{15.2}$$

Finally, the probability we wish to find, namely that of obtaining R runs or fewer, is given by $\sum_{k=2}^{R} P(k)$. The summation begins at $k=2$ since obviously there must be at least two runs. These probabilities have been tabulated by Swed and Eisenhart (1943) for all a, $b \leq 20$. In applying the test, therefore, we may simply consult the tables and there is no need to calculate and sum all the separate probabilities.

Outside the range of the tables use of this exact combinatorial test is laborious. For long sequences a different test is desirable. It is easy to devise one from the following considerations.

Each encounter with a plant, as we travel along the transect, may be thought of as a trial, the outcome of which must be A or B. If the species are unsegregated, the trials are independent, and the probabilities of the two possible outcomes are constant along the whole transect. Suppose these probabilities are p_A and $p_B = 1 - p_A$. Then the probability of encountering a run of r A's is $p_A^{r-1} p_B$; similarly, the probability of encountering a run of s B's is $p_B^{s-1} p_A$; that is to say, for each species the run lengths (numbers of individuals in the runs) are geometrically distributed. For the A's the expected run length is

$$E(l_A) = \sum_{j=1}^{\infty} j p_A^{j-1} p_B = \frac{p_B}{(1-p_A)^2} = \frac{1}{p_B},$$

and for the B's the expected run length is

$$E(l_B) = \frac{1}{p_A}.$$

Writing $\bar{l}_A$ and $\bar{l}_B$ for the observed mean run lengths of the species, we may therefore derive estimates $\hat{p}_A$ and $\hat{p}_B$ of the probabilities by putting $\hat{p}_A = 1/\bar{l}_B$ and $\hat{p}_B = 1/\bar{l}_A$. (These are, in fact, maximum likelihood estimators of the population probabilities.) As a test statistic we may therefore use $1/L = 1/\bar{l}_A + 1/\bar{l}_B$ and it is clear that, given the null hypothesis, $E(1/L) = 1$. We shall not explore the sampling variance of $1/L$ here. Details will be found in Pielou (1962b).

4. Distributions of Run Lengths for Three Model Populations

Given a sequence of species occurrences within a transect, we can do more than merely test the species for segregation. It is also interesting to devise models that might "explain" the patterns of the two species in relation to each other and to carry out tests to determine whether the models accord with the observations. A model that does not fit the observations is as instructive in some ways as one that does. The discrepancy between observation and expectation itself constitutes an observation, sometimes a more revealing one than that provided by the original raw data.

The Markov Chain Model

The simplest postulate that can be made is that the sequence of species-occurrences forms a realization of a simple two-state Markov chain. The Markov matrix may be written

$$\begin{pmatrix} p_{AA} & p_{AB} \\ p_{BA} & p_{BB} \end{pmatrix},$$

where p_{AB}, for example, denotes the probability that an A will be followed in the sequence by a B and the other three transition probabilities are defined analogously. Then the probability that a run of A's will be of length r is $p_{AA}^{r-1}p_{AB}$ and the mean run length of the A's is $E(l_A) = 1/p_{AB}$. Similarly, the mean run length of the B's is $E(l_B) = 1/p_{BA}$.

It is seen that the run lengths again have geometric distributions, as was true when the species were assumed to be randomly mingled, but if the sequence is a Markov chain we no longer have that $E(1/L) = 1$. If the species are positively segregated, $p_{AA} > p_{BA}$ and $p_{BB} > p_{AB}$. Therefore, since $p_{AA} + p_{AB} + p_{BA} + p_{BB} = 2$, $p_{AB} + p_{BA} < 1$ and consequently $E(1/L) = E(1/\bar{l}_A + 1/\bar{l}_B) < 1$.

In the few tests that have been done (Pielou, 1962b) on the relative patterns of pairs of plant species it was found that in all cases the observed distributions of run lengths had greater variances than had the fitted geometric distributions. On the scanty evidence that so far exists it seems unlikely that the Markov chain hypothesis will often be tenable. We must therefore search for a distribution whose variance is greater than that of a geometric distribution with the same mean.

The Geometric-Poisson Distribution

We now suppose that the population consists of nonoverlapping clumps of the two species and that the clumps are randomly mingled (see Figure

15.2). Within a belt transect, therefore, there are alternating runs of
A-clumps and B-clumps; (notice that we are now speaking of runs of
clumps). We do not postulate that there is any visible boundary between
adjacent clumps. Since the clumps are assumed to be randomly mingled,
their run-lengths must be geometrically distributed.

We next suppose that for each species the number of individuals per
clump (within the transect) is a Poisson variate.

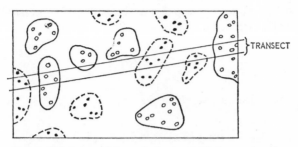

Figure 15.2. A model population in which the run-lengths
of the individuals of each species have geometric-Poisson
distributions. The clumps, shown with solid outlines for one
of the species and dashed outlines for the other, are randomly mingled. For each species the number of individuals
per clump within the transect is a Poisson variate.

For clarity of exposition we now confine attention to the run-lengths of
individuals of only one of the species, say species A. Our initial aim is to
find the relation between the mean and variance of the distribution, since
what is required is a distribution with variance greater than that of a
geometric distribution of the same mean. If a distribution has a known
probability generating function (pgf), its mean and variance are easily
determined directly; there is no need to derive an explicit expression for
the probability of encountering a run of r A's.

Now, since the run-lengths of the *clumps* of A's are geometrically
distributed, their pgf is

$$G(z) = qz + pqz^2 + p^2qz^3 + \cdots + p^rqz^{r+1} + \cdots$$

$$= \frac{qz}{1 - pz}.$$

For later reference we note here that the mean M and variance V of the
geometric distribution are $M = 1/q$ and $V = p/q^2$, so that

$$V = M^2 - M. \qquad (15.3)$$

The number of individual A's per A-clump is a Poisson variate and its pgf is therefore (see page 117)

$$g(z) = e^{\lambda(z-1)}, \qquad \text{with} \qquad \lambda > 0.$$

Thus the pgf of the run-lengths of *individual* A's is

$$H(z) = G[g(z)] = \frac{qe^{\lambda(z-1)}}{1 - pe^{\lambda(z-1)}}.$$

The distribution with this pgf may be called Geometric-Poisson. It is a generalized distribution (see page 118). Its mean is

$$\mu_1' = H'(1) = \frac{\lambda}{q} \tag{15.4}$$

and the second moment *about the origin* is

$$\mu_2' = H''(1) + H'(1) = \frac{\lambda^2}{q^2}(2-q) + \frac{\lambda}{q}. \tag{15.5}$$

Note, however, that some A-clumps (those with 0 individuals) will be unobservable. Some runs of A-clumps may consist entirely of these empty clumps and the whole run will then be unobservable. Thus the observed distribution will be truncated with the zero class missing. If the proportion of these empty runs is π_0, the first and second moments about the origin of the truncated distribution will be m_1' and m_2', say, where

$$m_1' = \frac{\mu_1'}{1 - \pi_0} \qquad \text{and} \qquad m_2' = \frac{\mu_2'}{1 - \pi_0}.$$

Now

$$\pi_0 = H(0) = \frac{qe^{-\lambda}}{1 - pe^{-\lambda}}$$

and

$$\frac{1}{1 - \pi_0} = 1 + \frac{qe^{-\lambda}}{1 - e^{-\lambda}}.$$

Therefore, using (15.4) and (15.5), the first two moments of the observed (truncated) distribution are

$$m_1' = \frac{\lambda}{q}\left(1 + \frac{qe^{-\lambda}}{1 - e^{-\lambda}}\right) \tag{15.6}$$

and

$$m_2' = \frac{\lambda}{q}(2-q)m_1' + m_1'. \tag{15.7}$$

We next eliminate q from these equations and find a relation between m_1'

and m_2' on the one hand, and λ on the other. From (15.6) we obtain

$$2m_1' = \frac{2\lambda}{q} + \frac{2\lambda e^{-\lambda}}{1 - e^{-\lambda}}$$

and from (15.7)

$$\frac{m_2'}{m_1'} = \frac{2\lambda}{q} - \lambda + 1,$$

whence

$$2m_1' - \frac{m_2'}{m_1'} + 1 = \frac{2\lambda e^{-\lambda}}{1 - e^{-\lambda}} + \lambda = \frac{\lambda(1 + e^{-\lambda})}{1 - e^{-\lambda}}. \qquad (15.8)$$

Recall from (15.3) that we require a distribution for which the first two moments are related by the inequality

$$m_2 = m_2' - m_1'^2 > m_1'^2 - m_1',$$

or, equivalently,

$$2m_1' - \frac{m_2'}{m_1'} + 1 < 2. \qquad (15.9)$$

We now ask whether this requirement is compatible with (15.8). The answer is obtained by considering the properties of the function

$$f(\lambda) = \frac{\lambda(1 + e^{-\lambda})}{1 - e^{-\lambda}}.$$

Using l'Hôpital's rule to evaluate $f(0)$ gives

$$f(0) = \lim_{\lambda \to 0} \frac{\lambda(1 + e^{-\lambda})}{1 - e^{-\lambda}} = \lim_{\lambda \to 0} \frac{1 + e^{-\lambda} - \lambda e^{-\lambda}}{e^{-\lambda}} = 2.$$

Further, by differentiating $f(\lambda)$ it is easy to show that $f'(\lambda) > 0$ for $\lambda > 0$, in other words, for $\lambda > 0$ the function is monotonically increasing.

It follows that the Geometric-Poisson distribution does *not* meet the requirement stated in (15.9). On the contrary, the Geometric-Poisson must always have a variance *less* than that of a simple geometric distribution of the same mean. It will therefore fit the observed frequency distributions less well than the simple geometric series.

In postulating that the run-lengths of the A-clumps had a geometric distribution, we took it for granted that the form of the distribution would be unaltered by the fact that some B-clumps, hence some runs of B individuals, would contain 0 individuals. The runs of A's preceding and following such an empty run of B's will, of course, appear to be a single unbroken run. This does not affect the form of the distribution of the run-lengths of A-clumps; it merely changes the parameter of the geometric distribution. Thus, if all B-clumps were observable and the probability of encountering one were p_B, we should have, as the probability of observing a run of r A-clumps, $p_A^{r-1}p_B$ (where $p_A = 1 - p_B$) (cf. page 231). Now let the number of individuals per B-clump be a Poisson variate with parameter λ_B. Then the probability of encountering an *observable* B-clump will be $p_B(1 - e^{-\lambda_B}) = p_B'$, say. Putting $p_A' = 1 - p_B'$, the probability of observing a run of r A-clumps now becomes $(p_A')^{r-1}p_B'$ and is seen to be still a geometric term.

The Compound Geometric Distribution

Consider again the run lengths of one of a pair of segregated species. As a third possible model, we shall suppose that the individuals of this species have run lengths that are geometrically distributed but that the parameter p of the geometric distribution is itself a random variable. It is assumed that p has a constant value within any one run but different values in different runs. The resulting distribution therefore is compound (see page 122).

We now choose a standard distribution that is likely to provide an approximation to the true distribution of p, whatever form the latter may take. The Beta distribution with range 0 to 1 is suitable because of its flexibility; it may be bellshaped or U-shaped and of any degree of skewness. We therefore assume that the pdf of p is

$$f(p) = \frac{1}{B(\alpha, \beta)} p^{\alpha-1}(1-p)^{\beta-1},$$

with $\alpha > 0$ and $\beta > 2$. The constraint $\beta > 2$ is necessary to ensure the convergence of the integral in (15.11).

We now wish to find the mean and variance of this compound geometric distribution to see how they are related.

For the simple geometric distribution the mean is $1/q$. Therefore the mean of the compound geometric distribution is

$$\mu_1' = \frac{1}{B(\alpha, \beta)} \int_0^1 \frac{1}{q} \cdot p^{\alpha-1}(1-p)^{\beta-1}\, dp = \frac{\alpha + \beta - 1}{\beta - 1}. \tag{15.10}$$

Next, for the simple geometric distribution the second moment about the origin is $(1+p)/q^2$. Hence the corresponding moment for the compound geometric is

$$\mu_2' = \frac{1}{B(\alpha, \beta)} \int_0^1 \frac{1+p}{q^2} p^{\alpha-1}(1-p)^{\beta-1}\, dp = \frac{(\alpha + \beta - 1)(2\alpha + \beta - 2)}{(\beta - 1)(\beta - 2)}. \tag{15.11}$$

(Notice that, as already remarked, we must have $\beta > 2$ if the variance of the distribution is to be finite.)

Then the variance of the compound geometric is

$$\mu_2 = \mu_2' - \mu_1'^2 = \frac{\alpha\beta(\alpha + \beta - 1)}{(\beta - 1)^2(\beta - 2)}. \tag{15.12}$$

Again recall that we are searching for a distribution with variance greater than that of a simple geometric distribution with the same mean. The compound geometric meets this requirement. From (15.10) and (15.12) it is seen that

$$\mu_2 = \frac{\alpha + \beta - 1}{\beta - 1} \cdot \frac{\alpha\beta}{(\beta - 1)(\beta - 2)} > \frac{\alpha + \beta - 1}{\beta - 1} \cdot \frac{\alpha}{\cdot\beta - 1}, \qquad \text{since } \beta > 2.$$

Therefore

$$\mu_2 > \frac{\alpha + \beta - 1}{\beta - 1}\left(\frac{\alpha + \beta - 1}{\beta - 1} - 1\right) = \mu_1'^2 - \mu_1'. \tag{15.13}$$

This is in contrast to the relation given in (15.3) for the moments of the simple geometric, namely, $\mu_2 = \mu_1'^2 - \mu_1'$.

The probability that a run of individuals will be of length r is given by

$$\pi_r = \int_0^1 p^{r-1}(1-p)f(p)\, dp = \frac{1}{B(\alpha, \beta)} \int_0^1 p^{\alpha + r - 2}(1-p)^\beta\, dp$$

$$= \frac{B(\alpha + r - 1, \beta + 1)}{B(\alpha, \beta)},$$

whence

$$\pi_1 = \frac{\beta}{\alpha + \beta}; \qquad \pi_2 = \frac{\alpha}{\alpha + \beta + 1}\pi_1; \ldots; \pi_r = \frac{\alpha + r - 2}{\alpha + \beta + r - 1}\pi_{r-1}; \ldots.$$

As shown by (15.13), it is always possible to find a compound geometric distribution with the same mean and variance as an observed distribution of run-lengths, however great the variance of the latter may be. This, of course, does not ensure a good fit of the theoretical distribution to the observed distribution. When the compound geometric was fitted to run lengths of plants observed in the field (Pielou, 1962b), it was found that the number of runs of unit length (i.e., consisting of a single individual) always exceeded expectation. This result suggests that even when two species of plant are unquestionably clumped relative to each other there is often considerable clump overlap. When this is so, it seems reasonable to conclude that the tolerance ranges of the species for microclimatic and edaphic factors must also overlap.

5. An Association Test Based on Transect Sequences

Regardless of whether two species are randomly mingled, one often wishes to judge whether they tend to occur with each other more frequently than with other species. If they do, they may be said to be positively associated even though we are not now concerned with how close (in space) their co-occurrences are. A test for association in this sense, using data on the sequence of occurrences of individuals in narrow transects has been devised by Knight (1974). It is free of the ambiguity introduced into the customary tests for association by the use of quadrats of arbitrary size.

The test is as follows. Suppose the two species whose association is to be tested, A and B, occur in a many-species community; let all the other species be treated as indistinguishable and be labeled, simply, X. An observed sequence of occurrences will then be of the form:

$$A\ XXX\ BA\ X\ AAB\ XX\ BAAABB\ XX\ A\ X\ BBBB\ XXXX\ AABB \ldots ,$$

for example. We wish to judge whether A and B are adjacent unexpectedly often.

Let the numbers of A's and B's in the sequence be a and b and let the total number of individuals of all species be n; thus $a + b < n$ unless A and B are the only species present in which case an association test would be meaningless. Denote the number of AB adjacencies by j; (in the sequence above, $j = 5$). Write $p(j; n, a, b)$ for the probability of there being j adjacencies if the A's, B's and Xs are randomly mingled. For $n > a + b$, j must be 0, 1, ... or min(a, b). A recursion formula for the probabilities may be arrived at as follows.

Suppose there are j AB adjacencies in a sequence of n individuals of which a are A's and b are ‘B's. Let a new X be inserted into the sequence at random. There are $n + 1$ equiprobable locations for the new individual since it may either go between two of the individuals already present or at one of the ends of the sequence. The probability that there are j AB adjacencies in this new random sequence is $p(j; n + 1, a, b)$ by definition. Now the newly added X could have been placed, with probability $j/(n + 1)$, between an adjacent A and B in the original sequence; this would reduce the number of AB adjacencies to $n - 1$. Or, with probability $(n + 1 - j)/(n + 1)$, it could have been so placed as to leave the number of AB adjacencies unchanged at j.

Therefore it is clear that

$$p(j; n+1, a, b) = \frac{j+1}{n+1} p(j+1; n, a, b) + \frac{n+1-j}{n+1} p(j; n, a, b) \quad (15.14)$$

When $a + b = n$, that is, in a sequence consisting only of A's and B's with no X's, the number of AB adjacencies is one less than the number of runs of the two species. Thus $p(j; n, a, b)$ when $n = a + b$ is the probability that there will be $j + 1$ runs in a random sequence of a A's and b B's and is given by (15.1) or (15.2) according as j is odd or even. Starting with this known probability, we can now use (15.14) recursively to determine $p(j; n, a, b)$ for $n = a + b + 1, a + b + 2, \ldots$.

Knight (1974) also showed that the distribution of the number of AB adjacencies is asymptotically normal with mean and variance given by

$$\mu_1' = \frac{2ab}{n} \quad \text{and} \quad \mu_2 = \frac{2ab}{n}\left[\frac{(n-a)(n-b) + ab - n}{n(n-1)}\right].$$

The normal approximation (with the usual continuity correction) is reasonably good even for short sequences. Knight found that the difference between the exact tail probability and that of the approximating normal curve never exceeded 0.012 in absolute value.

This "sequence test" for association between two species overcomes one of the chief defects of tests based on quadrat sampling in that it precludes the ambiguities arising when, as inevitably happens, quadrats of different sizes give discrepant results. However the sequence test is even more likely than a quadrat test to be influenced by chance association, that is the chance overlap, with no underlying biological cause, of patches of the two species (cf Section 14.2 and Figure 14.1).

16

Segregation Among Many Species: n-Phase Mosaics

1. Introduction

The methods described in Chapter 15 for studying what may be called segregation patterns in two-species populations can obviously be extended to many-species populations. These methods presupposed that the plants under study occurred as separate, easily distinguished individuals; they are therefore most likely to be applicable in investigations of the relative patterns of the different tree species in mixed forest. It is unrewarding to contemplate many-species segregation patterns in the abstract when one has no particular concrete problem in mind requiring solution, and no attempt is made here to generalize the preceding arguments. However, the concepts of relative pattern and spatial segregation should be borne in mind whenever the pattern of a many-species tree population poses some definite ecological problem (for an example see Pielou 1963b, 1965b).

Entirely different methods of studying many-species patterns are needed when the plants occur, not as distinct, discrete individuals, but as clumps or patches of appreciable area. The vegetation of swamps or of heaths or moors are examples. In drawing a map of such vegetation it would be impossible to represent individual plants by dimensionless dots, each marking a plant's center, as could be done with a map of a forest; instead the vegetation must necessarily be mapped as a many-phase mosaic, one phase of which might be bare ground.

There is an important contrast between dot maps and mosaic maps that should be emphasized at the outset. A dot map is made up of two kinds of entity: the dots, representing individual plants of various species and the undifferentiated background that constitutes the only continuous phase. (In practice the "background" may be bare ground or any form of low vegetation the ecologist is unconcerned with.) In a mosaic map every point in the mapped area is assigned to one phase or another and all the phases are present as patches of finite extent. Bare ground, if present, is most easily treated as an extra phase; the number of phases is then one

240

more than the number of species. It will be remembered that in considering two-phase mosaics (see Chapter 12) we were considering the pattern of a single plant species; the phases consisted of ground occupied, and not occupied, by the species concerned.

In this chapter we consider many-species mosaics from a theoretical point of view. The practical difficulties that may be encountered in field work can be dealt with only as they arise. It is assumed that there are no discrete "dimensionless" plants in the vegetation (or if there are they are to be ignored) and that every point in the area can be assigned to one and only one phase so that the patches of the various phases do not overlap. We are assuming, in short, that the many difficulties inherent in mapping vegetation have already been overcome and that the object of study is a finished mosaic map.

The arguments to be developed apply to all mosaic maps, not merely to those of small areas of ground in which each separate species of plant is distinguished and mapped. Other kinds of mosaic map also interest ecologists and a few examples were given on page 181. To that list may be added maps showing the outlines of recognizably distinct plant communities, or physiognomically defined vegetation types, or areas of ground differing among themselves in physical and chemical characteristics (e.g., see Pielou, 1965a).

2. n-Phase L-Mosaics

The first question to ask concerning an n-phase mosaic map is: is it random? As remarked on page 182, the question cannot be answered as it stands. The term "random," as applied to mosaics, may be variously defined and the question is unanswerable until the exact type of randomness contemplated has been stipulated. We have therefore to envisage, and specify, a random n-phase mosaic. How could one be constructed?

One obvious way is to draw a network of random cells, either by the random-lines process (page 182) or by the random-sets process (page 182). Now suppose that the mosaic is to have n colors, or phases, in proportions $a_1, a_2, \ldots, a_n$. The cells of the network are independent, and for each of them the probability that it will be of the ith color is a_i ($i = 1, 2, \ldots, n$). The cells are colored accordingly, and the resulting groups of one or more contiguous cells of the same color constitute the patches of the mosaic.

In what follows we consider only n-phase random lines mosaics (L-mosaics) because of their mathematical tractability. By generalizing the argument in Chapter 12 (page 184) it is easily seen that the sequence of phases encountered along a row of equidistant points across such a mosaic is a realization of a simple n-state Markov chain; but if an observed sequence is found to form a Markov chain it does not follow that the mosaic being sampled can be regarded as random. Another condition must also be met. For a mosaic to be considered random it is obviously necessary that the different kinds of patches should be randomly mingled, and this requires that the Markov matrix should be of the form

$$
\mathbf{P} = \begin{matrix} & \begin{matrix} x_1 & x_2 & & x_n \end{matrix} \\ \begin{matrix} x_1 \\ x_2 \\ \\ x_n \end{matrix} & \begin{pmatrix} P_1 & p_2 & \cdots & p_n \\ p_1 & P_2 & \cdots & p_n \\ \multicolumn{4}{c}{\dotfill} \\ p_1 & p_2 & \cdots & P_n \end{pmatrix} \end{matrix}. \tag{16.1}
$$

Here $x_1, x_2, \ldots, x_n$ denote the different phases (or species or colors) occurring along the sequence.

In this matrix all the off-diagonal elements in any one column are equal. In other words, the probability that any of the not-x_1's, say, will be succeeded by an x_1 is the same for all not-x_1's and is given by p_1.

Two properties of a Markov chain having a matrix of this form should be noted.

1. The chain is reversible; that is, if any two adjacent points in the sequence belong to phases x_j and x_k, the two possible orders $x_j x_k$ and $x_k x_j$ will have equal probabilities. Denoting the limiting vector of the chain by $\mathbf{a}' = (a_1 \, a_2 \cdots a_n)$, we find that the condition for reversibility is therefore $a_j p_k = a_k p_j$. That this is so may be seen from the following:

Since $\mathbf{a}'\mathbf{P} = \mathbf{a}'$, we may find the elements of $\mathbf{a}'$ in terms of the elements of $\mathbf{P}$ from the equations

$$
a_j P_j + P_j \sum_{k \neq j} a_k = a_j, \qquad j = 1, 2, \ldots, n,
$$

or

$$
p_j(1 - a_j) = a_j(1 - P_j),
$$

whence

$$
a_j = \frac{p_j}{1 - P_j + p_j}. \tag{16.2}
$$

Then the condition $a_j p_k = a_k p_j$ becomes

$$
\frac{p_j p_k}{1 - P_j + p_j} = \frac{p_k p_j}{1 - P_k + p_k}
$$

or

$$p_j + P_k = p_k + P_j. \tag{16.3}$$

Both members of (16.3) are equal to $1 - \sum_{l \neq j,k} p_l$ and the condition for reversibility is fulfilled.

Vegetation mosaics that might yield irreversible chains are easy to visualize. Suppose the phases of a mosaic consisted of different plant communities and their arrangement was controlled by the prevailing wind so that one type of community was often found downwind but seldom upwind of another type (e.g., see Watt, 1947). The sequence of phases observed along a row of points parallel with the wind direction would then depend on whether the sequence was observed upwind or downwind; that is, the chain would be irreversible. As another example, consider the sequence of phases likely to be observed along a north-south transect in temperate latitudes; the sequence would be irreversible if some small plants tended to grow only on the shaded side and others only on the sunlit side of clumps of taller vegetation. Mosaics such as these are obviously nonrandom.

2. A chain having a matrix of the form $\mathbf{P}$ in (16.1) is *lumpable* with respect to any partition of the phases (see Kemeny and Snell, 1960); that is, if the phases are grouped together in any manner whatever to give a smaller number of more broadly defined "lumped phases," the sequence will still form a Markov chain. To see this we may argue as follows: suppose the phases $x_1, x_2, \ldots, x_n$ are partitioned into sets $\{X_1, X_2, \ldots\}$. Each set consists of one or more phases lumped together. Now consider the sets X_u and X_v having s and t phases, respectively. Let the phases be so labeled that X_u consists of the phases $x_1, x_2, \ldots, x_s$, and let X_v consist of the phases $x_{s+1}, x_{s+2}, \ldots, x_{s+t}$. Denote by $P(x_i X_v)$ with $i = 1, 2, \ldots, s$ the probability of a transition from the phase x_i in the set X_u to any phase in the set X_v. Then the condition for lumpability of these phases is that

$$P(x_1 X_v) = P(x_2 X_v) = \cdots = P(x_s X_v);$$

that is, the probability of a transition from any phase in the set X_u to a phase in X_v is the same for all phases in X_u. It will be seen that

$$P(x_i X_v) = p_{s+1} + p_{s+2} + \cdots + p_{s+t} \qquad \text{for} \qquad i = 1, 2, \ldots, s,$$

and therefore the chain is lumpable with respect to any partition of the sets we care to make.

From this it follows that the two-phase mosaic formed of any one phase and all the others combined is a random two-phase L-mosaic of the sort described in Chapter 12. Therefore each phase, or species, considered alone has a mosaic pattern that is random in the sense already defined.

Random n-phase L-mosaics of vegetation are probably rare in nature. Presumably they are as uncommon as many-species populations of "discrete" plants in which every species has a random dot pattern.

It is interesting to compare the number of parameters needed to specify completely (a) the pattern of an n-species population of discrete plants in which all n species are at random and (b) a random n-phase L-mosaic. Clearly (a) is completely defined when the densities, or Poisson parameters, of all the species are known; that is, there are n parameters. Likewise, to specify (b), n parameters are needed as now shown. First it is necessary to know the relative areal proportions of the phases or equivalently the elements of the limiting vector $\mathbf{a}'$; the vector has n elements, but since they sum to unity only $n-1$ are independent. In addition one of the transition probabilities must also be known to complete the specification. This is equivalent to asserting that if $\mathbf{a}'$ is known only one of the elements in $\mathbf{P}$ is free to vary, a fact we shall now prove.

It has already been shown [see (16.2)] that a_j is a function of p_j and P_j only. Thus, when a_j is given, p_j is uniquely determined by P_j or vice versa. It was also shown in (16.3) that $P_j - p_j = P_k - p_k$ or that $P_j - p_j$ is constant for all j. It follows that if all the elements of $\mathbf{a}'$ and one of the elements of $\mathbf{P}$ are given all the other elements of $\mathbf{P}$ will be determined. Thus, of the n parameters needed to define the random mosaic, $n-1$ are determined by the areal proportions of the n phases and the nth is a measure of the "grain" of the mosaic. If $P_j - p_j$ (which is constant for all j) is large, the mosaic is coarse-grained; conversely, if $P_j - p_j$ is small, the mosaic is fine-grained.

3.　Unsegregated Mosaics

If, in studying a mosaic, we take as null hypothesis that it has the form of a random L-mosaic we are, of course, testing a very complicated hypothesis. We are assuming not only that the phases are randomly mingled but also that the pattern of each species considered by itself is a two-phase random mosaic. We shall next consider the properties of an *unsegregated mosaic*, or one in which the phases are randomly mingled, although no restrictions are placed on the patterns of the separate species. We are now concerned only with the *relative* arrangements of the different kinds of patches and not with their absolute patterns. In discussing the patterns of discrete "point" plants in Chapter 15, we emphasized the distinction between relative patterns (the patterns of species relative to one another) and absolute patterns (the spatial pattern of each species relative to the ground). We now take an analogous approach to mosaic maps. A random n-*phase* L-mosaic may be thought of as the analog of the pattern formed by the superposition of n independent random dot patterns. Similarly, an unsegregated mosaic, in which the different kinds of patches are randomly mingled, is the analog of a dot pattern of the kind shown in Figure 15.1b (see page 225), in which the different kinds of dots are randomly mingled, although their spatial patterns need not be random. Obviously the patches in a mosaic can be randomly mingled or unsegregated regardless of their shapes and sizes, and, if a vegetation mosaic is found to be unsegregated, we may conclude that there is no tendency for particular groups of species to grow adjacent to one another.

To investigate the properties of a mosaic that is random in this less restricted sense it is again convenient to consider the sequence of phases encounted at equidistant points along a line transect. Labeling the n different species with the letters $A, B, \ldots, N$, we obtain sequences such as the following:

$$AA\ CCCC\ N\ AA\ C\ BBB\ CC\ DDD\ A\ NNN \ldots.$$

The sequence is a succession of runs of the different letters. The lengths of these runs is now of no interest and the observations can be replaced by a "collapsed" sequence in which each run, whatever its length, is replaced by a single letter. The collapsed sequence corresponding to the observed sequence shown above is thus

$$A\ C\ N\ A\ C\ B\ C\ D\ A\ N \ldots.$$

We shall now show that if the species are randomly mingled this sequence constitutes a Markov chain and shall find its transition probabilities in terms of the elements of its limiting vector.

The argument is clarified by envisaging a balls-and-boxes model that generates a sequence of the desired kind. The following is such a model:

Take n boxes. Put a supply of balls of n different kinds, labeled $A, B, \ldots, N$, in each box; these balls are present in the same proportions in every box, namely $\pi_1, \pi_2, \ldots, \pi_n$. Now take the jth box and remove from it all balls bearing the jth letter, for $j = 1, 2, \ldots, n$. The jth box will now contain balls of all letters except the jth; in the jth box the proportion of balls bearing the kth letter becomes $\pi_k/(1 - \pi_j)$ for $k = 1, 2, \ldots, (j-1)$, $(j+1), \ldots, n$. The sequence we require can now be generated as follows: pick a box at random and from it take a randomly selected ball, observe its letter, and replace it. If it had the fth letter, pick the next ball from the fth box; observe the letter of this second ball and replace it; if the second ball bore the gth letter, pick next from the gth box; and so on. In the resulting sequence of letters clearly no two adjacent ones can be the same. It is also clear that at each step of the process the probability of picking a ball with a particular letter depends only on the letter of the preceding ball, since this ball determined which box the succeeding ball should be picked from. Hence the sequence is Markovian. The Markov matrix of the collapsed chain is given by

$$\mathbf{Q} = \{q_{jk}\} \quad \text{where} \quad q_{jk} = \begin{cases} \dfrac{\pi_k}{(1 - \pi_j)} & \text{when} \quad j \neq k, \\ 0 & \text{when} \quad j = k. \end{cases}$$

Denote the chain's limiting vector by

$$\mathbf{b}' = (b_1 \ b_2 \cdots b_n).$$

Then, since $\mathbf{b}'\mathbf{Q} = \mathbf{b}'$,

$$\sum_{r \neq j} \frac{b_r}{1 - \pi_r} = \frac{b_j}{\pi_j}$$

Therefore

$$\sum_{r \neq j} \frac{b_r}{1 - \pi_r} - \frac{b_j}{\pi_j} = 0 = \sum_{r \neq k} \frac{b_r}{1 - \pi_r} - \frac{b_k}{\pi_k},$$

whence

$$\frac{b_k}{1 - \pi_k} - \frac{b_j}{\pi_j} = \frac{b_j}{1 - \pi_j} - \frac{b_k}{\pi_k}$$

and thus

$$\frac{b_j}{b_k} = \frac{\pi_j(1 - \pi_j)}{\pi_k(1 - \pi_k)}.$$

It follows that $b_j = C\pi_j(1 - \pi_j)$, where C is a constant of proportionality.

The π's may now be found in terms of the b's; computational details will be found in Pielou (1967b). As estimates of the b's, we may take the observed proportions of A's, B's, C's, etc., in the collapsed chain. Notice that these proportions are not the same as the areal proportions of the different phases of the mosaic. They are the proportions of times that *runs* (of any length) of the various phases occur in the original uncollapsed sequence.

If some of the species (or phases) of the mosaic occur only as occasional, widely scattered patches, we may wish to pool two or more of these infrequent species into a single miscellaneous class. It can be shown (Pielou, 1967b) that such pooling will not affect the form of the chain. This is not the same as proving that the chain is lumpable (see page 243) which, in general, it is not. Thus suppose we are given the collapsed chain

$$A\,C\,B\,A\,X\,C\,A\,B\,C\,X\,Y\,X\,C\,B\,C\,A\,Y\,X\,A\,\ldots$$

and wish to pool the two uncommon species X and Y, calling the combined "species" Z. The sequence obtained by replacing all the X's and Y's by Z's is

$$A C B A Z C A B C Z Z Z C B C A Z Z A \ldots .$$

However, this is *not* the sequence we are interested in, since it no longer has the property that adjacent pairs of letters are always different; it contains runs of consecutive Z's.

Therefore let us collapse the last sequence by deleting all but one Z wherever a run of two or more of them occurs. The result may be called a twice-collapsed sequence, and in this example it is

$$A C B A Z C A B C Z C B C A Z A \ldots .$$

This twice-collapsed sequence is Markovian and its matrix of transition probabilities is of the same form as $\mathbf{Q}$. The proof entails showing that the twice-collapsed sequence is identical with the chain that would have been obtained had the pooling been done at the balls-in-boxes stage of constructing the model; that is, if we had used only $(n-1)$ boxes and had replaced the X-balls and Y-balls with Z-balls.

It follows that if an n-phase mosaic has randomly mingled patches the mosaic formed by combining some phases to form fewer, more broadly defined "pooled phases" also has randomly mingled patches. This fact allows us to combine rare species (or rare phases of any kind) into more inclusive phases in any way that may be convenient before doing a test for random mingling.

17

Patterns in Zoned Communities

1. Introduction

The mosaic patterns discussed in Chapter 16 were assumed to have irregularly shaped, though roughly isodiametric, patches. An entirely different kind of mosaic pattern is that exhibited by zoned communities on unidirectional environmental gradients. Both the geometrical regularity of zoned patterns, and the cause for it, are obvious. There are many ecological questions that can be investigated far more directly, and that can be answered with far greater clarity by studies in naturally occurring zoned communities than anywhere else. Zoned communities are common in nature: salt marshes, the altitudinally zoned communities of hillsides and mountainsides, rocky shore communities, and sublittoral benthic communities are some conspicuous examples. However, what may be called "zone mosaics" should not be sought only in these obvious sites where communities are always zoned to some extent. Because natural zonation (which can be thought of as the outcome of a natural experiment) is so ecologically informative, it is worth while to seek it out even in places where it would not automatically be expected. At any site where an environmental factor undergoes smooth, unidirectional change, so that its isopleths form (at least roughly) a set of parallel straight lines, some degree of community zonation is likely. Such places are far more deserving of study than the elusive "homogeneous community" traditionally beloved of ecologists.

2. Problems Posed by Zonation Patterns

The five panels in Figure 17.1 illustrate five of the problems that a statistical study of zone patterns can answer. The answers are statistical answers, of course, and consist in acceptance or rejection of a chosen null hypothesis; the ecological conclusions flowing from such statistical decisions depend, as always, on other circumstances.

The environmental gradient in Figure 17.1 descends from left to right. In the first four panels are possible zone maps in which the zones are

248

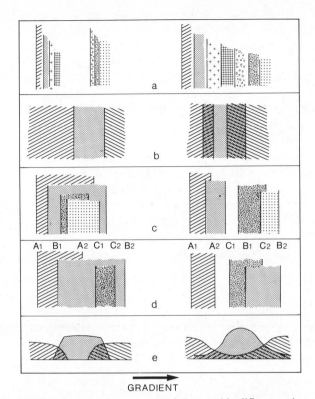

GRADIENT

Figure 17.1. To illustrate five of the testable differences between comparable zone patterns. The gradient descends from left to right across the page. (a, b, c, d) Maps in which the zones extend up and down the page (a shows positions of only the upslope zone boundaries). (e) Elevations in which the zones are perpendicular to the page. (See text for further explanation.)

shown as strips extending up and down the page; it is assumed, for clarity, that the zone boundaries are abrupt and obvious. The last panel shows an elevation (instead of a map) and the zones extend into the page. The contrasted pair of patterns in each panel illustrate some specific questions, raised by zonation patterns, that appropriate statistical tests will answer. The questions, labeled as in the figure, are as follows.

1. Are zone boundaries grouped, or evenly spaced out? The question is asked separately for upgradient (as shown in the figure) and downgradient boundaries. (The diagram shows only a narrow strip of each zone, adjacent to the upgradient boundary.)

2. Do the zones tend to abut on neighboring zones with little or no overlap, or do they overlap extensively?

3. For a chosen subset of the full species complement (e.g. a taxonomically related group of species), do the zones overlap one another more than, or less than, would be expected if they were mutually independent?

4. Consider two spatially separate realizations of a zoned community, having similar (in the limit, identical) species lists. Are the zone patterns (of any chosen subset of species that are present at both places) qualitatively congruent (i.e., concordant) in the two realizations? That is to say, do the several zone boundaries that make up the pattern (an upgradient and a downgradient boundary for each species) occur in the same order in the two realizations? In Figure 17.1d they do not: denoting the up- and down-gradient boundaries of species A by $A1$ and $A2$ (and correspondingly for species B, C, . . .), it is seen that the ordered lists of boundaries are on the left:

$$A1 \quad B1 \quad A2 \quad C1 \quad C2 \quad B2;$$

and on the right:

$$A1 \quad A2 \quad C1 \quad B1 \quad C2 \quad B2.$$

5. Are zone boundaries abrupt, or does the abundance of a species tend to diminish gradually above and below its optimum position on the gradient?

It should be noticed that the first four of these questions have a meaning only when several zones (or their boundaries) are considered; question 5 is the only one of the five that can be asked concerning each species separately.

These questions are dealt with in Sections 3 through 7 below. Before tackling them individually, it is necessary to consider how field observations on zonation patterns are best carried out. The boundaries of the species' zones are rarely clear enough (except, sometimes, for those of a few dominants) to be everywhere discernible with certainty and precision. An objective method for identifying boundary locations must therefore be devised.

Suppose the sampling unit used for examining a zone pattern is a row of contiguous quadrats forming a belt transect at right angles to the zones (see Figure 17.2). Assume that the gradient is monotonic. Then the zone of a given species may be taken to extend from the first quadrat in which it occurs to the last, as one traverses the transect in one direction.* Not all quadrats between these two extreme ones will contain representatives of

---

* For a modification of this definition of zone width, see Section 7 (page 265).

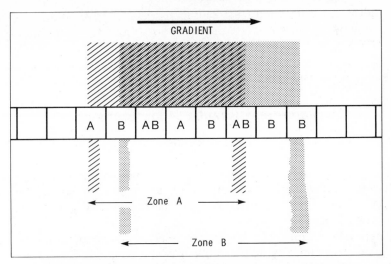

Figure 17.2. To illustrate the sampling of a zone pattern. The horizontal "ladder" is a row of contiguous quadrats. The letters A and B label quadrats containing species A and B. The estimated positions of their upgradient ($A1$ and $B1$) and downgradient ($A2$ and $B2$) boundaries are shown by the vertical strips. Species A's zone is hatched and species B's zone is stippled.

the species, especially if it is uncommon; but if the level of the factor(s) whose gradual variation is the cause of the zonation is the overriding cause of the species' success or failure, then it would presumably be possible for it to live anywhere between its observed uppermost and lowermost occurrences. Thus the positions of the uppermost and lowermost occurrences of a species within one transect provide estimates, yielded by that transect, of the boundaries between which the species can occur.

The estimates are subject to error of course. A species of low density will often chance to be absent from some of the high and low quadrats where it could occur, with the result that the width of its zone will be underestimated. Conversely, accidental "outliers" will sometimes lead to overestimation of the width of a zone. However, observation of a large enough sample, that is, of sufficiently many transects, will keep the risk of reaching erroneous conclusions acceptably low.

To estimate the location of a zone boundary as being somewhere within a quadrat yields only a coarse, discrete estimate of its position. In theory it is conceivably possible to measure accurately the distances (within a belt transect) separating the top of a gradient, a species' first and last occurrences, and the bottom of the gradient, and hence to treat boundary positions and zone widths as continuous variates. This procedure may occasionally be feasible, but probably not often. In what follows, we therefore treat zone widths and locations as discrete variates; when, as some tests require, boundaries have to be listed in order of occurrence, ties may occur and can be allowed for.

3. Clustered and Regular Zone Boundaries

To examine the pattern of zone boundaries, we take as null hypothesis that they are at random. The hypothesis is entertained, and the test performed, for upgradient and downgradient boundaries separately. The discussion below is limited, for clarity, to upgradient boundaries; the modifications needed in order to apply the test to downgradient boundaries are obvious.

To judge whether zone boundaries may be regarded as having been assigned independently and at random to the quadrats in a transect, we compare the observed and expected numbers of quadrats estimated to be cut through by boundaries, that is, containing species' "first occurrences" as one descends the transect starting from its upper end. Since all species in the first quadrat are, by definition, first occurrences in the transect, we discard this quadrat and treat its species as having zones whose upper boundaries lie above the start of the transect. Now assume (having discarded the first quadrat and its species) that the transect contains Q quadrats and s species. Write p_x for the probability that x quadrats are cut by one or more boundaries. Then

$$p_x = \frac{\left\{\begin{array}{c}\text{Number of ways of}\\\text{choosing } x \text{ of } Q\\\text{quadrats}\end{array}\right\}\left\{\begin{array}{c}\text{Number of ways of assigning } s\\\text{boundaries to these } x \text{ quadrats}\\\text{with none "empty"}\end{array}\right\}}{\left\{\begin{array}{c}\text{Number of ways of assigning } s\\\text{boundaries to } Q \text{ quadrats}\\\text{without restriction}\end{array}\right\}}$$

$$= \frac{\binom{Q}{x}\binom{s-1}{x-1}}{\binom{Q+s-1}{s}}, \qquad x = 1, \ldots, \min(s, Q).$$

Thus if the observed number of quadrats cut by boundaries is z we may calculate the probability that, under the null hypothesis, the s boundaries would have cut z or fewer (say) of the Q quadrats by summing the appropriate probabilities. This would yield the tail probability required for a one-tail test. If, for instance,

$$P = \sum_{x=1}^{z} p_x < \alpha \qquad (17.1)$$

where α is the chosen significance level, we should conclude that improbably few of the quadrats were cut by boundaries, and hence that the boundaries were clustered (as on the left in Figure 17.1a). Conversely, considering the other tail of the distribution, if $P > 1 - \alpha$ we should conclude that improbably many of the quadrats were cut by boundaries and hence that these were evenly or regularly spaced out (as on the right in Figure 17.1a).

The foregoing discussion assumes that only one transect is examined. Normally one would be reluctant to attach much weight to observations from a single transect and in any case the results would probably be inconclusive. Usually it is desirable to combine conclusions from several transects. This may be done by calculating P, from (17.1), for all the transects and judging whether $P - 0.5$ tends to have the same sign in an improbably high proportion of the cases. That is, significance of the departure of P from 0.5 is judged by a sign test.

If application of this test leads to acceptance of the null hypothesis it does not necessarily follow that the pattern of zone boundaries is random (that is, that the points at which they cut a transect constitute a Poisson point process). The number of quadrats cut by boundaries might accord with that predicted by the null hypothesis because, for instance, widely spaced groups of boundaries occurred and the pattern was regular within groups. Indeed, the difficulties of interpretation that arise in studying the patterns of zone boundaries parallel those that arise with dot patterns (cf. page 136). Results depend on the size of quadrat used, and this fact can of itself be useful if a judicious set of null hypotheses, and of tests for them, is chosen.

Examples of the application of this test to the pattern of zoned salt marsh vegetation have been given by Pielou (1975a,b) and Pielou and Routledge (1976).

4. Testing for Competition Between Species in Contiguous Zones

A species population on an environmental gradient may manage to establish itself wherever environmental conditions are within its tolerance limits, in which case we can say that it occupies the whole of its "fundamental zone." Or else its zone may be narrowed by the encroachments of competing species in zones above or below (or both) in which case it is confined to a "realized zone" that abuts against the realized zones of other species. Figure 17.1*b* shows the contrast diagrammatically.

Regardless of the pairwise interaction of particular pairs of species, it is interesting to ask, concerning any zoned community treated as a whole, whether its pattern is controlled at least to some extent by between-species competition. If it is, then upgradient and downgradient zone boundaries will not be independently located; instead, they will tend to occur close together and hence will be found in the same quadrat unexpectedly often.

To test the null hypothesis that boundaries are independently located, using data from one transect, we may set up the following 2×2 table. (For obvious reasons both the first and last quadrats in the transect are discarded; assume that Q quadrats remain.)

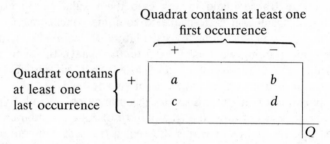

As before, a species' upgradient boundary is taken as crossing the transect through the quadrat in which the species first occurs as one traverses the gradient from top to bottom; likewise its downgradient boundary coincides with its last occurrence.

The observed proportion of the quadrats cut by both up- and downgradient boundaries (one or more of each) is a/Q, and under the null hypothesis the probability that this will happen is

$$p_a = \frac{\binom{a+b}{a}\binom{c+d}{c}}{\binom{Q}{a+c}}$$

where a can take the integer values $a(\min) = \mathrm{Max}(0, |a - d|), \ldots,$ $\mathrm{Min}(a + b, a + c) = a(\max)$. To test the null hypothesis against the alternative that a exceeds expectation, using a significance level α, it is now only necessary to calculate the tail probability

$$P = \sum_{x=a}^{a(\max)} p_x$$

[The required probabilities have been tabulated by Finney et al. (1963) and Bennett and Horst (1966); cf. page 207]. Then if $P < \alpha$, one may reject the null hypothesis and conclude that neighboring zones abut against each other (perhaps because of between-species competition) unexpectedly often.

As with the test described in Section 3 (page 252) one will usually wish to take account of observations on several transects. This may be done by carrying out a sign test to judge whether $P < 0.5$ in an improbably high proportion of the transects.

The test was applied to observations on the zoned vegetation of salt marshes by Pielou (1975a,b) and Pielou and Routledge (1976)

5.   The Overlap Score of Groups of Zones

The amount of overlap among the zones of a group of species is often of interest. In a biogeographical-evolutionary context, the gradient concerned might be thousands of kilometers long, a latitudinal gradient, say; in a more local, purely ecological context, the gradient might range from a few meters for shore communities to a few kilometers for montane communities. Regardless of the spatial scale of the enquiry, consider a group of related, perhaps congeneric, species and how they become sorted on a gradient. One might expect that because of their common ancestry, their environmental requirements will be similar and hence that their tolerance ranges will overlap strongly. Alternatively, their very similarity may have caused competitive exclusion followed by evolutionary divergence, with the result that they tend not to be found in the same place; on a gradient their zones will overlap slightly or not at all. To recognize which of these irreconcilable expectations has been realized in a particular case, we need to define the concept of random overlap; given a number of zones of unspecified width located independently (but, of course, parallel with one another) across a gradient, how much overlap is expected? Before this question can be answered, a method of measuring overlap must be devised.

Figure 17.3 demonstrates the way in which the overlap of two species' zones will be measured. If the positions of the zone boundaries can be determined so precisely that no ties occur, then the species-pair can be given a zone-overlap score of 0, 1, or 2 as shown on the left. If zone widths are discrete, and ties do occur, the scoring shown on the right ensures unbiased treatment of uncertainties. As a measure of the total overlap score for s zones, we take the sum of all $\binom{s}{2}$ possible pairwise scores. Let this total be L.

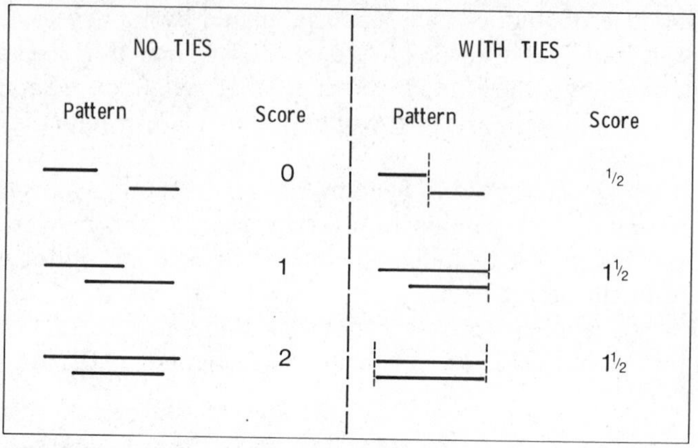

Figure 17.3. The possible overlap scores of a pattern of two zones.

Now let us derive the probability distribution of the score L given that the zones of all s species are independently located. Consider the order of occurrence of the zone boundaries, going from left to right down the gradient. We label the species with capital letters and use a 1 or a 2 following the letter to indicate the species' up- and down-gradient boundaries respectively. Then a random zone pattern yields a random ordering of the $2s$ zone boundaries, subject only to the constraint that a downgradient boundary must always fall to the right of the corresponding upgradient boundary. All permutations of the zone boundaries consistent with this constraint are equiprobable.

Suppose there are two species, B and C, and therefore four boundaries, $B1$, $B2$, $C1$, and $C2$. Let the species be so labeled that species B begins to the left of species C or, equivalently, that boundary $B1$ is the first in the ordered list. Then there are exactly three equiprobable orders for the boundaries:

$$
\begin{array}{llll}
B1 & B2 & C1 & C2 \quad \text{with score 0;} \\
B1 & C1 & B2 & C2 \quad \text{with score 1;} \\
B1 & C1 & C2 & B2 \quad \text{with score 2.}
\end{array}
$$

Put $f(L \mid s)$ for the number of ways in which s species can yield a score of L.

Then $f(L \mid 2) = 1$ for $L = 0$, 1, and 2.

Next, suppose that there are three zones, A, B, and C. Let them be so labeled that their upgradient boundaries are in the order $A1$, $B1$, $C1$. Thus $A1$ lies entirely to the left of zones B and C; these two zones must have one of the only three possible patterns listed above; and $A2$ must lie within one of five possible slots. The presence of the new zone (zone A)

will augment the overlap score of zones B and C (whatever it may be) by an amount, say a, that depends only on the location of $A2$ with respect to the other boundaries.

Thus let the boundaries of zones B and C be represented by indistinguishable asterisks. Then the five equiprobable possibilities, and the additions, a, to the total score that zone A's presence yields, are as follows:

$A1$	$A2$	*		*		*		*		with $a = 0$;
$A1$		*	$A2$	*		*		*		with $a = 1$;
$A1$		*		*	$A2$	*		*		with $a = 2$;
$A1$		*		*		*	$A2$	*		with $a = 3$;
$A1$		*		*		*		*	$A2$	with $a = 4$.

Therefore we may derive the values of $f(L \mid 3)$ in the way shown in Table 17.1.

The first row in the table lists the possible values of L, the total overlap score; these range from 0 to $s(s-1) = 6$. The next five rows give $f(L' + a)$, the conditional frequencies (conditional on a) for $a = 0, 1, \ldots, 4$ respectively; $f(L' + a)$ is defined as the number of ways (which is 1 in every case) in which a score of $L = L' + a$ can be obtained; the two components of L are L', the score that zones B and C alone would have if zone A were disregarded, and a, the additional score contributed to the total by zone A's presence; the rows match those in the schematic representation (of A's and asterisks) given above. Finally, $f(L \mid 3)$ for $L = 0, \ldots, 6$, the required unconditional frequencies, are found by summing each column of conditional frequencies. The frequencies $f(L \mid 3)$ are shown in the last row.

The same procedure yields $f(L \mid 4)$. The row of frequencies $f(L \mid 3)$ is written down $2s - 1$ times, with the ith row beginning in the column

TABLE 17.1. The derivation of $f(L \mid 3)$

$L = L' + a$	0	1	2	3	4	5	6
$a = 0$	1	1	1	·	·	·	·
$a = 1$	·	1	1	1	·	·	·
$a = 2$	·	·	1	1	1	·	·
$a = 3$	·	·	·	1	1	1	·
$a = 4$	·	·	·	·	1	1	1
$f(L \mid 3)$	1	2	3	3	3	2	1

$f(L' + a)$ when zone A's contribution is†:

† These rows match the corresponding rows in the scheme of A's and *'s in the text.

headed $L = i - 1$. The resultant columns are then summed to give $f(L \mid 4)$. And so on.

It will be found that, in general, $f(L \mid s)$ is given by the recurrence relation

$$f(L \mid s) = \sum_{j=L-2s+2}^{L} f(j \mid s-1)$$

with

$$f(j \mid s-1) = 0 \qquad \text{if } j \leq 0.$$

Also,

$$\sum_{L=0}^{s(s-1)} f(L \mid s) = \frac{(2s)!}{2^s s!} = T_s, \quad \text{say.}$$

Hence the probability that s zones with a "random" pattern will have an overlap score of L is

$$p(L \mid s) = \frac{f(L \mid s)}{T_s}, \qquad L = 0, 1, \ldots, s(s-1). \tag{17.2}$$

Since the distribution is symmetrical, the expectation of L given that there are s zones is

$$E(L \mid s) = \frac{s(s-1)}{2}. \tag{17.3}$$

We next derive the variance of the distribution.

First, write $V_x(L \mid s)$ for the second moment of L (with given s) about an arbitrary origin at x.

Then

$$V_x(L \mid s) = \sum_{L=0}^{s(s-1)} (L-x)^2 p(L \mid s)$$

$$= V_0(L \mid s) - xs(s-1) + x^2, \tag{17.4}$$

using the three facts that (i) $\sum_L L^2 p(L \mid s) = V_0(L \mid s)$, the second moment about zero; (ii) $\sum_L Lp(L \mid s) = s(s-1)/2$, the expectation of L given s; and (iii) $\sum_L p(L \mid s) = 1$.

Next, observe that

$$V_0(L \mid s+1) = \frac{T_s}{T_{s+1}} \sum_{x=-2s}^{0} V_x(L \mid s) \tag{17.5}$$

That this is so may be seen by inspecting the example in Table 17.1. The sum of squared deviations from zero of the distribution $p(L \mid 3)$ is the sum of the sums of squared deviations of the distribution $p(L|2)$ from, respectively, $0, -1, \ldots, -4$.

Substituting from (17.4) into (17.5),

$$V_0(L|s+1) = V_0(L|s) + s^2(s-1) + \frac{s(4s+1)}{3} \qquad (17.6)$$

which provides a recurrence relation for the second moment about the origin of $p(L|s)$.

A recurrence relation for the variance, $\text{Var}(L|s)$, may now be obtained. For brevity, put

$$s^2(s-1) + \frac{s(4s+1)}{3} = \Delta$$

so that (17.6) becomes

$$V_0(L|s+1) = V_0(L|s) + \Delta.$$

Then

$$\begin{aligned}
\text{Var}(L|s+1) &= V_0(L|s+1) - [E(L|s+1)]^2 \\
&= V_0(L|s) + \Delta - [E(L|s+1)]^2 \\
&= \text{Var}(L|s) + [E(L|s)]^2 + \Delta - [E(L|s+1)]^2.
\end{aligned}$$

Substituting for the expectations from (17.3), this is easily found to simplify to

$$\text{Var}(L|s+1) = \text{Var}(L|s) + \frac{s(s+1)}{3}. \qquad (17.7)$$

Then, using (17.7) repeatedly, we see that

$$\begin{aligned}
\text{Var}(L|s) &= \text{Var}(L|s-1) + \frac{(s-1)s}{3} \\
&= \text{Var}(L|s-2) + \frac{(s-2)(s-1)}{3} + \frac{(s-1)s}{3} \\
&\quad \cdot \ \cdot \ \cdot \ \cdot \ \cdot \ \cdot \ \cdot \ \cdot \ \cdot \ \cdot \ \cdot \ \cdot \ \cdot \ \cdot \\
&= \text{Var}(L|1) + \frac{1}{3} \sum_{j=1}^{s} (j-1)j.
\end{aligned}$$

Noting that $\text{Var}(L|1) = 0$, and evaluating the sum, finally shows that

$$\text{Var}(L|s) = \frac{s(s^2-1)}{9}. \qquad (17.8)$$

We now have, in (17.2), (17.3), and (17.8), expressions for $p(L|s)$, $E(L|s)$ and $\mathrm{Var}(L|s)$. These are, respectively, the general probability term, the mean, and the variance of the distribution of the overlap score, L, when there are s mutually independent zones. The distribution tends to normality with increasing s. The discrepancy between the true cumulative distribution of L and that of a normal distribution with the same mean and variance nowhere exceeds 1% when $s \geq 20$; hence, when there are 20 or more zones, a test based on the normal distribution may be used to judge whether the null hypothesis is acceptable. For $s < 20$ an exact test is desirable. A table of the critical values of L, for a two-tailed test with significance level $\alpha = 0.05$, when $s = 2, \ldots, 19$, is given in Pielou (1976); the paper also demonstrates application of the test to the zone patterns of congeneric species of benthic foraminifera found in the Gulf of Mexico.

The test need not be confined to examining the zone patterns of groups of taxonomically related species. It can also be used to test whether the classes yielded by applying statistical classification methods to data from zoned communities are ecologically "real." As we shall see in Chapter 20, the use of classification algorithms inevitably results in a classification regardless of whether the classes (of species or of quadrats) are true ecological entities or mere fortuitous assemblages. Now suppose the observations come from a zoned community. Then each set of species (or of quadrats or other sampling units) that a classification algorithm has singled out as a "class" occupies a set of zones whose overlap score may be determined. Only if this score is significantly higher than would be expected on the hypothesis that the zones are independent should we be willing to accept that the class is an ecological entity.

6. Concordance of Zone Patterns

We describe two zone patterns as completely concordant, or qualitatively congruent, if the sequence of zone boundaries is the same in both. Figure 17.1d illustrates patterns that are not completely concordant.

Concordance can only be assessed when both the zone patterns compared have identical species. Therefore it is necessary, when this aspect of two communities is to be measured, to exclude from consideration any species that are absent or uncommon in either community. Not only must the species whose zone patterns are to be compared be present in both communities, they must also be sufficiently abundant to be encountered in numerous transects. The locations of zone boundaries can then be sampled in the way described in Section 2 (see page 250).

Nonconcordance may result from three separate and unrelated causes, all of them ecologically interesting. They are as follows.

1. Suppose the zone patterns being compared are those of a subset of the species in the two communities; as remarked above, these subsets must have identical lists of member species. Suppose also, that the numbers of other species (i.e., not members of the sets being compared) is large in one community and small in the other. If competition is more intense in the species-rich community it may cause the zones to be narrower (that is, to have seemingly narrower tolerance ranges) than in the species-poor community; hence the zones will overlap one another less and have their boundaries differently ordered.

2. More than one factor may vary along a gradient. For instance, both salinity and frequency of inundation, among other things, vary across a salt marsh; hours of submergence, and daily and annual temperature ranges vary across a rocky intertidal region; amount of precipitation, duration of snow cover, temperatures, and wind speeds vary on a mountainside. Although the different factors obviously vary together and are interdependent, their rates of change relative to one another need not be the same in two separate communities. The possibility is shown diagrammatically in Figure 17.4 where only two factors are considered; if the axes are so scaled as to make the rate of change of factor 1 standard (so that, in each of the two graphs, the relation between factor 1 and distance across the community are straight lines of the same slope), the rate of change of factor 2 need not be the same in the two cases. Thus in the figure, the rate of change of factor 2 relative to that of factor 1 is lower in the right hand graph than in the left. Now suppose that, though all species are no doubt affected in greater or lesser degree by all environmental factors, the governing factors are different for different species; in the figure, the zone (on the abscissa) of species A is determined by its tolerance range (on the ordinate) for factor 1, and the zone of species B by its tolerance range for factor 2. The effect on the ordering of the zone boundaries is evident.

3. If a comparison is being made between communities in two geographically distant regions, the species populations whose zone patterns are observed may belong to different geographical races, with different physiological properties, in the two regions. One would not then expect an exact correspondence between zone boundary orders.

The three causes for nonconcordance listed above no doubt do not exhaust the possibilities. Nor is there any reason to suppose that, in a given case, only one cause is operative.

It should be emphasized, before we embark on a discussion of how nonconcordance may be diagnosed, that any comparative study of zone patterns conjures up a multitude of interrelated questions. Then, a

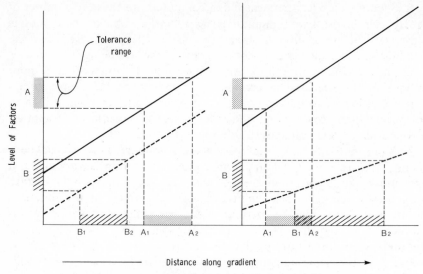

Figure 17.4. The relationship of factor 1 (solid line) and factor 2 (broken line) with distance down a gradient. The tolerance ranges for species *A* (stippled) and species *B* (hatched) are shown on the ordinates and their zones on the abscissa.

multitude of statistical tests can be devised and applied to the observational data. Many tests will give incomplete answers to more than one question, and many questions will need more than one test for their solution. For example, different intensities of between-species competition (the first of the causes listed above for nonconcordance of zone boundaries) will, simultaneously, cause changes in the overlap score of the species concerned (see Section 5); and may be accompanied by differing levels of interdependence between the positions of the zone boundaries (see Section 4). It is impossible in a brief treatment to envisage and describe more than a basic few of the possibilities that may arise. Endless ramifications will suggest themselves in each particular context, as will modifications of the various tests. Thus we are assuming in this section that both up- and down-gradient zone boundaries are included in the ordered lists of boundaries whose concordance is considered. Depending on the problem at hand, it will sometimes be desirable to treat up- and down-gradient boundaries separately.

The following is a way of judging whether two zone patterns are concordant. Suppose each has been sampled with t transects; (t should be even). Let the number of species (or zones) in the patterns to be compared be s. Then each transect yields an ordered list of $2s$ events, the sequence of occurrence of up- and down-gradient boundaries as a transect is traversed. Let us list the events in the order in which they occur in the first transect examined and assign to them the ranks 1, 2, ..., $2s$ respectively. The ranks of the events in all other transects (in either of the two patterns) can now be written down. For example, suppose the first three transects across a 4-zone pattern yielded the lists:

$$A1 \ A2 \ B1 \ C1 \ D1 \ C2 \ B2 \ D2$$
$$A1 \ C1 \ A2 \ D1 \ C2 \ B1 \ B2 \ D2$$
$$C1 \ B1 \ B2 \ C2 \ A1 \ D1 \ D2 \ A2.$$

Then the ranks of the events in these transect lists can be tabulated thus:

Event:	A1	A2	B1	C1	D1	C2	B2	D2
Rank in list $\begin{cases} \#1 \\ \#2 \\ \#3 \end{cases}$	1	2	3	4	5	6	7	8
	1	3	6	2	4	5	7	8
	5	8	2	1	6	4	3	7

We can now calculate Kendall's coefficient of rank concordance among the lists within each of the two zone patterns. The coefficient is defined as

$$W = \frac{6 \sum_{j=1}^{2s} (R_j - \bar{R})^2}{st^2 (4s^2 - 1)} \tag{17.9}$$

Here t is the number of lists in the set (i.e., transects in one community). R_j is the sum of the ranks of the jth event, summed over all t lists; thus for data tabulated as shown above, R_j is the sum of the ranks in the jth column. And $\bar{R} = (1/2s) \sum R_j = t(2s+1)/2$ is the mean of the R_j's.

The coefficient W is easily seen to be the ratio of the observed variance of the R_j's to the variance they would have if all t ordered lists were identical. Observations on a zone pattern whose boundaries were parallel lines visible on the ground would thus give $W = 1$ automatically. In practice it will almost always be found that the t transects across a single zoned community give $W < 1$ because of chance nonoccurrences of some species at places on a gradient where conditions are within their tolerance limits (cf. page 251). Apparent inconsistencies in boundary order are most likely to occur if several boundaries are nearly coincident.[†] However, in spite of some "wrong" orderings due to chance, the concordance within any one community will always be high.

---

† This will also result in ties in the ordered list of zone boundaries. The method of correcting for ties when calculating W is as follows (Siegel, 1956). From the denominator in (17.9) subtract tX where $X = \sum x(x^2 - 1)/12$. Here x is the number of events tied for a single rank position in any one list and the summation is over all t lists.

We now wish to test whether the concordance between two communities is as great as that within each community. If it is not, we can conclude that the zone patterns of the two communities are not qualitatively congruent. To perform a test we shall first obtain several estimates of between-community concordance, say W_b, each based on the same number, t, of lists as were used to calculate the within-community concordances W_1 and W_2. Recall that t is even and suppose $t = 2u$. Then a value of W_b may be obtained by selecting u transect lists at random from community 1, u lists at random from community 2, combining, or "splicing", them to form a single set, and then calculating the concordance, W_b within the set so formed. Since there are $\binom{t}{u}$ ways of partitioning each set of t lists into two equal subsets of u lists each, $\binom{t}{u}$ independent values of W_b may be calculated. One can then judge whether $W_b < \min(W_1, W_2)$ significantly often. If it is, the hypothesis that the two communities being compared are qualitatively congruent can be rejected. An example of the application of the test is given in Pielou (1975b).

Two further comments on the test should be made. (1) It can obviously be extended to permit judgment as to the congruence among several (as opposed to two) zone patterns. (2) The Friedman statistic (see, for example, Conover, 1971 or Lehmann, 1974) can be used in place of Kendall's coefficient of concordance and will lead, of course, to identical conclusions.

7. Within-Zone Density Variations

The tests described in Sections 3 through 6 are all concerned with interrelationships among zones; all require only that the presences and absences of species in quadrats be recorded; and in all cases it has been assumed that a species' zone, within a transect subdivided into contiguous quadrats, coincides with those quadrats between and including the uppermost and lowermost (along the gradient) in which it is found.

We have not inquired into the pattern of a species within its zone: in particular, whether a given species is of roughly constant abundance across the whole width of its zone or, alternatively, whether its density is greatest at or near the center of the zone and dwindles gradually to zero in either direction. The contrast is shown in Figure 17.1e. In what follows the two kinds of boundaries will be described as abrupt and tapered.

In studying any zone pattern, it is obviously interesting to know whether one or other of these two kinds of boundaries tends to predominate. There are two reasons for desiring an answer to the question. The first concerns the biology of the species: is their spread limited by between-species competition, or because their survival is impossible where threshold values of the gradient variables are transgressed? In either of these cases abrupt zone boundaries are to be expected. Or, does the success of a species vary continuously, in direct relation to a continuously varying environmental factor, in which case tapered boundaries are to be expected?

The second reason for investigating whether zone boundaries are abrupt or tapered is that the answer influences the method to be used to estimate zone widths and the positions of boundaries. We have assumed throughout the preceding sections of this chapter that it is satisfactory to treat the uppermost and lowermost occurrences of a species in a transect as marking the upper and lower limits of the species' zone. If zone boundaries are abrupt, this is undoubtedly the best method. It is intuitively evident, however, that the greater the tendency for boundaries to be tapered, the more frequent will be the sampling errors in estimating their positions by this method. Therefore, where boundaries tend to be tapered, more consistent data may be obtainable by disregarding ("trimming") the first and last of a species' occurrences (or even the first and last two occurrences) and then treating those remaining as, for practical purposes, "spanning" the species' zone. More research is needed on this topic. (A procedure sometimes attempted is to assume that the relationship between species density and position on the gradient has the form of a normal curve, and then to choose where the tails are to be truncated on each side of the mode so as to leave acceptably defined "zones"; however, this recipe demands far too much smoothness in field data for it to be practicable.)

We return now to the problem of testing whether a given species' zone boundaries are abrupt or tapered. To do a test requires that the quantity (e.g., number of individuals, biomass, or cover) of the species in each quadrat be observed; mere presence and absence records no longer suffice. Very occasionally, and then only for an abundant species, one can judge subjectively, without need of a test, whether these quantities show smooth trends, but usually this is not possible. Within-zone density variations of haphazard pattern, due either to environmental patchiness unrelated to the gradient variable, or to the species' mode of reproduction, usually obscure variations caused by the gradient variable itself. It is therefore useful to have a crude test for judging whether the density of a species tends to be greater in the middle than near the boundaries of its zone, in other words, whether the boundaries are tapered to an appreciable extent. (It is quite possible, of course, for a zone to have an abrupt boundary on one side and a tapered boundary on the other. If that is suspected, the test described below can be modified accordingly.)

The test is as follows. Consider a single transect of contiguous quadrats and suppose the species whose within-zone density variations are being considered "spans" n quadrats (i.e., the number of quadrats between and including those containing its first and last occurrences is n). Divide these n quadrats into three blocks: an upzone block consisting of the uppermost $[n/3]$ quadrats; ($[n/3]$ is the largest integer $\leq n/3$); a midzone block of the next $n - 2[n/3]$ quadrats; and a downzone block of the lowermost, and remaining, $[n/3]$ quadrats. Calculate the mean quantity per quadrat of the species within each of these blocks and denote the means by $\bar{u}$, $\bar{m}$, and $\bar{d}$ respectively; (these means are over all quadrats, empty as well as occupied, in their blocks). Under the null hypothesis that the density of the species is constant across the whole width of its zone, all the $3! = 6$ possible rankings of $\bar{u}$, $\bar{m}$, and $\bar{d}$, in order of magnitude, are equiprobable. In particular, the event $(\bar{u} < \bar{m} > \bar{d})$ has probability $1/3$. Therefore, suppose we examine t transects and (for the species concerned) record the number of transects, say x, in which this event occurs. Clearly, under the null hypothesis, x is a binomial variate with parameters $1/3$ and t, and hence one may judge whether an observed value of x significantly exceeds expectation; if it does, we reject the null hypothesis that the species' zone has abrupt boundaries in favor of the alternative that the boundaries are tapered.

This chapter summarizes some of the contrasts between zone patterns that are worth looking for, and some of the possible methods of testing for these contrasts. The analysis of zone patterns seems a promising field for further research.

IV
Many-Species Populations

18

Species Abundance Relations

1. Introduction

Most ecological communities contain many species of organisms, and the species may vary greatly in their abundance from very common to very rare. Therefore, as soon as one attempts to study whole communities, rather than the interrelations among a few chosen species, the question immediately arises: how are the abundances of the different species distributed? If there are N individuals belonging to s species and the numbers of individuals in the respective species are $N_1, N_2, \ldots, N_s$, have the N_j any consistent interrelationship, regardless of the type of community from which they come? Attempts to answer this question have led to the development of "species-abundance" curves. If it should turn out that one single form of probability distribution with a small number of parameters (say, two or three) fitted the data from the majority of observed communities, with only the parameter values varying from one community to another, interesting relationships might be discovered between the values of the parameters and the types of community they described.

Before delving into the mathematics, it is necessary to contemplate the word "community" and what is meant by it. In the present context it means all the organisms in a chosen area that belong to the taxonomic group the ecologist is studying. The chosen area is usually one that the ecologist regards as a convenient entity and is willing to consider as homogeneous in some intuitive sense. The reliance on intuition is necessary, since homogeneity cannot be precisely defined at present; exactly what meaning, if any, should be attached to the term "homogeneous community" has for many years been hotly debated and no end to the discussion is in sight. Community studies would have to be suspended indefinitely if ecologists refrained from investigating any community until a satisfactory definition of the word "homogeneous" had been attained. Delimitation of an area that the community under study is supposed to occupy is therefore nearly always a matter of common sense and convenience.

The same is true when it comes to defining the group of animals or plants that is to constitute the community. To take all the living things in the specified area will not do. It would be impracticable to consider every kind of living thing in, say, an acre of forest—the mammals, birds, reptiles, amphibians, arthropods, and soil microfauna, together with the trees, shrubs, herbs, ferns, mosses, and bacteria. A taxonomic group that the ecologist regards as an entity is usually chosen; often it is an entity only in the sense that it is a family, order, or class (or other taxon) that taxonomists are familiar with so that individuals can be fairly easily identified to species. The members of such a taxon that occur together at one place are designated a "taxocene."

Thus two subjective choices must be made in defining a community (more precisely, a collection): the taxocene, of whatever rank, whose individuals are to be collected and the area or volume within which the collecting is to be done. Examples (from Williams, 1964) are the breeding bird pairs in a tract of forest; the herbaceous vascular plants growing on a homogeneous area of ground, the fresh-water algae in small ponds, the beetles in flood refuse on river banks, the snakes in a stretch of tropical forest, and the moths caught in a light trap. In the last example the exact area involved is, of course, unknown; the collection studied consists of those moths that are phototactic, within range of the light, and happen to see it.

Without attempting either to justify or to disparage the types of collection examined in attempts to determine their species-abundance relations, we now proceed with the mathematical theory that has been developed to account for them. In many collections it is found that singleton species (those represented by one individual) are numerous, often the most numerous. Species with successively more representatives, doubletons with 2, trebletons with 3, . . . , and so on, are usually progressively less numerous. Roughly speaking, one often finds many rare species and a few abundant ones, although, of course, in terms of numbers of individuals those of the few common species far outnumber those of the many rare species. This frequently observed phenomenon has led to the method of tabulating species-abundance data customarily used: instead of listing the numbers of individuals in species 1, species 2, etc., we list the number of *species*, n_1, represented by one member, . . . , the number of species, n_r, represented by r members, . . . , and so on. The n_r are, in fact, frequencies of frequencies. Figure 18.1 is an example.

In all cases the collection at hand is treated as a random sample from some indefinitely large parent population. Assume further that each species is randomly dispersed; that is, the number of members the collection contains of, say, the jth species is a Poisson variate with

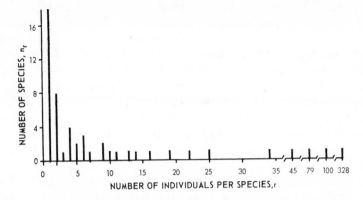

Figure 18.1. Numbers of species with 1, 2, 3,..., individuals in a collection of 822 individuals (52 species) of insects and mites. The individuals were adults emerging from a collection of fruiting bodies of the bracket fungus *Fomes fomentarius*. (Data from Pielou and Matthewman 1966.)

parameter λ_j. Then

$$\text{Pr(the } j\text{th species is represented by } r \text{ members)} = e^{-\lambda_j} \frac{\lambda_j^r}{r!}.$$

Now consider all the species in the community. Their densities vary from species to species over a wide range. If there are S^* species in the whole population, we may regard the several values of λ as constituting a sample of size S^* from some continuous distribution of λ values having pdf $f(\lambda)$. Then the probability that any species will be represented by r members is

$$p_r = \int_0^\infty \frac{\lambda^r e^{-\lambda}}{r!} f(\lambda) \, d\lambda \qquad \text{for} \qquad r = 0, 1, 2, \ldots; \qquad (18.1)$$

that is, the distribution of the different species frequencies $n_0, n_1, n_2, \ldots,$ where $n_r = S^* p_r$ is assumed to have the form of a compound Poisson distribution (see page 122).

The observed distribution is a truncated form of the theoretical distribution; the zero class is missing. We do not, in general, know the value of S^*, the number of species in the whole population, for presumably some of them will be missing from the collection, which is only a sample of the population. Suppose the observed number of species is S. Then $S^* - S = n_0$ is the number of species represented by zero members in the collection, which is to say unrepresented.

It is worth contrasting this situation with that obtaining when the spatial pattern of one species (e.g., of plant) is being investigated by quadrat sampling. In the latter case we count the numbers of individuals of the one species concerned in a known number of different quadrats located at different places; thus we can count the number of empty quadrats (those from which this species is absent) and so obtain an empirical value of n_0. In obtaining the empirical distribution of species-abundances in a collection, on the other hand, we are examining only a single area (equivalent to one quadrat) and counting the numbers of members it contains of each of S^* different species; since S^* is unknown, so also is n_0.

We now consider those members of the family of compound Poisson distributions that have been fitted to observed species-abundance data.

2. The Logarithmic Series or Logseries Distribution

Suppose the values of λ for the different species are assumed to have a Type III distribution; that is, $f(\lambda)$ in (18.1) is given by

$$f(\lambda) = \frac{P^{-k}\lambda^{k-1}e^{-\lambda/P}}{\Gamma(k)}, \qquad \lambda \geq 0 \qquad (18.2)$$

with

$$k, P > 0$$

Then (see page 122) p_r is a negative binomial variate, or

$$p_r = \frac{\Gamma(k+r)}{r!\Gamma(k)}\left(\frac{P}{1+P}\right)^r\left(\frac{1}{1+P}\right)^k \qquad \text{for} \qquad r = 0, 1, 2, \ldots. \qquad (18.3)$$

It is convenient to write $P/(1+P) = X$ so that

$$p_r = \frac{\Gamma(k+r)}{r!\Gamma(k)}(1-X)^k X^r \qquad \text{with} \qquad 0 < X < 1.$$

Next consider p'_r, the probability that a species will contain r individuals, given that the zero class is ignored; that is, p'_r is a term of a truncated negative binomial distribution.

Then, since

$$p_0 = (1-X)^k, \qquad (18.4)$$

$$p'_r = \frac{p_r}{1-p_0} = \frac{\Gamma(k+r)}{r!\Gamma(k)}\frac{(1-X)^k X^r}{[1-(1-X)^k]}. \qquad (18.5)$$

Collecting terms independent of r into a single constant C gives

$$p'_r = C \frac{\Gamma(k+r)}{r!} X^r \quad \text{where} \quad C = \frac{(1-X)^k}{[1-(1-X)^k]} \cdot \frac{1}{\Gamma(k)}.$$

Now, as already remarked (page 128), the parameter k measures (inversely) the variability of λ. A large value of k would be expected if the densities of the various species differed only slightly from one another, and a small value of k would be expected if there were pronounced differences in these densities. In natural communities it is usually found that the differences in abundance among the species is extremely great. This led Fisher (see Fisher, Corbet, and Williams, 1943) to propose that by letting $k \to 0$ in the formula for p'_r an approximation to species-abundance proportions might be obtained.

It should be noticed that mathematical difficulties are introduced if we allow k to tend to zero in (18.2). For then $f(\lambda)$ becomes $\lambda^{-1} e^{-\lambda/P}$ which is not integrable at the origin (Holgate, 1969); the result is that we cannot norm the function to ensure that $\int_0^\infty f(\lambda) \, d\lambda = 1$ as must be true of any legitimate pdf.

However, allowing $k \to 0$ in (18.5) to give the limiting form of p'_r, we see that

$$\pi_r = \lim_{k \to 0} p'_r = \gamma \frac{\Gamma(r)}{r!} X^r = \gamma \frac{X^r}{r} \quad \text{for} \quad r = 1, 2, \ldots,$$

where $\gamma = \lim_{k \to 0} C$. One may evaluate γ by noting that

$$\sum_{r=1}^{\infty} \pi_r = -\gamma \ln(1-X) = 1,$$

whence

$$\gamma = \frac{-1}{\ln(1-X)}.$$

The expression $\gamma X^r / r$ is the *probability* that a species will be represented in the collection by r individuals. The *expected frequency* of species with r individuals is then

$$n_r = S\gamma \frac{X^r}{r} \equiv \alpha \frac{X^r}{r}, \tag{18.6}$$

where S is the total number of species in the collection and $S\gamma = \alpha$. This is the form in which the logseries is usually given in the ecological literature; it should be emphasized that expressions of the form $\alpha X^r / r$ represent expected frequencies, not probabilities.

We may now express S, the number of species observed, and N, the number of individuals (of all species) in the collection, in terms of the parameters α and X. Thus

$$S = \sum_{r=1}^{\infty} n_r = \sum_{r=1}^{\infty} \frac{\alpha X^r}{r} = -\alpha \ln(1-X) \qquad (18.7)$$

and

$$N = \sum_{r=1}^{\infty} rn_r = \sum_{r=1}^{\infty} \frac{r\alpha X^r}{r} = \frac{\alpha X}{1-X}. \qquad (18.8)$$

These two equations may be solved (see Fisher, Corbet, and Williams, 1943) to give estimates $\hat{\alpha}$ and $\hat{X}$ of the true population values of these parameters, once S and N are known for a particular collection. Estimation of the parameters has also been discussed by Bliss (1965) and by Nelson and David (1967).

The magnitude of X depends only on the size of the sample taken from the parent community. Thus, if we increase the area from which a collection is taken or prolong the duration of operation of a light trap for insects, the only effect will be to change the value of X, *provided* the sample still comes from the same parent community.

The second parameter, α, is unaffected by sample size and is an intrinsic property of the community. It will be seen that for given X, α is proportional to S. Thus, in two collections having the same X and N, and in both of which the species abundances are graduated by a logseries distribution, the collection with the larger number of species will have the higher value of α; hence a high value of α corresponds with high diversity in the intuitive sense. When α is high, abundant species are very much less common than sparse ones, and when α is low, abundant species are relatively more common.

If a logseries is fitted to an observed frequency table of species-abundances, it is not possible to estimate S^*, the total number of species in the population. To see this note that in arriving at (18.6), namely $n_r = \alpha X^r / r$, we let $k \to 0$ in the negative binomial series; but $k = 0$ implies that the number n_0 of species unrepresented in the collection is infinite. This follows from the fact that [see (18.4)] $p_0 = (1-X)^k = 1$ when $k = 0$. What was, in fact, assumed was not that $k = 0$ but merely that k was very small indeed. This is tantamount to asserting that the reserve of uncollected species is indefinitely large.

Suppose, now, that a community being studied is enlarged, not by increasing the size of the area sampled, but by widening the range of species to be treated as community members; for example, in dealing with light-trap collections, we could treat only the moths as the community or study the larger community made up of all trapped insects. Similarly, in studying birds in a wooded area, we might consider only warblers or, alternatively, the larger community made up of all birds whatsoever. If the species-abundances in a narrowly defined community formed a logseries, those in a widely defined community might consist of a mixture of several logseries, each with its own parameters. Anscombe (1950) has shown that when this is so one would expect to find larger numbers of very sparse and very abundant species and smaller numbers of species of medium abundance than would be yielded by a single logseries with the same N and S.

As already remarked, when a logseries distribution is fitted to empirical data it is tacitly assumed that the community contains an inestimably large number of species over and above those that chance to have been included in the sample collection. This would prompt one to suppose that too small a sample had been taken from the parent community being investigated to permit any conclusions to be drawn concerning its species-abundance relations. An additional obstacle to deriving conclusions of ecological interest is that numerous other mechanisms that might account for a community's species-abundance distribution, besides the one described above, lead to the logseries distribution; an empirical species-abundance distribution cannot by itself give evidence on how to choose among them. These models have been summarized and compared by Boswell and Patil (1971) and by Watterson (1974).

3. The Discrete Lognormal Distribution

Consider (18.1) again. To derive the logseries distribution, we substituted the pdf of a Type III distribution for $f(\lambda)$. We shall now let $f(\lambda)$ be the pdf of the lognormal distribution; that is, we assume the λ-values are a sample of size S^* from a distribution having pdf

$$f(\lambda) = \frac{1}{\lambda\sigma\sqrt{2\pi}} \exp\left[-\frac{1}{2\sigma^2}\left(\ln\frac{\lambda}{m}\right)^2 \right]. \tag{18.9}$$

Equivalently, $\ln\lambda$ is assumed to be normally distributed with pdf

$$\phi(\ln\lambda) = \frac{1}{\sigma\sqrt{2\pi}} \exp\left[-\frac{1}{2\sigma^2}\left(\ln\frac{\lambda}{m}\right)^2 \right].$$

Then $E(\ln\lambda) = \ln m$ and $\text{var}(\ln\lambda) = \sigma^2$. Notice that $\ln m$ is the median as well as the mean of $\ln\lambda$. Therefore m is the median value of λ, or the median abundance.

Substitution of the formula for $f(\lambda)$ in (18.9) into (18.1) gives the

Poisson lognormal probability

$$p_r = \frac{1}{r!\sigma\sqrt{2\pi}} \int_0^\infty e^{-\lambda} \lambda^{r-1} \exp\left[-\frac{1}{2\sigma^2}\left(\ln\frac{\lambda}{m}\right)^2\right] d\lambda$$

$$= \frac{1}{r!\sigma\sqrt{2\pi}} \int_0^\infty \exp\left[-\lambda + r\ln\lambda - \frac{1}{2\sigma^2}(\ln\lambda - \ln m)^2\right]\frac{d\lambda}{\lambda}. \qquad (18.10)$$

This is the probability that a species in the collection (or sample) will be represented by r individuals, for $r = 0, 1, 2, \ldots$. The probability thus depends on the two parameters σ^2 and m; σ^2 is independent of the size of the sample but m, the median abundance, is a function of sample size.

An alternative form of (18.10), suggested by Grundy (1951), is obtained by putting $\ln\lambda = x$. Then p_r becomes

$$p_r = \frac{1}{r!\sigma\sqrt{2\pi}} \int_0^\infty \exp\left[e^{-x} + rx - \frac{1}{2\sigma^2}(x - \ln m)^2\right] dx.$$

There is no explicit expression for this integral but Bulmer (1974) has derived an approximate formula permitting evaluation of p_r when $r \geq 10$. For $r < 10$ it is necessary to use numerical integration. However, Preston (1948), who was the first to test this distribution against field observations, simply used theoretical lognormal frequencies to graduate the observed frequencies, which were, of course, grouped. This is equivalent to supposing that each species is represented by its expected number of individuals and that these numbers are not subject to sampling variation. According to Bliss (1965), such an approximation is quite adequate.

As with the logseries, so also with the discrete lognormal, the observed series is truncated; additional species over and above the S observed ones are assumed to be present in the population and to have escaped collection by chance.

We now consider Preston's (1948) original method of grouping the values of r and, as a consequence, what led him to suggest fitting the lognormal distribution in the first place. Since an observed species-abundance histogram is usually markedly J-shaped (as in Figure 18.1), with very few high frequencies for low values of r and a long tail representing the few abundant species, it is both natural and convenient to plot r on a logarithmic scale. Moreover, as Preston remarks, "Commonness is a *relative* matter"; one would say that a certain species was so many *times* as common as another. Preston therefore chose grouping into what he calls "octaves" as the most natural procedure; that is, the midpoint of each group is double that of the preceding group.

As group boundaries he takes $r = 1, 2, 4, 8, \ldots$, so that the midpoints of the groups are at $r = 1\frac{1}{2}, 3, 6, 12, \ldots$. A species that falls on a group boundary, for instance one containing 2^x individuals, is treated as contributing half a species to the octave (2^{x-1} to 2^x) and half a species to the octave (2^x to 2^{x+1}). When observations are grouped in this way and plotted (which is equivalent to plotting on semilogarithmic paper using logarithms to base 2) results are obtained of which Figure 18.2 is typical. The histogram looks as though it would be well fitted by a symmetrical normal curve truncated on the left. This leads to the idea that the values of λ for the several species might well be lognormally distributed.

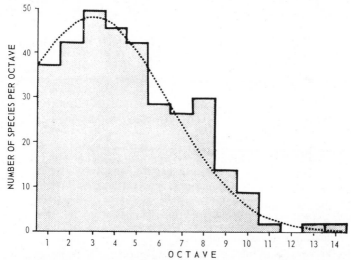

Figure 18.2.   Species abundances in a collection of moths caught in a light trap. Data from Dirks (1937) quoted in Preston (1948). The curve (fitted by Preston) is $n(R) = 48 \exp[-0.207(R - R_0)^2]$, where $n(R)$ is the number of species in the Rth octave and $R_0 = 3$ is the number of the modal octave.

Truncation of the curve on the left (at what Preston calls the "veil line") is inevitable. Those species so rare that the expected number of individuals in a sample of the size at hand is less than one are not expected to be found in the sample. Thus the relative abundances of these uncollected species cannot be observed. If the sample size were doubled, the expected number of individuals of every species would be doubled too. The whole curve would then shift one octave to the right or, equivalently, the veil line (or truncation point) would move one octave to the left, and some species that had hitherto been missed would now be found in the collection.

The salient feature of Preston's observations is this: when the species are grouped into octaves, the observed histogram often exhibits a maximum in some octave to the right of the first one. In other words, the observed octave frequencies first increase and then decrease. This could not happen if the species-abundance distribution conformed to a logarithmic series, as we now show.

Consider the limiting form of the logarithmic series distribution [see (18.6)], when $X = 1$. This is the harmonic series, $\alpha, \alpha/2, \alpha/3, \ldots$. Putting F_r for the number of species in the rth octave, we have

$$F_1 = \frac{n_1}{2} + \frac{n_2}{2} \qquad \propto \alpha(\tfrac{1}{2} + \tfrac{1}{4}) \qquad\qquad = 0.750\alpha,$$

$$F_2 = \frac{n_2}{2} + n_3 + \frac{n_4}{2} \qquad \propto \alpha(\tfrac{1}{4} + \tfrac{1}{3} + \tfrac{1}{8}) \qquad\qquad = 0.708\alpha,$$

$$F_3 = \frac{n_2}{2} + \sum_{i=5}^{7} n_i + \frac{n_8}{2} \propto \alpha(\tfrac{1}{8} + \tfrac{1}{5} + \tfrac{1}{6} + \tfrac{1}{7} + \tfrac{1}{16}) \qquad = 0.697\alpha,$$

$$F_4 = \frac{n_8}{2} + \sum_{i=9}^{15} n_i + \frac{n_{16}}{2} \propto \alpha(\tfrac{1}{16} + \tfrac{1}{9} + \cdots + \tfrac{1}{15} + \tfrac{1}{32}) = 0.694\alpha,$$

$$\cdots\cdots$$

The frequencies in successive octaves are seen to decrease monotonically. It follows a fortiori that when $X < 1$ the octave frequencies must also decrease monotonically and therefore there cannot be a maximum to the right of the veil line, which is what Preston so often observed. By postulating a lognormal distribution for λ he overcame this shortcoming of the logarithmic series distribution.

When the species abundance curve is lognormal, it becomes possible to estimate S^*, the total number of species in the population including those uncollected. Thus, if the area under the truncated lognormal curve fitted to the data is set equal to S, the number of species in the sample, an estimate of S^* is given by the area under the complete untruncated curve. Bliss (1965) gives an account of the calculations.

4.   The Negative Binomial Distribution

As already remarked, to assume that $\ln \lambda$ has a normal distribution ensures that the frequencies of the species-abundances when logarithmically grouped shall increase to some modal value before decreasing, provided the sample is large enough for the veil line to fall to the left of the mode. But the assumption also entails the hypothesis that λ itself has some modal value greater than zero; that is, very rare species as well as very abundant ones are assumed to be less frequent than species having some intermediate value of abundance. It is in this postulate that Preston's hypothesis differs most markedly from that of Fisher, Corbet, and Williams (1943). However, it is not a necessary postulate. Even if $n_1 > n_2 > \cdots > n_r > \cdots$, a semilogarithmic plot (i.e., a plot of species frequencies against octaves) may still show a humped distribution.

As an example we assume with Brian (1953) that the observed n_r can be graduated by a truncated negative binomial series for which we do *not* assume that $k \to 0$. It is then found that a plot of species frequencies against octave numbers may give a humped distribution in the same way that the lognormal series does. In this case, however, the curve is skewed instead of being symmetrical (see Figure 18.3). As before, unless the sample is large enough, the mode may be to the left of the veil line and

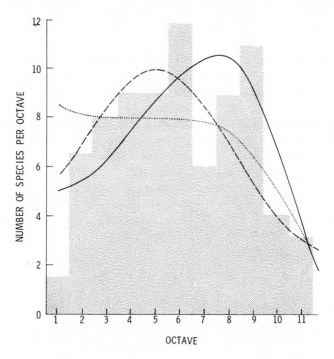

Figure 18.3. Three theoretical species-abundance curves and the empirical histogram to which they were fitted. Solid line: negative binomial with $k = 0.20$ and $P = 642$. Dashed line: lognormal with $n(R) = 10 \exp[-0.194(R - R_0)^2]$; fitted by Preston (1948). Dotted line: logseries with $\alpha = 11.18$. The curves were fitted to a single set of data (the stippled histogram) on a population of breeding birds. (Data from Saunders, 1936, quoted in Brian, 1953. Redrawn from Brian, 1953.)

therefore hidden. Only with large samples will the humped form of the histogram be manifest.

Suppose, then, that the n_r are assumed to form a negative binomial distribution. This is equivalent to assuming that the densities of the several species (the values of λ) have a Type III distribution. We must now enquire into the shape of this distribution which has pdf

$$f(\lambda) = \frac{1}{\Gamma(k)} P^{-k} \lambda^{k-1} e^{-\lambda/P}. \tag{18.2}$$

In particular, we wish to know whether $f(\lambda)$ decreases monotonically from its value at $\lambda = 0$ or whether, instead, it has a mode at some $\lambda > 0$.

Assume P to be constant. Then

$$\frac{df(\lambda)}{d\lambda} = \frac{1}{\Gamma(k)} P^{-k} \lambda^{k-2} e^{-\lambda/P} \left(k - 1 - \frac{\lambda}{P} \right).$$

It is seen that, if $k > 1$, $f(\lambda)$ has a maximum when $\lambda = P(k-1)$, whereas, if $0 \le k \le 1$, $df(\lambda)/d\lambda$ is negative for all values of λ.

It follows that if species-abundance data are fitted by a truncated negative binomial series for which $k > 1$ we shall be led to infer that species of intermediate abundance are commoner than rare species, but, if $k \le 1$, we infer that rare species are the most numerous. Then the *ungrouped* histogram of species frequencies will decrease monotonically with $n_1 > n_2 > \cdots$, whereas the histogram of grouped frequencies may have a mode in some octave to the right of the first.

When the negative binomial distribution, with $k > 0$, fits the data, it is possible to estimate S^* from observations on the sample. The probability that a species will contain r individuals is

$$p_r' = \frac{\Gamma(k+r)}{r!\Gamma(k)} \frac{P^r}{(1+P)^{k+r}[1-(1+P)^{-k}]}, \qquad r = 1, 2, \ldots. \tag{18.11}$$

The mean and variance of this distribution are

$$E(r) = \frac{kP}{1-(1+P)^{-k}}$$

and

$$\mathrm{var}(r) = (1 + P + kP)E(r) - [E(r)]^2.$$

The parameters P and k may therefore be estimated from the mean and variance of the empirical distribution. Then, since

$$S^* = \frac{S}{1-(1+P)^{-k}}, \tag{18.12}$$

we may estimate S^* from S and the estimates of P and k. The sampling variance of the estimate is unknown.

5. The Geometric Distribution

Suppose once more that the several values of λ, the Poisson parameters of a community's several species, have a type III distribution; but this time, instead of allowing $k \to 0$ (which yields the logseries distribution), or of letting k take a value controlled by the data (which yields the negative binomial distribution), we assume that $k = 1$.

One may put $k = 1$ in the pdf of $f(\lambda)$ in (18.2) so that it becomes

$$f(\lambda) = \frac{1}{P} e^{-\lambda/P}, \qquad \lambda \geq 0 \qquad (18.13)$$

which is the pdf of the exponential distribution (cf. page 30). Or one may put $k = 1$ in (18.11) to give p_r', the probability that the variate (the number of individuals in a species) takes the value r given that it cannot be zero. Then

$$p_r' = \left(\frac{P}{1+P} \right)^{r-1} \frac{1}{1+P}, \qquad r = 1, 2, \ldots, \qquad (18.14)$$

which is a geometric distribution, the discrete version of the exponential distribution. The mean of this distribution is $1 + P$. If the distribution fits, then S^* may be estimated. From (18.12) it is seen that $S^* = S(1+P)/P$.

Whenever negative binomial distributions have been fitted to field data, k has been found to be considerably less than 1 (Brian, 1953). As is clear from Figure 18.4 which shows examples of the three distributions (logseries, negative binomial, geometric) all with the same mean, the variance decreases with increasing k, and the proportion of low values of r (i.e., sparse species) becomes small. This suggests that the geometric distribution will seldom give a good fit to actual data since, as we have already seen, most collections have species-abundance distributions like that shown in Figure 18.1.

The distribution has been tested extensively, however, though not in the form given here. The occasions on which it has been used are those in which S, the number of species in the collection, has been low; one then tends to have $n_r = 1$ for S distinct values of r and $n_r = 0$ otherwise. It then becomes impracticable to group the n_r values to give an observed frequency distribution to which a theoretical distribution can be fitted; instead one merely has a list of the values of r for which $n_r = 1$. These

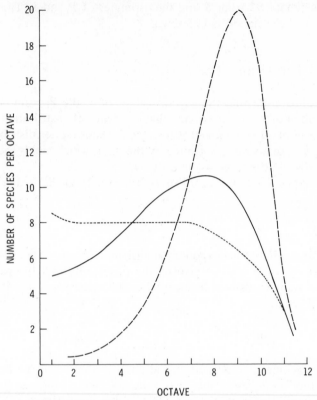

Figure 18.4. Three compound Poisson distributions. The solid and dotted curves are, respectively, the negative binomial and logseries distributions already given in Figure 18.3. The dashed curve is the geometric distribution with the same mean, $1 + P = 179.4$. (The distributions, though discrete, are shown as continuous curves for clarity.)

values are customarily denoted* by N_i with $i = 1, \ldots, S$ and the species are so labeled that $N_1 \le N_2 \le \cdots \le N_S$. Then N_1 is the number of individuals in the rarest species, $\ldots$, N_i the number in the ith rarest, $\ldots$, and N_S the number in the commonest. We now see that N_i is in fact the ith order statistic of the observed random variables. To judge the fit of any chosen theoretical distribution to the observations, we must therefore obtain the expected value of the distribution's order statistics. This is done as follows (David, 1970).

---

* We here use the traditional ecological symbols. Since the N_i are order statistics, it would conform with statistical usage, and perhaps add clarity at the expense of a more elaborate symbolism, to put $N_{(i)}$.

Consider first a sample of S random variables from any parent population, say with pdf $f(\lambda)$ and cumulative distribution function $F(\lambda)$. Writing $\mu_{a:S}$ for the expectation of the ath order statistic we have

$$\mu_{a:S} = S\binom{S-1}{a-1}\int \lambda[F(\lambda)]^{a-1}[1-F(\lambda)]^{S-a}f(\lambda)\,d\lambda,$$

where the integration is over the range of possible values of λ.

Now suppose the parent population is exponential so that $f(\lambda) = (1/P)e^{-\lambda/P}$ as in (18.13) and $F(\lambda) = 1 - e^{-\lambda/P}$. Since P is merely a scale factor we may, without loss of generality, put $P = 1$. We then have for the expected ath order statistic of the exponential distribution,

$$\mu_{a:S} = S\binom{S-1}{a-1}\int_0^\infty \lambda(1-e^{-\lambda})^{a-1}(e^{-\lambda})^{S-a}e^{-\lambda}\,d\lambda$$

$$= \frac{S^{(a)}}{(a-1)!}\int_0^\infty \lambda e^{-S\lambda}(e^\lambda - 1)^{a-1}\,d\lambda$$

$$= \frac{S^{(a)}}{(a-1)!}\sum_{j=0}^{a-1}(-1)^j\binom{a-1}{j}\int_0^\infty \lambda\exp([-(S-a+j+1)\lambda]\,d\lambda$$

$$= \frac{S^{(a)}}{(a-1)!}\sum_{j=0}^{a-1}(-1)^j\binom{a-1}{j}\frac{1}{(S-a+j+1)^2}$$

$$= \sum_{j=0}^{a-1}\frac{1}{S-j} \tag{18.15}$$

The hypothesis that the number of individuals in the ath species (ranked from smallest to largest) of an S-species community might be given by (18.15) has often been entertained. It is the value predicted by the well known "broken stick" model of species abundances first proposed by MacArthur (1957) who argued as follows. Suppose S species divide up the environment at random among themselves so that they occupy nonoverlapping niches; also, let the number of individuals in each species be proportional to the size of the species' niche. Then, on the analogy that the niche-space to be divided up may be likened to a line (or "stick") of unit length, we suppose the line to be broken at random into S segments; the lengths of these segments, ranked from smallest to largest, represent the abundances of the 1st, 2nd, ..., Sth species, ranked from rarest to commonest. The analogy is unsound if it is postulated that what the species divide among themselves is a multidimensional niche-space, since (as we have stressed in Chapters 12 and 16) there is no unique way of randomly partitioning a space of more than one dimension into nonoverlapping parts; the reasoning that applies when a one-dimensional space is randomly broken cannot be extrapolated to spaces of higher dimensionality. However, this objection to the model is easily countered by postulating that what is divided up by the species is some single abundance-limiting factor. If this were so, the broken-stick analogy might hold; the derivation of (18.15) then proceeds as follows. (The argument given here is a paraphrasis of that given by Whitworth (1934) in his Propositions LV and LVI.)

Imagine a line of unit length cut at $S-1$ points located at random on it; then S random segments are formed. Let the lengths of the segments, ranked from smallest to largest, be $l_1, l_2, \ldots, l_S$. Now put

$$l_2 - l_1 = d_1, \quad l_3 - l_2 = d_2, \quad \ldots, \quad l_S - l_{S-1} = d_{S-1};$$

That is, d_r is the difference in length between the rth and $(r+1)$th segments when they are ranked in order of increasing size. Clearly, the length of the original line is given by

$$1 = Sl_1 + (S-1)d_1 + (S-2)d_2 + \cdots + d_{S-1}.$$

Each of the S terms on the right-hand side has equal expectation, since the only condition to which all are subject is that they sum to unity. Therefore

$$E(Sl_1) = E[(S-1)d_1] = E[(S-2)d_2] = \cdots = E(d_{S-1}) = \frac{1}{S},$$

whence

$$E(l_1) = \frac{1}{S^2}, \qquad E(d_1) = \frac{1}{S(S-1)}, \qquad E(d_2) = \frac{1}{S(S-2)},$$

and, in general,

$$E(d_i) = \frac{1}{S(S-i)}.$$

We then see that

$$E(l_2) = E(l_1) + E(d_1) = \frac{1}{S^2} + \frac{1}{S(S-1)},$$

$$E(l_3) = E(l_1) + E(d_1) + E(d_2) = \frac{1}{S^2} + \frac{1}{S(S-1)} + \frac{1}{S(S-2)}.$$

$$E(l_a) = E(l_1) + \sum_{i=1}^{a-1} E(d_i) = \frac{1}{S} \sum_{j=0}^{a-1} \frac{1}{S-j}$$

which, except for a constant of proportionality, is the same as (18.15). The broken stick model now asserts that the expected number of individuals in the ath species, when species are ranked from rarest to commonest, is

$$E\left(\frac{N_a}{N}\right) = E(l_a) = \frac{1}{S}\sum_{j=0}^{a-1}\frac{1}{S-j} \qquad (18.16)$$

Thus the broken stick model is equivalent to the hypothesis that species-abundances are exponentially distributed, which is itself a special case of the hypothesis that they have a type III distribution; (and cf. Cohen, 1968, and Engen, 1974).

Testing the fit of the model to observations on a single community is impossible unless there are so many species that a geometric series can be fitted to, and compared with, a set of n_r values. For low values of S, field data consist merely of a list of values of N_i (the number of individuals in the ith species for $i = 1, \ldots, S$) and these are too few to be grouped. When this is so it is not legitimate to compare these observed values with their expectations as given by (18.16). This is because $E(l_a)$, say, is the expected *average* length of the ath smallest segment in infinitely many repetitions of the stick-breaking experiment. There is no reason whatever to expect the ath segment to be of this length in any one particular case (Pielou, 1975a). Indeed, the joint probability distribution of the S-valued variate $(l_1, l_2, \ldots, l_S)$, with $l_1 \le l_2 \le \cdots \le l_S$, and $\sum l_i = 1$ is

$$f(l_1, \ldots, l_S) = (S-1)!$$

inside the region

$$\{(l_1, \ldots, l_S) : 0 \le l_1 \le \cdots \le l_S \le 1\}$$

and zero elsewhere. [This is the ordered $(S-1)$-variate Dirichlet distribution (Wilks, 1962), with all parameters equal to unity.]

Equivalently, one may say that all possible values of the vector $(l_1, \ldots, l_S)$ are equiprobable. Therefore it is impossible to test the broken stick hypothesis by comparing single observed vectors of the form $(N_1, \ldots, N_S)$ with their expectations; and it can be shown also (Pielou, 1975a) that combining the data from a large number of communities does not permit a test either.

6. Species-Area Curves and Collector's Curves

Most of the work on species-abundance relations has been done by animal ecologists studying collections of animals. Equivalent studies on plant communities cannot be carried out in the same way, since it is seldom possible to count the numbers of individual plants, of each of the several species, in a sample of vegetation (see page 115). Thus data to which species-abundance distributions may be fitted are generally unobtainable.

An alternative approach consists in collecting data for species-area curves. A tract of vegetation is sampled repeatedly with quadrats of different sizes, and for every quadrat the number of species it contains is recorded. We can then investigate the relationship between quadrat area and the mean number of species per quadrat.

Let us postulate that the vegetation of all species in a given quadrat consists of a number of "plant units." These plant units are not identifiable as such and cannot be counted. Without attempting to define precisely what these units are, we merely assume that their number is proportional to the area of the quadrat. Denote by N the number of plant units in a quadrat of unit area.

Suppose now that we wish to judge whether the species abundances have a logseries distribution. If so, the expected number of species represented by r plant units must be given by a term of the form $\alpha X^r/r$. Writing S_1 for the number of different species in a quadrat of unit area it then follows [from (18.7) and (18.8) on page 274] that

$$N = \frac{\alpha X}{1 - X}$$

and

$$S_1 = -\alpha \ln(1 - X) = \alpha \ln\left(1 + \frac{N}{\alpha}\right). \tag{18.17}$$

Now let the sampling be repeated using quadrats of area q. The number of plant units per quadrat of all species combined becomes Nq and we write S_q for the expected number of species. From (18.17) it is seen that

$$S_q = \alpha \ln\left(1 + \frac{Nq}{\alpha}\right),$$

or, when q is large,

$$S_q \simeq \alpha \ln \frac{Nq}{\alpha} = \alpha \ln \frac{N}{\alpha} + \alpha \ln q. \tag{18.18}$$

It thus appears that if the species-abundance distribution has the form of a logarithmic series, and sampling is done with large enough quadrats, the number of species per quadrat will be proportional to the logarithm of quadrat area.

Hopkins (1955), who sampled several plant communities (including heaths, marshes, woods, and grassland) of area $400 \, \text{m}^2$ with quadrats ranging in area up to $100 \, \text{m}^2$, found that the observed species versus log area curve was roughly linear for large quadrats, though for small ones for which the expected curve flattens out, the observed numbers of species exceeded expectation (Figure 18.5 shows an example.)

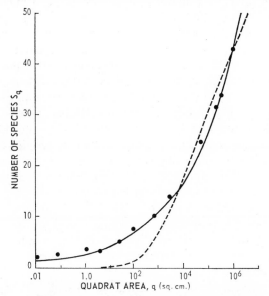

Figure 18.5. Species-area curves. Dots: data from Hopkin's (1955) Community II. Dashed line: the curve $S_q = 5.7 \ln(1 + q/420)$ fitted by Hopkins. Solid line: the curve $S_q = 2.82 q^{0.20}$ fitted by Kilburn (1966). (Redrawn from Kilburn, 1966, who gives the equation for his curve as $S_q = 17.8 q^{0.20}$, since he treated $1 \, \text{m}^2$ instead of $1 \, \text{cm}^2$ as the unit of area.)

Results such as these, however, do *not* permit one to draw conclusions, even approximate ones, about the species-abundance relations of these plants. In fitting the curve $S_q = \alpha \ln(1 + Nq/\alpha)$ to the observations in order to test the hypothesis that the species-abundance frequencies form a logarithmic series, we are tacitly assuming that the vegetation mosaic is so fine-grained that its patches correspond to "plant units" and that the patches of the various species are randomly mingled. Only if both assumptions were true would the contents of a quadrat constitute a random sample from the population. It is clear that to derive theoretical species-area curves we must make assumptions not only regarding the form of the species-abundance curve but also the grain of the vegetation mosaic and the way in which the species are mingled. Very elaborate hypotheses would be needed and to explore their consequences would be difficult.

Attempts to find empirical formulas to describe species-area relations seem worthwhile, however. How such formulas should be interpreted is admittedly not clear; they do not provide evidence bearing directly on the relative abundances of the species, since, as already remarked, the spatial pattern of the vegetation is at least as important as the species-abundance relation in determining the shape of a species-area curve.

Empirical studies of species-area curves were described by Kilburn (1966), who sampled communities of area $900 \, m^2$ with quadrats ranging in size up to $100 \, m^2$; the communities studied were prairie, deciduous forest, and jack-pine woodland. He found that curves of the form $S_q = kq^z$ (with parameters k and z estimated from the data) fitted the observations better than did the curve tested by Hopkins. He inferred that in general it is the logarithm of the number of species rather than the number itself that is proportional to log area. Figure 18.5 shows examples of the two curves fitted to data from Hopkins (1955); the vegetation was grassland heavily grazed by sheep.

In constructing a true species-area curve it is important to ensure that the quadrats (or larger plots) of different areas are independent of one another; they must not be nested. If they are nested, the result is a "collector's curve."

Collector's curves are often constructed casually; if one is trying to compile an exhaustive list of the species in a circumscribed area, one repeatedly wishes to inspect a graph of the number of species collected so far versus the area so far examined, in order to judge whether the curve is leveling off. Clearly, since the number of species is indubitably finite, the curve must ultimately reach the value S^* and then stop; but the way in which it does so depends on the form of the species abundance distribution. Indeed, provided the community mosaic is fine-grained, the shape of the species versus log area curve (i.e., the collector's curve) is of considerable interest, and so also is the proportion of singleton species (those represented by a single individual) in the collection.

It has been shown by Holgate (1969) that if the species abundances follow a logseries distribution, then S increases with log area without leveling off (as in Figure 18.5) and at the same time n_1/S, the proportion of the collected species that are singletons, rises to and remains at a constant level. By contrast, if the underlying species abundance distribution is negative binomial, the rate of increase of S gradually decreases so that S approaches S^* asymptotically; and n_1/S, after rising at first, reaches a maximum and then decreases to zero.

Let us look at the further consequences of these findings. Obviously if there is still an appreciable proportion of singletons even when the whole of the area concerned has been examined exhaustively, we must conclude either: (1) that the community concerned occupies a larger area than that chosen for study and that we have examined only a part of it; or (2) that there is no truly autonomous community, that is a community in the strict sense, closed to immigration, but rather that the chosen area contains a chance assemblage of species among which are immigrant singletons from other areas and/or individuals of species whose range of occurrence just overlaps the chosen area at its periphery.

An attempt to distinguish among these various possibilities may be made as follows. Suppose the area under investigation has been sampled with an array of systematically located quadrats in order to provide the most representative sample possible of the community, or pseudocommunity, it contains. There are two ways in which a collector's curve can be constructed. One may count the number of species accumulated in a sample that grows either (1) by the adding in of quadrats picked at random from the whole array; or (2) by starting with a quadrat at some point in the interior of the area and progressively adding those within a steadily expanding circle centered on the point. Recall now that there are two shapes that the collector's curve can assume: it may rise continuously, or it may level off.

Thus there are four possibilities to consider and they may be displayed in a 2×2 table and labeled with identifying letters.

		Collector's Curve	
		Levels off	Rises continuously
Quadrats accumulated:	At random	A	B
	Over expanding area	C	D

Now suppose we do in fact construct collector's curves in the two different ways; four pairs of results can be envisaged, namely AC, AD, BC, and BD; only three of these are possible in practice. The conclusions they lead to are as follows.

AC: The sample is large enough for S^* to be estimated. The area contains a homogeneous community. The logseries is not appropriate for graduating the species abundances.

AD: S^* for the area is estimable but the community is not homogeneous and it would be futile to try fitting a theoretical distribution to the observed species abundances.

BC: Impossible.

BD: S^* is not estimable. No decision can be reached on community homogeneity. Perhaps it is inhomogeneous. Or perhaps it is homogeneous and the species abundances have a logseries distribution.

Notice that we can never judge decisively whether the fitting of a logseries distribution is appropriate. Even if it gives a fairly good fit in fact, the "community" concerned may not be a true, homogeneous community at all (case *AD*). Or (case *BD*) its appropriateness may be undecidable; if a sample is too small for the collector's curve to level off, one cannot convincingly distinguish between a true logseries distribution and the descending arm of some other distribution (possibly, but not necessarily, humped) with its truncation point (or "veil line") far to the right.

The contrast between *AD* and *BD* is worth emphasizing. In both cases it is likely to be found that n_1/S, the proportion of singletons, increases with log area as does S itself. Then case *AD* would suggest that these singletons were chiefly at the periphery of the area under investigation; whereas, given case *BD*, it would be impossible to decide whether the singletons were concentrated at the periphery of the area or scattered throughout.

In any case, when species abundance distributions are compiled, or collector's curves drawn, it is always important to observe how many, and which, of the species that are sparsely represented in the collection are globally rare (i.e., rare members of the whole world's flora or fauna) and how many are only locally rare. This must obviously have an enormous influence on any conclusions suggested by formal analysis of the data.

19

Ecological Diversity and Its Measurement

1. Introduction

In Chapter 18 we considered species-abundance curves whose form depends on two things: the number of different species present in a community or population and the relative proportions of their abundances. It is worthwhile to inquire how we might sum up, in one or two descriptive statistics, the properties of these curves.

When the species-abundance frequencies in an actual collection are well fitted by one or another of the theoretical distributions already described, the parameters of the fitted distribution are obviously suitable as descriptive statistics. If the distribution is lognormal, the appropriate statistics are the estimates of S^*, the total number of species in the population, and σ^2, the variance of the lognormal curve. If the distribution is negative binomial with $k \neq 0$, the appropriate statistics are the estimates of S^* and k (the parameter P depends on sample size). The logarithmic series distribution has two parameters, α and X, but since X is a function of sample size and S^* is assumed to be indefinitely large this leaves only α as a statistic that describes an intrinsic property of the population being sampled. The geometric distribution's only parameter, P in (18.13) and (18.14), is the mean number of individuals per species in the sample; thus it is largely determined by sample size and is without clear biological significance. The only remaining parameter of interest for communities whose species abundances are fitted by the geometric distribution is $S^* = S(1 + P)/P$, the estimated total number of species in the community.

The statistics described in the preceding paragraph all suffer from the defect that they are not sufficiently widely applicable. What is needed are descriptive statistics that can be used for any community, no matter what the form of its species-abundance distribution and even when no theoretical series can be found to fit the data.

We begin by considering the properties of any collection, regardless of whether it is to be treated as a population in its own right or as a sample from some larger parent population. Two statistics are clearly needed to

291

describe a collection, of which the first and most obvious is S, the number of species it contains. Now suppose we are dealing with data consisting of a list of the numbers of individuals, $N_1, N_2, \ldots, N_S$, in each of the S species. If the data are portrayed in histogram form, S is the range of the data or the width of the histogram. As a second statistic, to describe the shape of the histogram, we require something analogous to variance. If the N_j were frequencies of some discrete *quantitative* variate, variance as ordinarily calculated would, of course, be the obvious statistic to use, but we are now considering an unordered *qualitative* variate; the individuals are classified according to the species to which they belong and there is no a priori reason for listing them in any particular order. The shape of the histogram is therefore best described in terms of what may be called its "evenness." Thus the distribution has maximum evenness if all the species abundances (the N_j) are equal; and the greater the disparities among the different species abundances, the smaller the evenness.

Before considering evenness per se, however (in Section 5), it is necessary to discuss what has come to be called the "diversity" of a collection.* Various ways of defining and measuring diversity have been proposed and some are discussed here in detail, but it should first be emphasized that diversity, however defined, is a single statistic in which the number of species and the evenness are confounded. A collection is said to have high diversity if it has many species and their abundances are fairly even. Conversely, diversity is low when the species are few and their abundances uneven. It will be seen that since diversity depends on two independent properties of a collection ambiguity is inevitable; thus a collection with few species and high evenness could have the same diversity as another collection with many species and low evenness.

This difficulty did not arise when the notion of diversity was first introduced by Williams (see Fisher, Corbet, and Williams, 1943). On the assumption that most species-abundance distributions would be well fitted by logarithmic series distributions, he proposed that the parameter α of that distribution be used as an index of diversity (see page 274). This index can be applied only if the logarithmic series does indeed fit the species-abundance data, but to determine whether it does may be impossible if there are only a few species and each is represented by a different number of individuals. Thus α is suitable as an index of diversity only if the collection at hand has many species and even then only if its species abundances form a logarithmic series. Some other measure of diversity is needed. In particular, we require one that will be applicable to small as well as large collections.

* "Diversity" is sometimes used merely as a synonym for "number of species." That is not the sense in which it is used here.

It should not be (but it is) necessary to emphasize that the object of calculating indices of diversity is to solve, not to create, problems. The indices are merely numbers, useful in some circumstances but not in all. Much futile effort has been devoted to calculating diversity indices and then trying to explain the results as though the indices were worthy of study in their own right. Indices should be calculated for the light (not the shadow) they cast on genuine ecological problems.

2.   The Information Measure of Diversity

A useful measure of diversity may be arrived at as follows (see Khinchin, 1957). It is important to stress at the outset that we are considering here an indefinitely large (effectively infinite) population. The arguments about to be given must be modified when the diversity of a finite population is being measured and a discussion of these modifications is deferred to a later section.

Imagine, then, an infinite population of individuals that can be classified into s classes (or species) $A_1, A_2, \ldots, A_s$. Every individual belongs to one and only one class, and the probability that a randomly selected individual will belong to the class A_j is p_j. Thus $\sum_{j=1}^{s} p_j = 1$. As a measure of the diversity of the population, we wish to find a function of the p_j, $H'(p_1, p_2, \ldots, p_s)$, say, that meets the following conditions:

1. For a given s the function takes its greatest value when $p_j = 1/s$ for all j. Denote this greatest value by $L(s)$. Then

$$L(s) = H'\left(\frac{1}{s}, \frac{1}{s}, \ldots, \frac{1}{s}\right).$$

2. The diversity of the population is unchanged if we postulate the existence of $(s+1)$th, $(s+2)$th, $\ldots$, classes to which no individuals belong; that is,

$$H'(p_1, p_2, \ldots, p_s, 0, \ldots, 0) = H'(p_1, p_2, \ldots, p_s).$$

3. Suppose the population is subjected to an additional separate classification process that divides it into t classes, $B_1, B_2, \ldots, B_t$. Each individual belongs to exactly one B-class, and the probability that it will belong to class B_k is q_k with $\sum_{k=1}^{t} q_k = 1$. Then the double classification yields st different classes, $A_j B_k (j = 1, \ldots, s; k = 1, \ldots, t)$, and the probability that a randomly chosen individual will belong to the class $A_j B_k$ may be written π_{jk}. Clearly, if the A-classification and the B-classification are independent, $\pi_{jk} = p_j q_k$, but if the attributes on which the classifications are based are mutually dependent, then $\pi_{jk} = p_j q_{jk}$, where q_{jk} is the

conditional probability that an individual will belong to B_k, given that it belongs to A_j. For the diversity of the doubly classified population we may write

$$H'(AB) = H'(\pi_{11}, \pi_{12}, \ldots, \pi_{st}).$$

Next, put $H'_j(B) = H'(q_{j1}, q_{j2}, \ldots, q_{jt})$ for the diversity under the B-classification *within* the class A_j, and put

$$H'_A(B) = \sum_j p_j H'_j(B)$$

for the mean diversity under the B-classification within all the A classes. Condition 3 then is that we should have

$$H'(AB) = H'(A) + H'_A(B). \tag{19.1}$$

If the classifications are independent so that $q_{jk} = q_k$ for all j, then $H'(AB) = H'(A) + H'(B)$.

Having specified the three conditions that H' is to satisfy, we now show that the only function with these properties is

$$H'(p_1, p_2, \ldots, p_s) = -C \sum_j p_j \log p_j, \tag{19.2}$$

where C is a positive constant.

Before doing so, however, it is desirable to show how the conditions are to be interpreted in an ecological context.

Condition 1 merely ensures that for a population with a given number of species the measure of diversity will be a maximum when all the species are present in equal proportions (or with maximum evenness).

Condition 2 ensures that, given two populations in which the species are evenly represented, the population with the larger number of species will have the higher diversity. This will become clear in the proof.

To understand the implications of condition 3 imagine a forest of s different species of tree and suppose that the trees are also classified according to height. For the purpose of the argument assume height to be a discrete variate; there are t possible heights. The A-classification is the species classification and the B-classification is the height classification. Thus we could measure separately the species diversity of the trees, $H'(A)$, and their height diversity, $H'(B)$. Also, we could calculate their "species-height" diversity, $H'(AB)$, taking both classifications into account. Condition 3 stipulates that if species and height are wholly independent (as in Figure 19.1a) then $H'(AB) = H'(A) + H'(B)$. In this

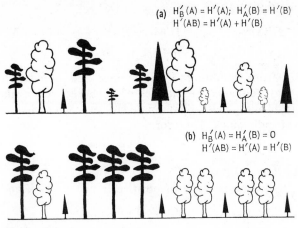

(a) $H'_B(A) = H'(A);\ H'_A(B) = H'(B)$
$H'(AB) = H'(A) + H'(B)$

(b) $H'_B(A) = H'_A(B) = 0$
$H'(AB) = H'(A) = H'(B)$

Figure 19.1

case knowledge of a tree's species gives no information about its height and vice versa. But, if trees of one species are all of one height, peculiar to the species (or, equivalently, if trees of a given height are all of the same species), as in Figure 19.1b, then $H'_B(A) = H'_A(B) = 0$ and condition 3 now requires that $H'(AB) = H'(A) = H'(B)$. In this case knowledge of a tree's species immediately reveals its height and vice versa; neither classification adds anything to the information yielded by the other.   The reasons why it is desirable for a diversity measure to meet condition 3 will become clear later (see Section 4).

In deriving (19.1), we begin by dealing with the special case in which $p_j = 1/s$ for all j.

From conditions 2 and 1

$$L(s) = H'\left(\frac{1}{s}, \frac{1}{s}, \ldots, \frac{1}{s}\right) = H'\left(\frac{1}{s}, \frac{1}{s}, \ldots, \frac{1}{s}, 0\right)$$
$$\leq H'\left(\frac{1}{s+1}, \frac{1}{s+1}, \ldots, \frac{1}{s+1}\right)$$
$$= L(s+1).$$

This shows that $L(s)$ is a nondecreasing function of s.

Next, consider m mutually independent classifications of the population, $S_1, S_2, \ldots, S_m$, under each of which the individuals are assigned in equal proportions to r different classes. Then under the kth classification, S_k, for instance, the diversity is $H'(S_k) = L(r)$. This relation holds for all k. Since the classifications are independent, the diversity of the population under the multiple classification is given by

$$H'(S_1 S_2 \cdots S_m) = H'(S_1) + H'(S_2) + \cdots + H'(S_m)$$
$$= m L(r).$$

This follows from condition 3. But, since the population has been subdivided into r equal classes at each of m successive classifications, the multiple classification yields r^m ultimate classes, each containing the same proportion of individuals. Therefore

$$H'(S_1 S_2 \cdots S_m) = L(r^m).$$

Thus we see that $m L(r) = L(r^m)$.

Likewise, for any other pair of positive integers, say n and s,

$$n L(s) = L(s^n).$$

Now, for arbitrary values of r, s, and n let m be chosen so that $r^m \leq s^n \leq r^{m+1}$ or $m \log r \leq n \log s \leq (m+1) \log r$.

Then

$$\frac{m}{n} \leq \frac{\log s}{\log r} \leq \frac{m}{n} + \frac{1}{n}. \tag{19.3}$$

Recall that $L(s)$ is a nondecreasing function of s, as has already been shown. It then follows that

$$L(r^m) \leq L(s^n) \leq L(r^{m+1})$$

or

$$m L(r) \leq n L(s) \leq (m+1) L(r),$$

whence

$$\frac{m}{n} \leq \frac{L(s)}{L(r)} \leq \frac{m}{n} + \frac{1}{n}. \tag{19.4}$$

From (19.3) and (19.4) it now follows that

$$\left| \frac{L(s)}{L(r)} - \frac{\log s}{\log r} \right| \leq \frac{1}{n}.$$

Therefore, since n may be taken as large as we like,

$$\frac{L(s)}{L(r)} = \frac{\log s}{\log r} \quad \text{or} \quad L(s) \propto \log s.$$

Then, writing C for the constant of proportionality, we have

$$H'\left(\frac{1}{s}, \frac{1}{s}, \ldots, \frac{1}{s}\right) = L(s) = C \log s = -C \sum \frac{1}{s} \log \frac{1}{s}$$

as the diversity of an s-species population in which all the species are equally represented.

Now consider the general case in which the p_j are not all equal. As before, we wish to define the diversity of a population classified into s classes, $A_1, A_2, \ldots, A_s$, with a proportion p_j of the individuals in the class A_j. Put $p_j = g_j/g$, where g_j and g are integers. Now perform a second, dependent classification of the population, the B-classification, as follows: B has g classes, all of the same size and labeled $B_1, B_2, \ldots, B_g$. These classes are divided into s groups, with g_j classes in the jth group. If an individual is in the class A_j under the A-classification, it is *defined* as belonging to the jth *group* of B-classes under the B-classification. The probability that it belongs to any given class in this group is the same for all classes in the group and is $1/g_j$.

The double classification process should become clear from an actual example which is shown schematically here. There are three A-classes ($s = 3$) in proportions $p_1 = \frac{6}{10}$, $p_2 = \frac{3}{10}$, and $p_3 = \frac{1}{10}$. There are $g = 10$ B-classes, with $g_1 = 6$ B-classes in group 1 (equivalent to A_1), $g_2 = 3$ B-classes in group 2 (equivalent to A_2), and $g_3 = 1$ B-classes in group 3 (equivalent to A_3).

A-classification $\qquad A_1 \qquad\qquad A_2 \qquad\qquad A_3 \quad (s = 3)$

B-classification $\quad B_1\ B_2\ B_3\ B_4\ B_5\ B_6 \qquad B_7\ B_8\ B_9 \qquad B_{10}$

$\qquad\qquad$ first B-group $\quad$ second B-group $\quad$ third B-group

$\qquad\qquad\quad g_1 = 6 \qquad\qquad g_2 = 3 \qquad\qquad g_3 = 1$

$$\sum_j g_j = g = 10.$$

We now see that *within* class A_j the diversity under the B-classification is

$$H_j'(B) = L(g_j) = C \log g_j.$$

It follows that $H_A'(B)$, the *mean* diversity under the B-classification within the A-classes, is

$$H_A'(B) = \sum_j p_j H_j'(B) = C \sum_j p_j \log g_j = C \sum_j p_j \log p_j + C \log g,$$

since $g_j = g p_j$.

Now consider the doubly classified population with classes $A_j B_k$ ($j = 1, \ldots, s$; $k = 1, \ldots, g$). (In the example shown schematically these are the classes in the last line. Giving them their full labels under both classifications, they are $A_1B_1, \ldots, A_1B_6, A_2B_7, \ldots, A_2B_9$ and A_3B_{10}.) The total number of classes under the double classification is $\sum g_j = g$ and these classes are of equal size. The diversity of the population under the double classification is thus

$$H'(AB) = L(g) = C \log g.$$

Now applying condition 3, we see that

$$H'(A) = H'(AB) - H'_A(B)$$

$$= C \log g - \left\{ C \sum_j p_j \log p_j + C \log g \right\}$$

$$= -C \sum_j p_j \log p_j.$$

We have now proved that $-C \sum_j p_j \log p_j$ is the only function of the p_j's (when they are rational) that meets the conditions given earlier. The assertion is also true for real p_j's and a proof will be found in Khinchin (1957).

This formula is the one proposed by Shannon (see Shannon and Weaver, 1949) as a measure of the information content per symbol of a code composed of s kinds of discrete symbols whose probabilities of occurrence are $p_1, p_2, \ldots, p_s$. In the ecological context H' measures the diversity per individual in a many-species population. It remains to choose the units in which diversity shall be measured. The size of the unit depends on the value assigned to the constant C and on the base used for the logarithms. It is customary to put $C = 1$. Commonly chosen logarithmic bases are the numbers 2, e, and 10. Information theorists use logarithms to base 2 and the information units are then called "binary digits" or "bits." When natural logarithms are used, the unit has been called a "natural bel" (Good, 1950) or a "nat" (McIntosh, 1967b); and with logarithms to base 10 the unit becomes a "bel" (Good, 1950), a "decimal digit" (Good, 1953), or a "decit" (Pielou, 1966a). In ecology the unit to be used and the name to be given it have not yet become standardized.

There has been much debate on whether H' is a suitable measure of ecological diversity. The fact that it measures "information" and "entropy" is beside the point; these fashionable words have been bandied about out of their proper context (the mathematical theory of information) and have led to false analogies that produced no noticeable advance in ecological understanding. It is more illuminating to regard H' as a measure of "uncertainty." If an individual is picked at random from a many-species population, we are uncertain which species it will belong to, and the greater the population's diversity (in an intuitive sense), the greater our uncertainty. The three conditions that H' was required to meet are those appropriate to a measure of uncertainty (Shannon and Weaver, 1949; Khinchin 1957). It is therefore reasonable to equate diversity to uncertainty and use the same measure for both.

3. Measuring Diversity in Sampled and Censused Communities

Recall that in Section 2 we assumed that the community whose diversity was to be measured was indefinitely large; in other words, it was assumed to be too large to be censused, that is, too large for all its individuals to be examined and identified. When this is so, diversity must perforce be estimated from a sample and the estimate will be subject to sampling error. However, if one chooses to define, as the community under study, an assemblage sufficiently small for it to be censused, its diversity can be exactly determined without sampling error. Observe that it is rarely legitimate to treat a censused collection as a sample from a larger (conceptually infinite) parent community. This is permissible only if the boundaries of the postulated parent community can be precisely specified and the collection at hand is a truly random sample from it, which is seldom the case. Usually a censused collection is best treated as an entity in its own right.

The procedures for estimating the information measure of diversity in an indefinitely large community, and for determining it in a small, fully censused community are entirely different, and in this section we consider them in turn. As we shall see, the information function is defined differently in the two cases and for this reason it is desirable to use different symbols. As in Section 2, we use H' (with a prime) for the diversity of "large," sampled communities; and H (with no prime) for "small," censused communities. For reasons that will become clear we consider censused communities first.

Censused Communities

The Shannon function

$$H' = -\sum_j p_j \log p_j \tag{19.5}$$

as used in information theory is strictly defined only for an infinite population; it measures the information content of a *code* as distinct from a *particular message* in the code (Goldman, 1953).

For a particular message, containing N symbols of s different kinds with N_j of the jth kind ($\sum N_j = N$), the analogous measure is Brillouin's (1962) function, defined as

$$H = \frac{1}{N} \log \frac{N!}{N_1! N_2! \cdots N_s!}. \tag{19.6}$$

Margalef (1958) was the first to use this function to measure ecological diversity.

Because of the analogy between codes and messages on the one hand, and "infinite" and "small" communities on the other, it is desirable to use, as measures of diversity, H' as defined in (19.5) for "infinite" communities, and H as defined in (19.6) for "small" ones. It was shown in Section 2 that H' is the only function of $(p_1, \ldots, p_s)$ with real p's and $\sum p_j = 1$ that meets the three conditions listed at the beginning of the section. In the same way, H is a function of $(N_1, \ldots, N_s)$ with integer N's and $\sum N_j = N$ that meets analogous conditions. In particular, condition 3 as expressed by (19.1) is true of Brillouin's function as well as of Shannon's; in other words, the equation may be written with or without the primes.

Two further reasons for using H as a measure of diversity in censused communities are the following.

1. As $\min(N_j) \to \infty$, $H \to H'$. To see this use natural logarithms and replace the factorials in (19.6) with the approximation $\ln n! = n(\ln n - 1)$. Then since

$$H = \frac{1}{N} \left\{ \ln N! - \sum_j \ln N_j! \right\},$$

$$\lim_{\min(N_j) \to \infty} H = \frac{1}{N} \left\{ N \ln N - \sum_j N_j \ln N_j \right\}$$

$$= -\sum_j \frac{N_j}{N} \ln \frac{N_j}{N},$$

which is formally the same as the maximum likelihood estimator of H'. It is not a good approximation to H in practice, for unless *all* the N_j are very large (which seldom happens) the approximation to $\ln n!$ used in deriving it is not sufficiently close. For purposes of evaluating (19.6) it is better to put

$$\ln n! \sim n(\ln n - 1) + \tfrac{1}{2} \ln 2\pi n,$$

which is the commonly used form of Stirling's approximation to the factorial. Alternatively, we may obtain log factorials from tables.

2. The use of different formulae for the diversities of censused and sampled collections emphasizes the contrast between them, a contrast that should not be overlooked. It is especially important to keep in mind that H is determined, not estimated, hence (provided there have been no mistakes in identifying and counting community members) it is free of error. On the other hand, H' must always be estimated from a sample. Thus estimates of H' are always subject to sampling error and a particular numerical estimate of H' should always be accompanied by an estimate of its sampling variance (or of its standard error).

The fact that H depends on community size is sometimes regarded as a drawback. Thus, given two s-species communities one with M and the other with N individuals but with identical relative abundances (i.e., with $N_1:N_2:\ldots:N_s = M_1:M_2\ldots:M_s$), then if $N > M$, we have $H_N > H_M$. However, except for communities so small that it would be absurd to compute their diversities, the discrepancy is negligible. Moreover, to many people it is intuitively reasonable to regard a small community as less diverse than a large one, other things being equal; not everyone agrees, and intuition is an unreliable guide on this point.

Sampled Communities

If the diversity of a large community is to be estimated from a sample, and if we now write N_j for the number of individuals of the jth species in the sample ($j = 1, \ldots, s; \sum N_j = N$), then

$$\hat{H}' = -\sum \frac{N_j}{N} \log \frac{N_j}{N}$$

is the maximum likelihood estimator of H'. However, it is biased; if natural logs are used, $\hat{H}'$ underestimates the true community value of H' by an amount approximately equal to $s^*/2N$, where s^* is the number of species in the whole community (Basharin, 1959; Pielou, 1966b); thus no correction can be made for the bias unless s^* is known, which it rarely is.

Another, practical, difficulty that arises when a community is sampled is that the individuals in the separate sampling units (quadrats, say) are, in general, not randomly and independently drawn from the parent community. On the contrary, because of the universal patchiness of ecological communities, they are closely dependent. Thus the diversity of the contents of a single small sampling unit is nearly always very much less than that of the community from which it comes, and only trial and error can show how many sampling units must be drawn from the community to give an adequate representation of it.

A way of estimating H' that overcomes these difficulties is as follows. Suppose a sample of n sampling units (or quadrats) has been examined and their contents listed. These quadrats are now to be taken in random order and added, one after another, to a growing pool of quadrats.

Let N_{xj} be the number of individuals of the jth species in the xth quadrat.

Let $M_{kj} = \sum_{x=1}^{k} N_{xj}$ be the number of individuals of the jth species in the pool of the first k quadrats; and let $M_k = \sum_{j=1}^{s} M_{kj}$ be the number of individuals of all species in this pool.

Also, let

$$\mathcal{H}_k = \frac{1}{M_k} \log \frac{M_k!}{M_{k1}! \cdots M_{ks}!}$$

be the diversity, as measured by the Brillouin index, of the contents of this pool. If $\mathcal{H}_k$ is plotted against k, and if n, the total number of quadrats, is large enough, then it will be found that $\mathcal{H}_k$ increases (not necessarily monotonically) at first, and then levels off when the pool of quadrats has become large enough for its contents to provide an adequate representation of the community as a whole; Figure 19.2 shows two examples.

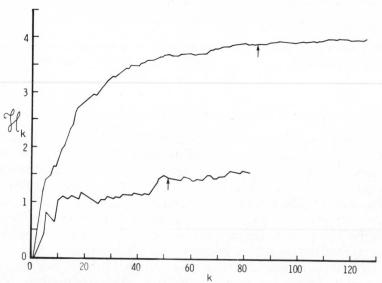

Figure 19.2 Plots of $\mathcal{H}_k$ versus k for communities of amphibians and reptiles in moist tropical forest in Ecuador (upper curve) and dry evergreen forest in Thailand (lower curve). The arrows show the positions of t. (Redrawn from Heyer and Berven, 1973.) (From Pielou, 1975a.)

Suppose the curve shows no upward trend once $k > t$. Calculate

$$h_k = \frac{M_k \mathcal{H}_k - M_{k-1} \mathcal{H}_{k-1}}{M_k - M_{k-1}}$$

for $k = t+1,\ t+2, \ldots,\ n$.

It can be shown that an estimate of H' is given by

$$\tilde{H}' = \frac{1}{n-t} \sum_{k=t+1}^{n} h_k = \bar{h}, \tag{19.7}$$

say; the sampling variance of this estimate is estimated by

$$\mathrm{var}(\tilde{H}') = \frac{1}{n(n-1)} \left(\sum h_k^2 - n\bar{h}^2 \right).$$

This method of estimating H' is discussed more fully in Pielou (1975).

4. Hierarchical and Habitat Components of Diversity

The advantages of using a diversity index that meets condition 3 of the conditions listed on page 293 is that if the community under study is subdivisible in any way, then the diversity index can be subdivided into additive components in a corresponding way. This property, which is shared by Shannon's H' and Brillouin's H but by no other diversity measures, is useful in at least two contexts.

Hierarchical Diversity

In the first place, the property permits us to take into account the hierachical nature of biological classification. Suppose we were comparing two communities and that both had the same number of species in the same relative proportions. Whatever function of these proportions we use as a measure of diversity, the diversities of the two communities must be equal; but, if in one community all the species belonged to a single genus and in the other every species belonged to a different genus, it would be reasonable to regard the latter community as the more diverse of the two. This suggests that it would be desirable to be able to split the diversity measure into two components, a generic component and a specific component. For simplicity we consider only two taxonomic levels, but it is straightforward to extend the argument to as many levels as desired and to split the diversity measure into a corresponding number of components.

Consider a fully censused community whose individual members have been classified as to genus and species. The Brillouin index, H, is to be used to measure the community's total diversity, and also its genus diversity and the species diversities within the separate genera: these latter diversities are to be treated as components of the total diversity.

Let classification of the individuals into their genera be called the G-classification and suppose there are a total of g genera; the number of individuals in the ith genus is N_i ($i = 1, \ldots, g$; $\sum_{i=1}^{g} N_i = N$).

The classification of the individuals into species will be called the S-classification. There are s_i species in the ith genus, and N_{ij} individuals in the jth species of the ith genus ($j = 1, \ldots, s_i$; $\sum_{j=1}^{s_i} N_{ij} = N_i$).

Now put:

$H(G)$ for the genus diversity of the community;

$H(GS)$ for the species diversity of the community, that is, the "total" diversity;

$H_i(S)$ for the species diversity within the ith genus; and

$$H_G(S) = \sum_{i=1}^{g} \frac{N_i}{N} H_i(S)$$

for the weighted mean of the species diversities in all g genera. Clearly,

$$H(GS) = \frac{1}{N} \log \frac{N!}{\prod_{i=1}^{g} \prod_{j=1}^{s_i} N_{ij}!}$$

$$= \frac{1}{N} \log \frac{N!}{N_1! \cdots N_g!} \frac{N_1!}{N_{11}! \cdots N_{1s_1}!} \cdots \frac{N_g!}{N_{1g}! \cdots N_{gs_g}!}$$

$$= H(G) + \sum_{i=1}^{g} \frac{N_i}{N} \frac{1}{N_i} \log \frac{N_i!}{N_{i1}! \cdots N_{is_i}!}$$

$$= H(G) + H_G(S)$$

It is seen that the total species diversity, $H(GS)$, has now been split, as required, into two components: the genus diversity, $H(G)$, and the species diversity within a genus averaged over all genera, $H_G(S)$.

Likewise, for a large community whose diversity is measured by the Shannon index H', we have

$$H'(GS) = H'(G) + H'_G(S). \tag{19.8}$$

Putting $G = A$ and $S = B$, this is seen to be identical with (19.1).

Diversity Components in Different Habitats

There are obviously other ways of classifying the members of a community besides that of assigning them to their places in the taxonomic hierarchy. For instance, they may be classified according to the habitats they occupy. If community members are classified twice, first taxonomically and then by habitat, the result is a double classification. Indeed, multiple classifications (into more than two classes) are easy to envisage and it is possible, in addition, to have hierarchical subdivision of each primary class. For brevity, we here consider only simple double classification; the arguments are easily elaborated in principle to cover as complicated a classification as desired. The object is to split a community's diversity into additive components that accord with various ecologically interesting classifications; these may be dependent or independent.

A simple double classification is involved if we consider a narrowly defined community (narrowly defined in the taxonomic sense) made up of a group of related species that occupy a group of related habitats. Examples are: a congeneric group of phytophagous insect species that attack a congeneric group of plant species; a group of wood warbler species that nest in trees forming a group of species whose individuals together form a forest; a group of intertidal organisms that occupy tide pools belonging to a group of qualitatively distinguishable classes.

We shall suppose that the M individuals comprising the community concerned have been fully censused so that the Brillouin index is the appropriate measure of diversity. With a simple double classification, the observations are easily tabulated as an $r \times c$ table in which the r rows represent r different species, say, and the c columns the c different habitats in which these species are found.

Let M_{ij} ($i = 1, \ldots, r; j = 1, \ldots, c$) be the number of individuals of the ith species found in the jth habitat. Also, let $\sum_j M_{ij} = M_i.$ (the ith row total) be the total number of members of the ith species in the community, in all habitats; and let $\sum_i M_{ij} = M_{.j}$ (the jth column total) be the total number of community members, of all species, found in the jth habitat.

Now let the rowwise classification (by species) be called the A-classification and the columnwise classification (by habitats) be called the B-classification. Clearly

$$H(AB) = \frac{1}{M} \log \frac{M!}{\prod_i \prod_j M_{ij}!}$$

$$= \frac{1}{M} \log \frac{M!}{\prod_i M_i.!} + \sum_{i=1}^{r} \frac{M_i.}{M} \cdot \frac{1}{M_i.} \log \frac{M_i.!}{\prod_j M_{ij}!}$$

$$= H(A) + H_A(B), \tag{19.9}$$

or, alternatively

$$H(AB) = \frac{1}{M} \log \frac{M!}{\prod_j M_{.j}!} + \sum_{j=1}^{c} \frac{M_{.j}}{M} \frac{1}{M_{.j}} \log \frac{M_{.j}!}{\prod_i M_{ij}!}$$
$$= H(B) + H_B(A). \tag{19.10}$$

Here $H(AB)$ is the total diversity of the doubly classified observations; $H(A)$ and $H(B)$ are, respectively, the species diversity and habitat diversity in the whole community; $H_A(B)$ and $H_B(A)$ are, respectively, the weighted mean habitat diversity per species, and the weighted mean species diversity per habitat. Indeed, both (19.9) and (19.10) are formally identical with (19.1); the fact that $H(GS)$ in (19.8) can be decomposed in only one way, unlike $H(AB)$ which can be decomposed in two ways, is merely because a purely taxonomic classification is hierarchical: members of one genus can belong to several species but members of one species must belong to one genus.

Consider how the various diversity measures in (19.9) and (19.10) are to be interpreted. Observe first that if the A- and B-classifications are independent, then

$$E(M_{ij}) = \frac{M_i . M_{.j}}{M},$$

whence, if the M_{ij} are large, it can be shown (Pielou, 1972) that

$$H_A(B) \simeq H(B) \qquad \text{and} \qquad H_B(A) \simeq H(A).$$

Conversely, if the two classifications are wholly dependent, that is, if each species occurs in exactly one habitat and each habitat contains exactly one species, then $M_{ii} = M_{i.} = M_{.i}$ and $M_{ij} = 0$ when $i \neq j$. In this case, therefore,

$$H_A(B) = H_B(A) = 0 \qquad \text{and} \qquad H(AB) = H(A) = H(B).$$

Thus the range of possible values of the mean habitat diversity per species and of the mean species diversity per habitat are given by the inequalities

$$0 \leq H_A(B) \leq H(B) \qquad \text{and} \qquad 0 \leq H_B(A) \leq H(A).$$

As a standardized measure of $H_A(B)$ we may therefore use $H_A(B)/H(B) = W$, say, with $0 \leq W \leq 1$. A low value of W is obtained when each species in the community tends to be confined to a restricted subset of the habitats available to the community as a whole; a high value is obtained when each species tends to occur indifferently in many or all of the available habitats. Thus W can be thought of as proportional to the average "niche width" of the community's species.

The corresponding standardized measure of $H_B(A)$ is $H_B(A)/H(A) = L$, say, with $0 \leq L \leq 1$. A low value of L is obtained when each habitat contains a comparatively small subset, of low species diversity, of the community's species complement; whereas a high value of L is obtained when the species diversity within each habitat approaches that of the species diversity of the community treated as an undivided whole. Thus L, in effect, measures the average "niche overlap" within the community.

5. The Measurement of Evenness

As remarked on page 292, diversity depends on both the number of species in a community and on the evenness with which they are represented, and it is sometimes desired to treat these two things separately. It is therefore worth deciding how evenness may be measured.

Consider a fully censused community first, for which the Brillouin index, H, is the appropriate diversity measure. For such a community, the total numbers of individuals, N, and of species, s, are known, and the minimum and maximum values that H could take, given these values of N and s, are easily determined. Let us put $N = sX + r$ where $X = [N/s]$ is the integer part of N/s; also, put $Y = X + 1$ so that $N = (s-r)X + rY$. Then

$$H(\text{max}) = \frac{1}{N} \log \frac{N!}{(X!)^{s-r}(Y!)^r}$$

and

$$H(\text{min}) = \frac{1}{N} \log \frac{N!}{(1!)^{s-1}(N-s+1)!}.$$

A convenient measure of evenness is now given (Hurlbert, 1971) by

$$V = \frac{H - H(\text{min})}{H(\text{max}) - H(\text{min})}. \tag{19.11}$$

As with H itself, V pertains to a fully censused collection and is therefore free of sampling error.

Now consider a large community whose characteristics can only be inferred from a sample and for which the Shannon index, H', is the appropriate diversity measure. Clearly

$$H'(\text{min}) = \lim_{N \to \infty} H(\text{min}) = 0.$$

Further, from condition 1 that H' is required to satisfy (see page 293), we have that

$$H'(\text{max}) = -\sum \frac{1}{s^*} \log \frac{1}{s^*} = \log s^*$$

where s^* is the total number of species in the community. To estimate community evenness from a sample, let us therefore consider whether V', defined on analogy with (19.11) as

$$V' = \frac{H' - H'(\min)}{H'(\max) - H'(\min)} = \frac{H'}{\log s^*}$$

would be useful.

Suppose we obtain an estimate, $\tilde{H}'$, of H', by the method described in Section 3. Suppose, also, we happen to *know* s^*, the number of species in the whole community; it may exceed s, the number in the sample used to estimate H'. Then, as estimator of V', we have

$$\tilde{V}' = \frac{\tilde{H}'}{\log s^*} \qquad (19.12)$$

with

$$\text{var}(\tilde{V}') = \frac{\text{var}(\tilde{H}')}{(\log s^*)^2}.$$

But now suppose, as is far more likely to be true in practice, that s^* is *not* known. Then V' can obviously not be estimated; we can say only that $\tilde{H}'/\log s$ estimates, with unknown error, an upper bound of the community value of V', and it is thus not a very useful descriptive statistic.

The inestimability of evenness when s^* is unknown is hardly surprising. It is obvious on intuitive grounds. For if a community contains an unknown number of undetected species of unknown (but presumably low) abundances, one can say only that the evenness of the whole community is less than meets the eye. Unfortunately V' cannot be estimated by a method analogous to that usable for H' (see Section 3). To see this, note (1) that the number of species in a sample, so long as it is less than s^*, must increase monotonically as sample size is increased; and (2) that values of $\tilde{H}'$ as defined by (19.7) become trendfree once a growing sample has become large enough to be "representative" of the community. Now consider a sequence of values of $\tilde{V}' = \tilde{H}'/\log \tilde{s}^*$ obtained by progressively augmenting sample size; here $\tilde{H}'$ and $\tilde{s}^*$ are the corresponding sequence of estimates, obtained as sample size is steadily augmented, of the community parameters H' and s^*. Clearly $\tilde{V}'$ will tend to decrease as sample size is increased, until all s^* of the community's species have found their way into the growing sample, and there is no means of knowing (unless s^* is independently known) when this stage has been reached.

The foregoing arguments lead to the following two conclusions.

1. Evenness as defined by V in (19.11) is exactly determinable for fully censused collections. The evennesses of different communities (provided

all are censused) can therefore be compared with one another, though such comparisons are unlikely to mean much if the communities have arbitrary boundaries and are not natural ecological entities.

2. Evenness as defined by V' for large communities is inestimable unless s^* is known, which it seldom is. Hence it is usually impossible to intercompare large communities in respect of their evennesses.

It is worth explaining the difficulty, indeed, impossibility of estimating the evenness of a large, sampled, community in somewhat different terms. To estimate evenness entails giving as much weight to observations on rare species as to observations on common ones, and observations on rare species are highly susceptible to error. Indeed, it is only because rare species are as important as common ones when evenness is the focus of interest that observations on the rarest species are not treated as "outliers" in the statistical sense and excluded from consideration.

A possible way around the difficulty would be to choose some wholly arbitrary measure of evenness, such as the ratio of the abundances of the commonest to the rth commonest species for some low r (say 2, 3 or, 4). The usefulness, and sampling properties, of such a measure have not been investigated.

6.   Simpson's Measure of Diversity

Thus far we have considered only the Shannon–Brillouin pair of related diversity indices. There are many others that have found varying degrees of favor with community ecologists and we shall here consider the best known of them, the Simpson index.

Simpson's (1949) original index was not designed to measure diversity but its converse, *concentration,* now more commonly called *dominance.* To obtain a measure of diversity, we therefore require a function of the index that increases as the index itself decreases. We return to this point later and first consider the index of concentration itself.

Suppose two individuals are drawn at random and without replacement from an s-species community containing N individuals of which N_j belong to the jth species ($j = 1, \ldots, s$; $\sum N_j = N$). If the probability is great that both the individuals drawn belong to the same species, we can say that the community exhibits a high degree of concentration and use the probability itself as an index of concentration, say C.

Thus

$$C = \sum_j \frac{N_j(N_j - 1)}{N(N - 1)}. \tag{19.13}$$

Given a fully censused community, C is determinable exactly, without sampling error.

Now consider an indefinitely large community for which C is to be estimated from a sample. Assume that a random sample of the community's individuals is available even though (see page 301), owing to the patchiness of natural communities, such a sample may be hard to obtain. Assume also, that all the community's species are represented in the sample, that is, that $s^* = s$. Let p_j be the proportion, in the parent community, of individuals of the jth species ($j = 1, \ldots, s$). The true value of p_j is, of course, unknown; its maximum likelihood estimator is $\hat{p}_j = N_j/N$. The true value of C is $C = \sum_j p_j^2$ and we wish to estimate it. An unbiased estimator is given by

$$\tilde{C} = \sum_j \frac{N_j(N_j - 1)}{N(N - 1)} \tag{19.14}$$

as we now show (see Good, 1953; Herdan, 1958).

Denote by n_r the number of species represented in the sample by r individuals; that is, n_r is a "frequency of a frequency" (see page 270) and $\sum_r r n_r = N$. Next, put $M = \sum_r r^2 n_r$; it is seen that $M = \sum_j N_j^2$. It remains to prove that $(M - N)/[N(N - 1)]$ is an unbiased estimator of $\sum p_j^2 = C$. Note first that

$$\frac{M - N}{N(N - 1)} = \frac{1}{N(N - 1)} \left\{ \sum_r r^2 n_r - \sum_r r n_r \right\} = \frac{1}{N(N - 1)} \sum_r r(r - 1) n_r,$$

whence

$$E\left[\frac{M - N}{N(N - 1)}\right] = \frac{1}{N(N - 1)} \sum_r r(r - 1) E(n_r). \tag{19.15}$$

Now, each time an individual is drawn from the population, the probability that it will belong to the jth species is p_j. Therefore the probability that in a sample of size N there will be r representatives of the jth species is

$$\binom{N}{r} p_j^r (1 - p_j)^{N-r}.$$

Taking all s species into account, the expected number that will be represented by exactly r individuals is thus

$$E(n_r) = \sum_{j=1}^{s} \binom{N}{r} p_j^r (1 - p_j)^{N-r}..$$

Then, from (19.15),

$$E\left[\frac{M - N}{N(N - 1)}\right] = \sum_j \sum_r \frac{r(r - 1)}{N(N - 1)} \cdot \frac{N!}{r!(N - r)!} \cdot p_j^r (1 - p_j)^{N-r}$$

$$= \sum_j p_j^2 \sum_r \binom{N - 2}{r - 2} p_j^{r-2} (1 - p_j)^{N-r}.$$

The second sum on the right is 1 for all values of j; hence

$$E\left[\frac{M-N}{N(N-1)}\right] = E\left[\frac{1}{N(N-1)}\sum_{\cdot j} N_j(N_j-1)\right] = \sum_j p_j^2.$$

We have therefore shown that (19.14) provides an unbiased estimator of C. It is formally identical with (19.13), the formula for the index of concentration of a fully censused community.

Next recall that C measures concentration or dominance and that what we require is an index of diversity. As will now be shown, the best function of C for this purpose is $-\log C = D$, say. The reason for using D rather than any other function of C that increases with decreasing C is that both D and the Shannon function, H', are special cases of a more general function (Pielou, 1975). The general function (Renyi, 1961) is used in the mathematical theory of communication and is known as the entropy of order α of an s-symbol code of which a proportion p_j of the symbols are of the jth kind. It is defined as

$$H_\alpha = \frac{\log \sum p_j^\alpha}{1-\alpha}.$$

It will be seen that

$$H_1 = -\sum p_j \log p_j = H',$$

the Shannon index; and

$$H_2 = -\log \sum p_j^2 = -\log C = D,$$

as defined above. Thus the Simpson index and the Shannon index are closely related.

The Simpson index does not share with H and H' the merit of being decomposable into additive components and therefore it cannot be adapted for measuring hierarchical diversity. One of the most interesting topics of current community ecology is that of determining, for various communities, at what level in the taxonomic hierarchy diversity is most strongly manifested, and why the level of greatest within-taxon diversity should vary from one community to another. The evolutionary implications of the problem are obvious. Hence H and H' are likely to be much more ecologically useful than D. However, since evenness has turned out to be such an elusive property of communities, there is perhaps room for both H (or H') and D in summarizing some of a many-species community's quantitative characteristics.

20

The Classification of Communities

1. Introduction

Whenever a many-species population—an ecological community—is sampled, the data obtained consist of (a) lists of the species present in each sample unit or (b) records of the amount of each species in each sample unit. Whichever form the data take, that is, whether they are qualitative or quantitative, it is interesting to inquire whether the units are naturally classifiable into distinct groups.

In ecology most of the work on classification has been done by students of vegetation, and the classification problem is considered here in this context. The sample units are quadrats or larger stands of vegetation. When arbitrarily delimited quadrats are used, there is always a risk that the classification obtained may be markedly affected by quadrat size; if stands of vegetation with natural boundaries form the sample units, they may differ greatly in area. These difficulties are merely pointed out here; they will not be discussed further. For convenience we shall speak consistently of the sample units as quadrats, although units much larger than those we should ordinarily describe as quadrats are often used.

Classification methods have also been much studied by marine biologists. The benthic fauna in deep water, and the infauna (in the sediment) at any depth, are invisible; they can be studied only by taking grab samples at a number of scattered points. Classification of these samples ("quadrats") is an obvious way of attempting to gain an insight into the ecology of the community concerned.

A major problem faced by the classifier is whether classification of his data is even appropriate. Thus consider vegetation, and suppose we have a record of the contents of a collection of sample quadrats. Clearly it is always possible to subdivide them in one way or another (i.e., classify them), but it does not follow that the vegetation they represent is classifiable into well-defined separate parts. Apparent discontinuities in the data may range from the clearcut and conspicuous to the dubious and barely perceptible. The act of classification does not of itself answer the

312

question: does the vegetation consist of a number of distinct communities or do the communities merge imperceptibly into one another because the vegetation varies continuously? Regardless of the answer to this question, we may still wish to classify as a matter of convenience; an unnatural classification has been called a "dissection" (Kendall and Stuart, 1966). As an analogy, it is often convenient to show the relief of an area in the form of a map with contour layer coloring. Nobody looking at such a map would suppose that the boundaries between colors represented stepwise discontinuities in elevation on the ground. Thus classification may be done *either* as a convenient system of cataloguing the data, *or* to enable us to perceive real discontinuities.

This brings us to the two well-known theories concerning the nature of vegetation: the *community concept* and the *continuum concept*. Reviews of these theories have been given by McIntosh (1967) and by Langford and Buell (1969). McIntosh's definitions of the two concepts are quoted below. It should first be remarked that probably few ecologists now hold either theory in pure form.

According to the community concept, vegetation is composed of "well-defined, discrete, integrated unis which can be combined to form abstract classes or types reflective of natural entities in the 'real world.'" If this is true, it is natural to attempt a classification of vegetation. Although transition zones between adjacent communities undoubtedly occur in nature, supporters of the community concept exclude them from consideration on the grounds that they form only a negligible proportion of the total area in any large tract of vegetation.

According to the continuum concept [again quoting McIntosh (1967)], "vegetation changes continuously and is not differentiated, except arbitrarily, into sociological entities." If this statement describes the true state of affairs, classification is not "natural," although it may still be convenient, for example, when we wish to map vegetation. The fact that obvious clear-cut vegetational discontinuities do occur in nature does not invalidate the continuum concept. A vegetational discontinuity may be due solely to a discontinuity in some abiotic environmental factor (e.g., a sudden change in the bedrock from which the soil is derived) or to a historical accident (e.g., it may be the boundary of an area that was once burned over). The problem therefore is, do abrupt discontinuities often or habitually occur that can*not* be accounted for by extrinsic causes and must therefore be attributed to interactions among the plants themselves?

For any one particular area this question could, at least in principle, be answered as follows. Suppose the area were known to contain no abrupt environmental discontinuities; then, if there were s plant species, each quadrat could be represented by a point in an s-dimensional coordinate frame, any one point having as coordinates the quantity of each species in the quadrat. If the points formed a single hypersphere, we would infer that the vegetation sampled constituted a single homogeneous community. If the points were clustered into two or more separate hyperspheres, it would follow that the vegetation contained as many discrete natural communities as there were clusters of points. If the points formed a single cluster that was not isodiametric, a hyperellipse, say, we would infer that the vegetation was neither homogeneous nor divisible into separate communities, but instead was continuously variable (see Goodall, 1954).

The interpretation of empirical scatter diagrams is not at all straightforward, however. Thus, if the points fell into separate clusters, a disciple of the continuum hypothesis could always argue that a sufficiently diligent search might reveal hitherto undetected environmental discontinuities. Conversely, an apparent hyperellipse might be regarded (by supporters of the community concept) as two hyperspheres linked by intermediate points representing those quadrats in a transition zone that chanced to be unusually wide. Statistical tests may, of course, be done to judge objectively whether a swarm of points should or should not be regarded as forming separate clusters, but they appear to have been used rarely. As Goodall (1966) says, classification techniques explain how classification may be done but not whether it should be done. If the community or, alternatively, some other realization of the same community type, can be found at a site where there is a unidirectional environmental gradient, the test described in Chapter 17 (see page 252) may be useful. It enables one to test objectively whether the member species of each of the several species groups, that a classification algorithm recognizes as forming distinct classes, have zones of occurrence (on the gradient) that tend to coincide.

If classification is regarded as inappropriate or undesirable, we may still wish to find some method of condensing (and, with luck, clarifying) a mass of field observations. An "ordination" of the data may then be attempted, the aim being to arrange the quadrats in some coordinate frame to display their interrelationships, using as few dimensions as will suffice. Ordination methods are discussed in Chapter 21. Often it is useful to ordinate the data before classifying them, in the hope that a two- or three-dimensional ordinated graph may show more clearly than can a direct visual examination of the community on the ground whether clusters are present. Further, after classification has been done, the interrelationships of the resulting clusters can be portrayed and examined by ordinating the cluster centers.

2. The Different Kinds of Classifications

We now proceed to a consideration of methods of classification. Before one can begin, five decisions must be reached. We list them before considering each in detail.

1. Should classification be hierarchical or reticulate?
2. Should the method be divisive or agglomerative?
3. Should monothetic or polythetic criteria be used to separate classes?
4. Should the data be qualitative or quantitative?
5. How is similarity between classes to be measured?

1. In a hierarchical classification the classes at any level are subclasses of classes at a higher level. Ordinary taxonomic classification with such levels as orders, families, genera, and species, is an example. In a reticulate classification the clusters are defined separately and the links between them have the form of a network rather than a tree. Here we consider only hierarchical systems for the reasons given by Williams and Lambert (1966), namely, that they are "better known, less cumbersome and more widely used in ecological work." These three reasons are admittedly not compelling. If vegetation is truly classifiable, that is, if discrete, nonintergrading communities are the rule rather than the exception, we ought, of course, to determine whether their relationships are reticulate or hierarchical and classify accordingly. To use a hierarchical method is to make untested assumptions about the true relationships among the communities. A hierarchical system is certainly easier to understand and we can only hope that it represents the true state of affairs.

A hierarchical procedure does not yield a classification directly but leads in the first place to construction of a tree diagram or dendrogram showing how the entities being classified are assumed (by the method in use) to be interrelated; Figure 20.2 on page 321 shows examples. The classes to be recognized are then decided by examining the dendrogram. We return to this point in Section 7.

2. In a divisive classification we begin with the whole quadrat collection and divide and redivide it to arrive at the ultimate classes. In an agglomerative classification we start at the bottom and work upward, beginning with the individual quadrats and combining and recombining them to form successively more inclusive groups. Divisive methods have two great advantages: (a) The computations are generally much quicker, since we do not usually continue the subdivision process down to the point at which individual quadrats are recognized as classes. When an agglomerative method is used, we must begin with individual quadrats.

(b) Divisive methods are free from the following difficulty that may often arise with agglomerative methods: in the latter the combining process is begun with the smallest units (the quadrats themselves) and these are the ones in which chance anomalies are most likely to obscure the true affinities. The result is that bad combinations may be made at an early stage in the agglomerative process and they will affect all subsequent combinations.

3. In a monothetic classification two "sister" groups are distinguished by the fact that one has and one lacks a single attribute: for instance, the possession of a particular species. In a polythetic classification two groups are combined or separated on the basis of their overall similarity. A polythetic method thus has the obvious advantage that it can be made to take account of as many properties of the vegetation as we wish to measure or record. A monothetic method is wasteful of information and can lead to meaningless subdivisions if the attribute chosen as distinctive is ecologically unimportant. However, monothetic methods have the great merit that each division is usually made into quadrats that do and quadrats that do not contain a particular species; the ultimate groups are defined by the set of species each contains. The recognition of these "species-sets"—groups of co-occurring species—may be a worthwhile objective quite apart from the classifier's primary purpose, namely, classifying the quadrats in a collection.

4. Whether to use qualitative or quantitative data is often decided by circumstances. If some of the species are so ubiquitous that they are present in every quadrat, obviously quantitative data are essential, at least for these species. When we have a choice, the decision hinges on whether small plants, those that do not contribute much mass or volume to the vegetation, are to be treated as important or unimportant to the classification. The use of presence-or-absence data ensures the small species an effect out of all proportion to their quantity, which may be negligibly small. Whether this is desirable or not is a matter for the ecologist's judgment. Many classification methods can be used with either qualitative or quantitative data or, if not, could be modified for use with data of the kind for which they were not originally devised.

5. All hierarchical methods except those based on information analysis (see Section 6) require that some coefficient of similarity (or its complement, dissimilarity) between pairs of the entities being classified be calculated. The entities may be groups of quadrats or individual quadrats which may be thought of as one-member groups. There is a bewildering variety of ways of measuring intergroup similarity and the whole subject has been clearly reviewed by Cormack (1971) and by Goodall (1973a). Here we mention only four of the commonly used methods.

If the data are qualitative, the coefficient of community (Sorensen's index) is a useful measure. It is defined as $CC = 2a/(2a + b + c)$ where a is the number of species common to the two entities concerned and b and c are, respectively, the numbers of species found in only the first, and only the second, entity. Clearly $0 \leq CC \leq 1$.

With quantitative data, one may measure the similarity between entities i and j, say, by their proportional similarity, PS, defined as

$$PS = 2 \sum_{v=1}^{s} \min \left[\frac{x_{iv}}{z}, \frac{x_{jv}}{z} \right].$$

Here x_{iv} and x_{jv} denote the amounts of species v in entities i and j; s is the total number of species in the two entities combined; and

$$z = \sum_{v=1}^{s} (x_{iv} + x_{jv}).$$

It is seen that $0 \leq PS \leq 1$.

As a measure of the dissimilarity between entities i and j one can use the Euclidean distance d_{ij} (or its square) between the points representing them in s-space. Thus

$$d_{ij}^2 = \sum_{v=1}^{s} (x_{iv} - x_{jv})^2.$$

Examples of methods using this measure are given in Sections 3, 4, and 5. A less frequently used measure of dissimilarity is the city-block metric $\sum_{v=1}^{s} |x_{iv} - x_{jv}|$; it is the distance measured along steps parallel with the axes (rather than the shortest distance) between two points in s-space whose coordinates represent the amounts of the s species in the entities being compared.

It is now apparent that an ecologist must make a number of choices before he can decide on a classification method. At present no strategy exists for making the choices objectively, and the result is that a multitude of methods has been devised and there is continual controversy over their pros and cons. The nature of the problem makes it unlikely that any one method will be voted "best" by a majority of ecologists. Comprehensive comparative reviews of the whole subject of classification have been given by Cormack (1971), Williams (1971), and Goodall (1973b). We here consider in detail a few strongly contrasted methods in order to give some idea of the range of possibilities.

Section 3 describes two "naive" methods; I describe as naive, methods in which the similarity between two groups of quadrats is set equal to the similarity between a single quadrat in one group and a single quadrat in the other. In contrast are the many "centroid" methods; these are methods in which, when a new group is formed, it is assigned a center of its own for the purpose of future similarity measurements instead of being represented by a particular one of its member quadrats as in the naive methods. We must therefore consider the number of similarity coefficients that have to be calculated in carrying out an agglomerative centroid classification.

Beginning with the individual quadrats (which are to be thought of as the initial groups), we determine the similarity coefficient for every pair. Thus, if there are N quadrats, $N(N-1)/2$ similarity coefficients are calculated. The two most similar quadrats are then united to form the first combined group. There are now $N-1$ groups in the collection, $N-2$ of them being single quadrats and the $(N-1)$th a combined pair of quadrats. Next, the similarity coefficients are compared for all the $(N-1)(N-2)/2$ possible pairs of these groups and again the two that are most similar are combined. The process is continued until in the end all the original quadrats have been combined into a single group.

The number of similarity coefficients to be computed is $(N-1)^2$. To see this note that at the first stage $N(N-1)/2$ coefficients must be computed. Once the first pair of quadrats has been combined, the similarities between it and all the remaining groups (single quadrats) have to be computed anew so that $N-2$ new coefficients are needed. After another pair of groups has been combined the similarities of the newly formed group to each of the other $N-3$ groups are computed, and so on. When the process is complete, the number of coefficients that have been obtained is therefore

$$\frac{N(N-1)}{2} + (N-2) + (N-3) + \cdots + 1 = \frac{N(N-1)}{2} + \frac{(N-1)(N-2)}{2}$$
$$= (N-1)^2.$$

In divisive classifications the number of comparisons required may be much larger than this as we shall see in Section 5.

Agglomerative methods are described in Sections 3, 4, and 6, and divisive methods in Sections 5 and 6. The two methods described in Section 6 use measures of within-group information content to decide whether groups should be united (in the agglomerative method) or split (in the divisive method); methods based on information analysis differ from all others in that measurements of between-group similarity is indirect.

3. Naive Classification Methods

Suppose we represent each quadrat by a data point in s-space and use as a measure of the dissimilarity of any pair of quadrats the Euclidean distance between the corresponding data points. We may now do an agglomerative polythetic classification by combining pairs of points to form groups, pairs of groups to form supergroups, and so on. The combinations are carried out one after another, each being between the two groups that are "closest" at that stage; (a single point is treated as a one-member group). It remains to decide how we are to measure the distance between one set of (possibly scattered) points and another such set.

The simplest way is to take as intergroup distance the shortest of all the distances between a point in one group and a point in the other. This procedure is known as nearest neighbor, or single linkage, classification. It is almost as simple, and gives quite different results, to take as intergroup distance the longest of all pairwise distances between a point in one group and a point in the other; this is farthest neighbor, or complete linkage, classification.

An artificial example is shown in Figures 20.1 and 20.2. The object is to classify 15 quadrats from two-species vegetation. The species compositions of the quadrats are represented by the points in the two-dimensional coordinate frame of Figure 20.1, and the distance between points is taken as a measure of interpoint dissimilarity. The dendrograms yielded by single linkage and complete linkage classifications respectively are shown in the upper and lower parts of Figure 20.2. If we choose to classify the points (and hence the quadrats they represent) into four classes by each method, the resultant classes are as shown (in Figure 20.1) by solid outlines (for complete linkage) and dashed outlines (for single linkage). As this example shows, the results can be conspicuously different. In general, single linkage tends to cause excessive "chaining," that is, the addition of single points one after another to the earliest clusters. In contrast, complete linkage may lead to the formation of unnatural "miscellaneous" classes whose members' only resemblance to one another is that they are leftovers from other classes.

Naive classifications need not use Euclidean distance as the (inverse) measure of interquadrat similarity. Thus Field (1971), in classifying the fauna of marine sediments, used proportional similarity as a measure of between-group resemblance and carried out single linkage classification.

4. Orloci's Sums of Squares Method

This is an agglomerative polythetic method in which (in contrast to the naive methods) each group's location in s-space is taken as the centroid of the points comprising it. A group's centroid necessarily shifts whenever a new point is added to it; (Orloci, 1967).

The points, and subsequently the groups formed from them, are to be

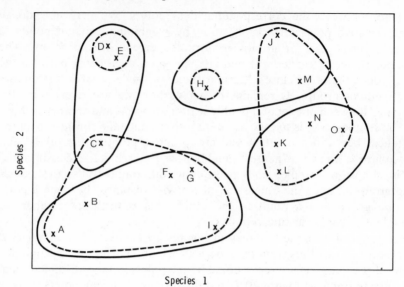

Figure 20.1 A plot of the species-composition of 15 quadrats in two-species vegetation. The dashed outlines show classes defined by single-linkage classification and the solid outlines show classes defined by complete-linkage classification. Euclidean distance was used as a measure of similarity. The number of classes to be recognized was set at four in each case. The dendrograms are shown in Figure 20.2.

combined in a sequence of stages. At any stage two groups are to be combined, provided the increase in within-group dispersion resulting from their union is less than it would be had either of the component groups been joined with some other group.

The within-group dispersion q_n of a group of n points is defined as the sum of squared distances between every point and the group's centroid. The centroid is the point representing the average quadrat of the group and has coordinates

$$\left(\frac{1}{n}\sum_{j=1}^{n} x_{1j}, \frac{1}{n}\sum_{j=1}^{n} x_{2j}, \ldots, \frac{1}{n}\sum_{j=1}^{n} x_{sj}\right) = (\bar{x}_1, \bar{x}_2, \ldots, \bar{x}_s),$$

where x_{aj} denotes the quantity of the ath species in the jth quadrat and $\bar{x}_a$ is the mean quantity of this species in all quadrats. Then

$$Q_n = \sum_{a=1}^{s}\left[\sum_{j=1}^{n}(x_{aj}-\bar{x}_a)^2\right] = \sum_{a=1}^{s}\left[\sum_{j=1}^{n} x_{aj}^2 - n\bar{x}_a^2\right]$$

$$= \sum_{a=1}^{s}\left[\frac{n-1}{n}\sum_{j=1}^{n} x_{aj}^2 - \frac{2}{n}\sum_{i<j} x_{ai}x_{aj}\right] = \sum_{a=1}^{s}\left[\frac{1}{n}\sum_{i<j}(x_{ai}-x_{aj})^2\right].$$

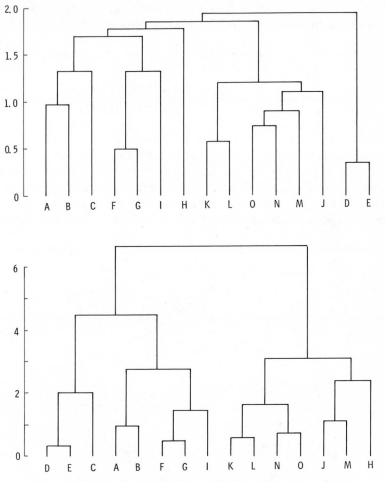

Figure 20.2 The dendrograms obtained by classifying the quadrats plotted in Figure 20.1. (*a*) Single linkage; (*b*) complete linkage. The scales on the left show the Euclidean distance between classes.

Here $\sum_{i<j}$ denotes summation over all pairs of points, counting each pair only once. Thus

$$Q_n = \frac{1}{n} \sum_{i<j} \left[\sum_{a=1}^{s} (x_{ai} - x_{aj})^2 \right] = \frac{1}{n} \sum_{i<j} d_{ij}^2,$$

where d_{ij} is the distance between the ith and jth points.

We therefore see that the within-group dispersion may be directly calculated from the interpoint distances.

Groups are now combined in pairs according to the following rule. Consider groups u and v, which have within-group dispersions of Q_u and Q_v, respectively, and denote by Q_{uv} the dispersion of the group formed by combining them. The combination of u and v is "permissible" if

$$Q_{uv} - (Q_u + Q_v) < \begin{cases} Q_{uw} - (Q_u + Q_w), \\ Q_{vw} - (Q_v + Q_w), \end{cases}$$

for all w (i.e., for all other groups). Note that groups u and v are combined only if *neither* of them would combine better with some other group. At each stage of the agglomeration *all* permissible pairwise combinations are made before another round of computations begins.

The method just described is based on the absolute distances between pairs of points. Orloci also proposed an alternative method based on

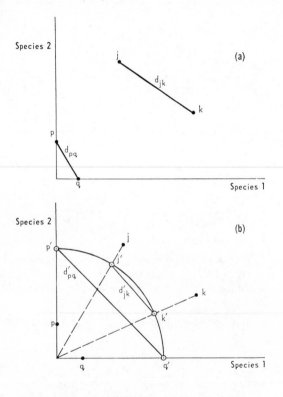

Figure 20.3 The relation between absolute distances [d_{pq} and d_{jk} in (a)] and standardized distances [d'_{pq} and d'_{jk} in (b)] (After Orloci 1967a).

"standardized distances." The argument is best illustrated diagrammatically (see Figure 20.3), in which it is assumed that there are only two species.

In Figure 20.3a d_{pq} and d_{jk} show the absolute distances between quadrats p and q and between quadrats j and k. It is seen that $d_{pq} < d_{jk}$; that is, quadrats p and q, which have no species in common but in which the amount of plant material is small, are close together; whereas quadrats j and k, both of which contain large amounts of both species (though in very different proportions), are far apart. If this consequence of using absolute distances is regarded as undesirable, we may use standardized distances instead, as shown in Figure 20.3b. The points representing the quadrats are projected onto a unit circle centered on the origin and the standardized distances between the two pairs of quadrats are then given by d'_{pq} and d'_{jk}, the lengths of the chords $p'q'$ and $j'k'$, respectively. Then $d'_{pq} > d'_{jk}$. The extension to s-dimensions is straightforward and details may be found in Orloci (1967a).

5.   Divisive Polythetic Methods

Probably the most satisfactory ecological classifications would be yielded by a method that was both divisive and polythetic and therefore combined the advantages of both (see Section 2). Unfortunately however, there is at present no direct divisive-polythetic method that is computationally feasible. This is because if, at each stage, every quadrat group present is subjected to all possible dichotomous partitions so that they may be compared and the appropriate one chosen, the number of possibilities quickly becomes very large indeed, as shown below. If it were not for this drawback the method of Edwards and Cavalli-Sforza (1965) would have much to recommend it and it is worth describing for this reason. It is inapplicable in nearly all ecological contexts, however, since its inventors do not recommend its use if the number of entities to be classified exceeds 16.

The method is as follows. Beginning with a swarm of n points, which represent n quadrats, we wish to divide them into two groups in such a way that the within-groups sums of squares will be a minimum and consequently the between-groups sum of squares a maximum. In order to choose which partition has the desired property all possible partitions must be tested. If there are n quadrats in the collection, they can be divided into two groups in $2^{n-1} - 1$ different ways. To see this assume n is even; note that there are $\binom{n}{1}$ ways of dividing the n points into groups of sizes $n - 1$ and 1; $\binom{n}{2}$ ways of dividing them into groups of sizes $n - 2$ and 2, ..., and $\binom{n}{n/2}$ ways of dividing them into two equal groups of size $n/2$.

The total number of possibilities is thus

$$\binom{n}{1} + \binom{n}{2} + \cdots + \binom{n}{n/2}$$

$$= \frac{1}{2}\left\{ \left[\binom{n}{0} + \binom{n}{1} + \cdots + \binom{n}{n-1} + \binom{n}{n} \right] - \binom{n}{0} - \binom{n}{n} \right\}$$

$$= \frac{1}{2}[(1+1)^n - 2] = 2^{n-1} - 1.$$

Similar arguments yield the same result when n is odd.

For every possible partition we now perform what is in effect a one-way analysis of variance in which the variates are vectors in s dimensions. Let

Q = the sum of squared distances from all points to the centroid of the whole swarm.

Also, with $i = 1, 2$, let

q_i = the sum of squared distances from the points in group i to the group's own centroid.

Q_i = the squared distance from the centroid of group i to the centroid of the whole swarm.

Then, if there are n_i points in the ith group,

$$Q = \sum_{i=1}^{2} q_i + \sum_{i=1}^{2} n_i Q_i.$$

As already proved (page 320), Q, q_i, and Q_i may all be found by summing the squares of the interpoint distances (taking each distance only once) and dividing by the number of points concerned. We may therefore tabulate these squared interpoint distances in the form of a half-matrix:

Points	α	β	γ	$\cdots$
α		$d^2_{\alpha\beta}$	$d^2_{\alpha\gamma}$	$\cdots$
β			$d^2_{\beta\gamma}$	$\cdots$
γ				
.				
.				
.				

Then $Q = (1/n)\sum d^2$, where the summation is over all elements in the half-matrix and there are n points.

Now consider how a particular partition is to be tested. For concreteness assume that $n = 7$ and let the quadrats be labeled $\alpha, \beta, \ldots, \eta$. Let the partition to be tested be that into groups $(\alpha, \beta, \gamma, \delta)$ and (ε, ξ, η) so that $n_1 = 4$ and $n_2 = 3$.

Then

$$q_1 = \frac{1}{n_1}(d_{\alpha\beta}^2 + d_{\alpha\gamma}^2 + \cdots + d_{\gamma\delta}^2)$$

and

$$q_2 = \frac{1}{n_2}(d_{\varepsilon\xi}^2 + d_{\varepsilon\eta}^2 + d_{\xi\eta}^2).$$

The within-groups sum of squares sought is $q_1 + q_2$. To decide which of the $2^{7-1} - 1 = 63$ possible partitions of the seven quadrats is to be made we must calculate all 63 values of $(q_1 + q_2)$ and choose that partition for which $(q_1 + q_2)$ is smallest. In this manner the original collection of quadrats is divided and the resultant groups redivided until we obtain a classification as fine as desired.

The computing time for this method may be enormously long. Thus at the first stage $2^{n-1} - 1$ partitions have to be tested. Suppose the chosen partition yields groups with n_1 and $n_2 = n - n_1$ points, respectively. Then at the second stage there are $2^{n_1-1} + 2^{n_2-1} - 2$ partitions to test and so on. According to Edwards and Cavalli-Sforza (1965), an initial collection of "16 points can be treated in a reasonable time." For 41 points, however, Gower (1967a) states that using a computer with 5 μsec access time the process would require more than 54,000 years. Some form of short cut seems called for, perhaps along the lines suggested by Dagnelie (1966), who comments on the possibility of ruling out inadmissible partitions. As an example of an inadmissible partition, consider a swarm of four points in a plane; if one of them were inside a triangle formed by the other three, it would be inadmissible to treat the interior point as one group and those at the vertices of the surrounding triangle as the other. Scott and Symons (1971) have also devised a time-saving modification.

An indirect method of performing a divisive-polythetic classification has been proposed by McNaughton-Smith et al. (1964). It consists in first identifying which quadrat is most dissimilar from all the others and using it as the founding member of a new "accreting" class to which is added, one by one, all other quadrats that resemble it more closely than they resemble the original "source" class. Thus, let $f(M, N)$ denote any convenient function that measures the dissimilarity between groups M and N. Denote the n quadrats in the collection to be classified by X_i with $i = 1, \ldots, n$. Denote by A_j and B_j respectively the source and accreting classes after j classes have been added to the latter.

The first step in the classification consists in determining $f(X_i, A_0 - X_i)$ for all i. The X_i for which this dissimilarity is greatest becomes B_1 and

$A_0 - B_1$ becomes A_1. Next, for each X in A_1, calculate $f(X_i, A_1 - X_i) - f(X_i, B_1) = d_{i1}$, say. Suppose $d_{i1} > 0$ for at least one i. Then that X_i for which d_{i1} has the largest (positive) value is now removed from A_1 (which becomes A_2) and added to B_1 (which becomes B_2). The process is continued until after k steps, say, it is found that $f(X_i, A_k - X_i) - f(X_i, B_k) = d_{ik} < 0$ for all i such that X_i is in A_k. This shows that the first division of the original collection, into the two groups A_k and B_k, is complete. These groups themselves are now divided and redivided by repeated application of the same method. The procedure has the virtues of being divisive, polythetic, and computationally feasible; it deserves more attention from ecologists than it has so far received.

6. Classification Methods Based on Information Analysis

Classification using information analysis can be agglomerative or divisive. In the agglomerative method, judgment as to which of two groups of quadrats in a collection of groups is most similar, and should therefore be united, is made as follows. Each group has a certain "information content" (in the sense of Chapter 19), and if two groups are united, the combined group so formed contains more information than either of its component groups. But the more similar the two groups are, the smaller the increase in information when the two are combined. Thus at any stage, in an agglomerative process, those two groups are to be combined whose union produces the smallest gain in information content. Conversely, in the divisive method, groups are to be subdivided into subgroups in such a way as to maximize the reduction in information content.

The agglomerative method, which we describe first, was devised by Williams and Lambert (1966) and uses qualitative data. For each quadrat we know only whether or not it contains a given species. Let the group consist of n quadrats with a total of s species and let a_j of the quadrats contain the jth species. Then the proportion of quadrats containing this species is a_j/n and the proportion lacking it is $(n - a_j)/n$. Treating the group as made up of n individuals (i.e., quadrats) of two kinds (with and without the jth species) we may then use Shannon's formula (see page 298) to define the information *per quadrat* in respect of the jth species as

$$H'_j = -\frac{a_j}{n} \log \frac{a_j}{n} - \frac{(n - a_j)}{n} \log \frac{(n - a_j)}{n}.$$

(In other words H'_j now measures the "diversity" of the *quadrats*.)

The total information in the group (still with respect to the jth species) is then nH'_j. Summing over all s species, we obtain I, the total information in ·the group with respect to all the species; that is

$$I = \sum_{j=1}^{s} nH'_j = - \sum_{j=1}^{s} \left[a_j \log \frac{a_j}{n} + (n - a_j)\log \frac{(n - a_j)}{n} \right]$$

$$= sn \log n - \sum_{j=1}^{s} [a_j \log a_j + (n - a_j)\log(n - a_j)].$$

Now consider the first stage of the agglomeration process, when a pair of single quadrats is to be combined into a group for which $n = 2$. The information content of a single quadrat is zero by definition. Let two quadrats be combined and let the total number of species in the resulting group be s. Suppose, also, that of these s species t are common to both quadrats and $u = s - t$ were present in only one quadrat. The species may now be labeled so that

$$a_j = 2 \quad \text{for} \quad j = 1, 2, \ldots, t,$$
$$a_j = 1 \quad \text{for} \quad j = t + 1, t + 2, \ldots, s.$$

Then the information content of the group of $n = 2$ quadrats is

$$I = 2s \log 2 - \sum_{j=1}^{t} 2 \log 2 = 2(s - t)\log 2 = 2u \log 2.$$

We see that if the two quadrats have identical species lists (i.e., if $u = 0$) the group formed by combining them again has zero information content. The agglomeration then proceeds by uniting, at each stage, that pair of groups (with one or more members), say groups G and H, for which

$$\Delta I = I(G) + I(H) - I(G + H)$$

is a minimum [here $I(G)$ denotes the information content of group G and correspondingly for the other symbols]. Thus the groups united are the two resembling each other most closely.

Although Williams and Lambert (1966) originally proposed the method for use with qualitative data, it is also usable when the data are quantitative. This has been done by Orloci (1968, 1969) in classifications of the vegetation of grassland and hardwood forest. When quantitative data are available, one records each quadrat as containing so many units (e.g., of weight, volume, or cover-area) of species 1, so many of species 2, and so on. The total information content of the quadrat can then be defined either as vH' (H' is Shannon's formula for information per unit; see page 298) or, better, as vH (H is Brillouin's formula; see page 300). Here v is the total number of units, of whatever quantity is measured, of all species taken together.

A divisive-monothetic method of classification based on information analysis has been proposed by Lance and Williams (1968). To carry it out, first calculate I, the total information content in the whole collection of quadrats to be classified. Then subdivide the collection into two groups, those with, and those without, species i; label these groups G_i and g_i respectively and calculate

$$\Delta I_i = I - I(G_i) - I(g_i).$$

Do this for all species (i.e., for all values of i) in turn in order to determine for which species ΔI_i is greatest. The first division of the classification is then into those quadrats containing, and those lacking, this species. Each of these groups is now treated as a collection in itself, and the division process is repeated until as fine a classification as desired has been attained.

The method has the advantages of being both divisive and (unlike the method of Edwards and Cavalli-Sforza) computationally feasible. The computational load depends, of course, on the number of species in the collection. The method also has the advantages and disadvantages of being monothetic. And it may prove to be free of the tendency shown by the agglomerative information method of combining into groups miscellaneous unrelated quadrats whose only resemblance to one another is that they differ from clearly identifiable natural groups (Goodall, 1973).

Wallace and Boulton (1968; and see Boulton and Wallace, 1970) have devised a divisive-polythetic method of classification based on information analysis. Replicates of the entities to be classified are required and the procedure is rather complicated.

7.   Hierarchical Levels and Stopping Rules

Classification of a quadrat collection by any of the methods described, or indeed by any hierarchical method, permits construction of a tree diagram or dendrogram to show the sequence in which the divisions or unions of the groups were made. A dendrogram thus has a number of nodes (shown by horizontal lines) each of which represents an intermediate group that was divided into two subgroups (in a divisive classification) or formed by the union of two subgroups (in an agglomerative classification). These intermediate groups are of different status and the status of each can be represented by the position of the corresponding node in the dendrogram. There are two ways in which this can be done.

1. Suppose the entities being classified (quadrats, say) are arranged along a horizontal base line. Then the internal heterogeneity of any group may be represented by the height of its node above the base line. Within-group heterogeneity may be measured in various ways and the

measure will normally (though not necessarily) depend on the classification criterion. Thus for complete linkage classification (Section 3) the obvious measure of intergroup heterogeneity is the greatest of all the between-member dissimilarities in the group, measuring dissimilarity by whichever of the numerous possible methods was used in performing the classification. For single-linkage classification, the corresponding measure is the sum of all the linkage distances that led to the linking up of the group. The dendrograms shown in Figures 20.2b and 20.2a respectively were constructed in these ways. When the within-group sum of squares is used as the classification criterion, as in Orloci's method (Section 4), within-group dispersion provides a measure of heterogeneity; and for information analytic methods, a group's total information content is the appropriate measure.

2. An alternative way of deciding the heights of the nodes in a dendrogram, when classification is divisive, is to require that the distance between any node and the next one below it should be proportional to the reduction in heterogeneity (however measured) brought about by performing the division that the lower node represents. The dendrogram is constructed from the top downward and the ultimate branch tips (the entities being classified) need not be on the same level. A dendrogram can also be drawn in this way, of course, when classification is agglomerative, but the agglomeration must be carried to completion before construction of the dendrogram can begin.

For many classification procedures the two ways of drawing a dendrogram yield identical results; this is obviously true of single linkage classification for example. It is not true of the information analytic methods, however. For these methods the second way described above is especially appropriate; the length of each of the dendrogram's internodes portrays ΔI, the difference in information content between the united group represented by the internode itself and that of its two constituent subgroups.

It remains to decide how a classification is to be obtained by cutting across a dendrogram. When a divisive classification is done, this decision amounts to a stopping rule (Lambert and Williams, 1966). There are three possibilities:

1. One may choose the number of ultimate classes to be recognized. The dendrogram is then read from the top downwards until the number of branches corresponds with the chosen number of classes.

2. One may choose the amount of heterogeneity to be permitted in the recognized classes. This is equivalent to choosing where to cut across a

dendrogram whose nodes are at levels representing intragroup heterogeneity.

3. Given a dendrogram in which the lengths of the internodes show changes in heterogeneity, one may read the dendrogram from the top downwards and terminate each branch as soon as an internode is reached that is shorter than some chosen length. The cutoff points will be at various levels. In effect, one is deciding to stop further subdivision as soon as the resultant reduction in heterogeneity becomes too small to be worth while.

It should be noticed that in each of the three ways, outlined above, of deriving a classification from a dendrogram, a choice is called for. Choice 1, of the number of classes to be recognized, is usually guided by expedience and is wholly arbitrary.

In the other two cases an attempt is often made to set up a so-called "probabilistic" stopping rule. Thus one may stipulate that the amount of heterogeneity to be allowed in each of the ultimate classes (choice 2) is to be just small enough to make acceptable, using an appropriate test and a chosen significance level, the null hypothesis that the classes are internally homogeneous. Or (for choice 3) one may stipulate that a division is to be made only if the resultant groups differ sufficiently for one to reject, again using an appropriate test and a chosen significance level, the null hypothesis that they are homogeneous.

Use of these probabilistic stopping rules requires the classifier to carry out a sequence of significance tests on successively smaller subsets of a single body of data. No allowance is made for the fact that the calculated probabilities are conditional, and that therefore no great faith can be placed in the significance, or lack of it, of within-group homogeneity and between-group heterogeneity. Null hypotheses, tests, and significance levels, are all the outcome of arbitrary choices. However, a probabilistic rule, though arbitrary, is objective. It can be made part of the whole list of instructions for performing classifications by a designated method and to that extent enables comparable classifications to be made of different bodies of data and by different people.

8. Concluding Remarks

There is now a huge literature on ecological classification; here we have described in detail only a few of the many different methods that have been proposed. Each method consists of a set of rules for operations to be carried out on the raw data. The vexing problem of the method that is best must be left open; it is not even possible to define "best" in this context. As Williams and Lambert (1966) have said, "the difficulty . . . is to find objective criteria in an essentially subjective situation." All that we can reasonably demand of a method is that "the major groupings which arise shall not be fewer than, or markedly different from, those recognized intuitively as distinct ecological entities . . . ." Work is continuing but much of it amounts to no more than pointless exercises in the devising of new methods and the intercomparison of old ones. This could (and perhaps will) continue without end. If classificatory procedures are to be used rather than played with, ecologists must choose one or a very few methods and use them consistently. The choice will unavoidably be somewhat arbitrary.

21

The Ordination of Continuously Varying Communities

1. The Purpose of Ordination

As already remarked (page 314), we can, conceptually, represent the data obtained by measuring the amounts of s species in each of n sample units by a scatter diagram of n points in an s-dimensional coordinate frame. Classification consists in subdividing the swarm of points into a number of disjoint sets or groups in what we belive is a natural manner. If the points chance to fall into several compact, widely separated groups no difficulty arises and formal rules for effecting a classification are scarcely needed. This ideal result is hardly ever obtained when natural communities are sampled. More often than not the points (representing the quadrats) are diffusively scattered and any classification procedure is largely arbitrary.

A way out of this difficulty is to ordinate the quadrats rather than to classify them. The .purpose, as in classification, is still to simplify and condense the mass of raw data yielded by sampling in the hope that relationships among the species and between them and the environmental variables will be manifested. Ordination consists in plotting the n points in a space of fewer than s dimensions in such a way that none of the important features of the original s-dimensional pattern is lost. Ideally we would plot them in two or three dimensions so that they might be easily visualized. As a method of summarizing the results of a survey, ordination has two great advantages over classification; it obviates the need for setting up arbitrary criteria for defining the classes and there is no need to assume that distinct classes (if there are any) are hierarchically related.

2. Principal Component Analysis

Many possible methods of ordination have been devised. The most straightforward is to project the original s-space onto a space of fewer dimensions in such a way that the arrangement of the points suffers the least possible distortion. To take the simplest conceivable case, suppose we wished to project a swarm of points in the plane onto a line to obtain

332

a one-dimensional ordination. The distortion will be a minimum if the line is oriented so as to preserve as far as possible the spacing of the points. An artificial example is given in Figure 21.1a which shows results that might be obtained by sampling vegetation made up of only two species of plants. The points represent the quantities of species 1 (measured along the x_1-axis) and of species 2 (on the x_2-axis) in each of $n = 27$ quadrats. The origin of the coordinates is at $(\bar{x}_1, \bar{x}_2)$, the mean quantities of the species averaged over all quadrats.

We now wish to subject the axes to a rigid rotation through an angle θ, say, to form new orthogonal (perpendicular) axes shown in the figure as the y_1- and y_2-axes. Then the y_1-axis is the line $x_1/\cos \theta = x_2/\cos \phi$, where $\phi = \pi/2 - \theta$. (We write $\cos \phi$ in place of $\sin \theta$ to emphasize that $\cos \theta$ and $\cos \phi$ are the direction cosines of the line.) A point with coordinates (x_{1r}, x_{2r}) relative to the old axes now has coordinates (y_{1r}, y_{2r}) relative to the new axes. A one-dimensional ordination of the points may be obtained by

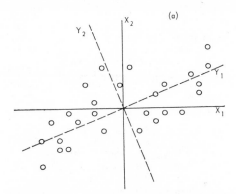

Figure 21.1a. The principal component axes y_1 and y_2 for the swarm of points whose original coordinates were given as (x_1, x_2).

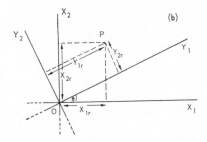

Figure 21.1b. To illustrate the rotation of coordinate axes (see text).

projecting them onto the y_1-axis, and we may say that the distortion produced by the ordination is least when the sum of squares of the y_1 values, $\sum_{r=1}^{n} y_{1r}^2$, is a maximum. Our object now is to determine what value of θ will give this result.

Notice first that

$$\sum_{1}^{n} x_{1r}^2 + \sum_{1}^{n} x_{2r}^2 = \sum_{1}^{n} y_{1r}^2 + \sum_{1}^{n} y_{2r}^2 .$$

That this is so may be seen from Figure 21.1b in which P is the point (x_{1r}, x_{2r}) in the original coordinates and (y_{1r}, y_{2r}) in the new coordinates. Clearly $(OP)^2 = x_{1r}^2 + x_{2r}^2 = y_{1r}^2 + y_{2r}^2$. It follows also that since the origin is at the centroid of the swarm of points (in other words since $\bar{x}_1 = \bar{x}_2 = \bar{y}_1 = \bar{y}_2 = 0$)

$$\mathrm{var}(x_1) + \mathrm{var}(x_2) = \mathrm{var}(y_1) + \mathrm{var}(y_2) \tag{21.1}$$

Figure 21.1b also shows that

$$y_{1r} = x_{1r} \cos \theta + x_{2r} \sin \theta \quad \text{and} \quad y_{2r} = -x_{1r} \sin \theta + x_{2r} \cos \theta.$$

This may be written in matrix form as

$$\begin{pmatrix} y_{1r} \\ y_{2r} \end{pmatrix} = \begin{pmatrix} \cos \theta & \sin \theta \\ -\sin \theta & \cos \theta \end{pmatrix} \begin{pmatrix} x_{1r} \\ x_{2r} \end{pmatrix}, \qquad r = 1, 2, \ldots, n,$$

or, more compactly, as

$$\mathbf{Y} = \mathbf{UX}, \tag{21.2}$$

where

$$\mathbf{U} = \begin{pmatrix} \cos \theta & \sin \theta \\ -\sin \theta & \cos \theta \end{pmatrix}$$

and $\mathbf{X}$ is the $(2 \times n)$ data matrix

$$\begin{pmatrix} x_{11} & x_{12} & \cdots & x_{1n} \\ x_{21} & x_{22} & \cdots & x_{2n} \end{pmatrix};$$

$\mathbf{Y}$ is defined similarly. It is seen that $\mathbf{U}$ is orthogonal; that is, $\mathbf{UU'} = \mathbf{U'U} = \mathbf{I}$. We also see that the covariance matrix of the x's is

$$\boldsymbol{\Sigma}_x = \frac{1}{n} \mathbf{XX'} = \frac{1}{n} \begin{pmatrix} \sum_{1}^{n} x_{1r}^2 & \sum_{1}^{n} x_{1r} x_{2r} \\ \sum_{1}^{n} x_{1r} x_{2r} & \sum_{1}^{n} x_{2r}^2 \end{pmatrix} = \begin{pmatrix} \mathrm{var}(x_1) & \mathrm{cov}(x_1, x_2) \\ \mathrm{cov}(x_1, x_2) & \mathrm{var}(x_2) \end{pmatrix}. \tag{21.3}$$

We return now to the problem of determining the value of θ that will make $\sum y_{1r}^2 = n \, \mathrm{var}(y_1)$ a maximum. It is clear from (21.1) that maximizing $\mathrm{var}(y_1)$ is equivalent to minimizing $\mathrm{var}(y_2)$.

From (21.2) we have

$$\mathbf{YY'} = \mathbf{UXX'U'} \tag{21.4}$$

or

$$\begin{pmatrix} \text{var}(y_1) & \text{cov}(y_1, y_2) \\ \text{cov}(y_1, y_2) & \text{var}(y_2) \end{pmatrix} = \begin{pmatrix} \cos\theta & \sin\theta \\ -\sin\theta & \cos\theta \end{pmatrix} \begin{pmatrix} x_1 \\ x_2 \end{pmatrix} (x_1 x_2) \begin{pmatrix} \cos\theta & -\sin\theta \\ \sin\theta & \cos\theta \end{pmatrix}, \tag{21.5}$$

where for conciseness we have replaced the $(2 \times n)$ data matrix with the vector $(x_1\ x_2)'$.

From (21.5)

$$\text{var}(y_1) = x_1^2 \cos^2\theta + x_1 x_2 \sin 2\theta + x_2^2 \sin^2\theta.$$

Obviously var(y_1) is maximized when

$$\frac{d}{d\theta}\text{var}(y_1) = 0;$$

that is, when

$$(-x_1^2 + x_2^2)\cos\theta \sin\theta + x_1 x_2 \cos 2\theta = 0.$$

From (21.5) it is also seen that

$$\text{cov}(y_1, y_2) = (-x_1^2 + x_2^2)\cos\theta \sin\theta + x_1 x_2 (\cos^2\theta - \sin^2\theta)$$
$$= (-x_1^2 + x_2^2)\cos\theta \sin\theta + x_1 x_2 \cos 2\theta.$$

Thus var(y_1) is a maximum, and var(y_2) a minimum, when cov(y_1, y_2) = 0; the new variates y_1 and y_2 are uncorrelated.

Therefore, when var(y_1) is a maximum, (21.4) becomes

$$\mathbf{U\Sigma_x U'} = \begin{pmatrix} \lambda_1 & 0 \\ 0 & \lambda_2 \end{pmatrix} = \mathbf{\Lambda} \quad \text{or} \quad \mathbf{U\Sigma_x} = \mathbf{\Lambda U}, \tag{21.6}$$

where $\lambda_i = \text{var}(y_i)$ for $i = 1, 2$.

It is now clear that λ_1 and λ_2 are the latent roots of the symmetric matrix $\mathbf{\Sigma_x}$; they are the roots of the determinantal equation $|\mathbf{\Sigma_x} - \lambda\mathbf{I}| = 0$, and the rows of $\mathbf{U}$ are the latent vectors of $\mathbf{\Sigma_x}$.

Therefore θ may be found by solving the equation

$$\text{var}(x_1)\cos\theta + \text{cov}(x_1, x_2)\sin\theta = \lambda_1 \cos\theta.$$

We are now in a position to do the following. Instead of defining any point by the coordinates (x_1, x_2), we can treat it as the point (y_1, y_2), where

$$y_1 = x_1 \cos\theta + x_2 \sin\theta, \qquad y_2 = -x_1 \sin\theta + x_2 \cos\theta.$$

Each of the two new variates is thus a linear combination of the original variates (the measured quantities of the species). The new variates are so defined that y_1, which is known as the first principal component, has maximum possible variance. The best one-dimensional ordination of the points is obtained by projecting them onto the y_1-axis, the first principal axis. In this simple two-dimensional case the direction of the y_2-axis (the second principal axis) is given once that of the first is known, since it must be orthogonal to it. We have also shown above that the two new variates y_1 and y_2 are uncorrelated.

Now consider the general s-dimensional case. Our objective is to find the rigid rotation of the original axes, or equivalently the linear combination of the original variate values (the x's), that will yield derived variates (the y's) with the following properties:

(1). The variance of the y_1's is to be as great as possible.

(2). The variance of the y_2's is to be as great as possible, subject to the restriction that the y_2-axis must be orthogonal to the y_1-axis. The variates y_1 and y_2 are uncorrelated.

(3). The variance of the y_3's is to be as great as possible, subject to the restriction that the y_3-axis must be orthogonal to the y_1- and y_2-axes. There are no correlations among the variates.

. . . .

(s) The final axis, the y_s-axis, is to be orthogonal to all the $(s-1)$ axes already fixed.

This process is principal component analysis.

What we wish to find are the s direction cosines of each of the s principal axes, which are given by the elements of the $s \times s$ matrix $\mathbf{U} = \{u_{ij}\}$ with $i, j = 1, \ldots, s$. They are obtained by solving the matrix equation

$$\mathbf{U}\mathbf{\Sigma}_x\mathbf{U}' = \mathbf{\Lambda} \qquad \text{[this is the same as (21.6)].}$$

Here $\mathbf{\Sigma}_x$ is the $s \times s$ covariance matrix of the x's and $\mathbf{\Lambda}$ is the diagonal matrix whose elements (the latent roots of $\mathbf{\Sigma}_x$) are $\lambda_j = \text{var}(y_j)$ ($j = 1, \ldots, s$). The direction cosines of the jth principal axis are the elements of the jth row of $\mathbf{U}$ (which is the jth latent vector of $\mathbf{\Sigma}_x$).

In terms of the original coordinates, the y_j-axis (the jth principal axis) is the line

$$\frac{x_1}{u_{j1}} = \frac{x_2}{u_{j2}} = \cdots = \frac{x_s}{u_{js}}.$$

The jth derived variate (the jth principal component) is

$$y_j = u_{j1}x_1 + u_{j2}x_2 + \cdots + u_{js}x_s.$$

Thus the coordinates of the rth point $(r = 1, \ldots, n)$ are given by

$$y_{jr} = u_{j1}x_{1r} + u_{j2}x_{2r} + \cdots + u_{js}x_{sr} \qquad (j = 1, \ldots, s).$$

These assertions about the general s-dimensional case are not proved here. Proofs may be found in books on multivariate analysis; for example, Kendall (1957), Anderson (1958), Kendall and Stuart (1966), and Morrison (1967).

We have ranked the transformed variates so that

$$\mathrm{var}(y_1) > \mathrm{var}(y_2) > \cdots > \mathrm{var}(y_s).$$

Then, depending on how much of the information in the original data we are willing to sacrifice, we may disregard the variates $y_{k+1}, y_{k+2}, \ldots, y_s$ (those with smallest variances) for some chosen k and, retaining only the k variates with largest variances, ordinate the data in a space of k dimensions. It is, in fact, often found that the first few latent roots of Σ_x account for a large proportion of the total variance; for example, Orloci (1966), analyzing data on the vegetation of sand dunes and dune slacks and considering the 101 most frequent species, found that the first three principal components accounted for more than 40% of the total variance. Likewise, Greig-Smith, Austin, and Whitmore (1967) present a three-dimensional ordination of forest types in the British Solomon Islands Protectorate. In both the examples quoted it was found that clusters of points recognizable in two- or three-dimensional plots of the vegetation samples could be associated with environmental differences.

3. Practical Considerations in Principal Component Analysis

When we wish to ordinate vegetation by means of a principal component analysis, there are four decisions to make.

1. A method of measuring the amount of each species in each sample unit must be chosen. If the sample units are quadrats, there are several ways of measuring the amount of each species; for instance, by counting individuals or by determining fresh weight, dry weight, basal area, or cover. When larger sample units are used—whole plots or stands of vegetation—a convenient method (Orloci, 1966; Gittins, 1965) is to sample each stand with a number of quadrats and take as a measure of the quantity of a species in the stand the frequency of that species in the quadrats; that is, the number of quadrats that contained the species.

2. Whether to standardize the original variates must be decided. If they are standardized the covariance matrix Σ_x in (21.6) is replaced by a

correlation matrix but the analysis is otherwise unaltered. Standardization is commonly done when the x variates are measured in several different units: for instance, in psychological studies when measurements on people may be a mixture of test scores, ages, incomes, and other dissimilar quantities measured on entirely different scales. In botanical work, in which the amounts of all the species in a sample are measured by the same method and in the same units, standardization is not necessary. If the raw data are standardized before analysis, those species whose abundances vary only slightly (usually the less abundant species) will have a greater influence than if unstandardized variates are used. Whether this is desirable is debatable. It should be noticed that standardizing the original data and then finding the principal components does not lead to the same result as finding principal components first and then standardizing the derived variates.

3. The number of principal components to extract—or of axes to use in the final ordination—must be chosen. If the number is to be small, a compromise is necessary between simplicity and precision; this is a matter of subjective judgment. In this connection we must also consider the following. Up to this point we have treated the observations as a population and not as a sample, subject to error, from some larger parent population. If the observations are to be treated as a sample, what we have obtained are the *sample* principal components. It may well happen that even though the s roots of (21.6) (when the elements of Σ_x are sample values) are all different, so that the s principal components of the sample can be found unambiguously, the population values of the $s-t$ smallest roots are equal. If this is so, there is no point in extracting more than t components; the remaining $s-t$ can have any direction as long as they are orthogonal to one another and to the previously extracted components. Methods exist for testing whether the $s-t$ smallest sample roots could have come from a parent population in which these roots are equal (see Kendall, 1957; Lawley and Maxwell, 1963; and references therein). These tests require that the parent distribution be s-variate normal, which is probably seldom true in the context we are considering—that of plant species on units of ground. Transformation of the data to normalize it is necessary before any test can be performed. In any case, for the purpose of ordinating vegetation we usually wish to use comparatively few components, those we can confidently regard as "real."

4. An entirely different form of component analysis, known as Q-type analysis, is sometimes performed. (The usual analysis described in the preceding pages is called R-type analysis.) Formally, the two analyses are identical, but whereas in R-type analysis we began by obtaining the $s \times s$

matrix $\mathbf{\Sigma}_x = (1/n)\mathbf{XX}'$ [see (21.3)], a Q-type analysis starts from the $n \times n$ matrix $(1/s)\mathbf{X}'\mathbf{X}$. The rth diagonal element of this matrix, $(1/s)\sum_{j=1}^{s} x_{jr}^2$, is the variance of the species quantities (all species) within the rth quadrat; the (r, t)th element is the covariance of the species quantities in the rth and tth quadrats. We are, in effect, contemplating an n-dimensional scatter diagram in which there are s points that represent species. Instead of treating the quantity of a species as an attribute of a quadrat, we treat quantity in a quadrat as an attribute of a species. The result is an ordination of species, not quadrats. Its interpretation in ecological contexts is not so intuitively clear as the usual R-type quadrat ordination. Moreover, as Sokal and Sneath (1963) have remarked, "the Q-matrix ... does not have a sampling distribution expected of ordinary correlations." This is because the species within a quadrat are (presumably) not independent, whereas in ordinary R-type analysis we generally assume that the observed quadrats are independent. If $n < s$, that is, if there are fewer quadrats than species, the Q-type matrix is of lower order than the R-type and less unwieldy from a computational point of view.

We can, in any case, perform an R-type analysis by means of a "Q-technique" (Gower, 1966; Orloci, 1967b). This is possible, since a swarm of n points occupies a space of $n-1$ dimensions at most, even when they are plotted in an s-dimensional frame with $s > n$. It follows that, regardless of whether we are doing an R- or Q-type analysis, we can perform the computations with an $n \times n$ or an $s \times s$ matrix. Usually we choose the smaller.

As on page 334, let $\mathbf{X}$ denote the $s \times n$ data matrix;
Now write

$$\mathbf{R} = \mathbf{XX}' \qquad (\mathbf{R} \text{ is an } s \times s \text{ matrix})$$

and

$$\mathbf{Q} = \mathbf{X}'\mathbf{X} \qquad (\mathbf{Q} \text{ is an } n \times n \text{ matrix}).$$

and suppose that $s > n$. Assume that $\mathbf{R}$ and $\mathbf{Q}$ are of full rank.

Let α be a latent root of $\mathbf{R}$ and $\mathbf{a}$, the corresponding latent vector so that

$$\mathbf{Ra} = \alpha\mathbf{a} \qquad \text{or} \qquad \mathbf{XX}'\mathbf{a} = \alpha\mathbf{a}. \qquad (21.7)$$

Similarly, let β and $\mathbf{b}$ be a root and vector of $\mathbf{Q}$ so that

$$\mathbf{Qb} = \beta\mathbf{b} \qquad \text{or} \qquad \mathbf{X}'\mathbf{Xb} = \beta\mathbf{b}. \qquad (21.8)$$

Premultiplying (21.7) by $\mathbf{X}'$ gives

$$\mathbf{X}'\mathbf{X}(\mathbf{X}'\mathbf{a}) = \alpha(\mathbf{X}'\mathbf{a}) \qquad \text{or} \qquad \mathbf{Q}(\mathbf{X}'\mathbf{a}) = \alpha(\mathbf{X}'\mathbf{a}).$$

Comparing this with (21.8), we see that $\alpha = \beta$ and $\mathbf{X}'\mathbf{a} = \mathbf{b}$. Thus the latent roots and vectors of $\mathbf{R}$ (the larger matrix) are directly obtainable from those of $\mathbf{Q}$ (the smaller matrix).

4. Principal Coordinate Analysis

In principal component analysis every quadrat is represented by a point whose coordinates are the quantities of each of the s species in that quadrat. As a result, the distance d_{ij} between the points representing quadrats i and j is

$$d_{ij} = \left[\sum_{t=1}^{s} (x_{ti} - x_{tj})^2 \right]^{1/2},$$

where x_{ti} and x_{tj} denote the quantities of species t in quadrats i and j.

In effect, we treat the Euclidean distance between the plotted points as a measure of the difference between the two quadrats. However, this distance is not necessarily the best measure of a *difference*. If we were given only two quadrats, the ith and jth, say, and were asked to propose a suitable measure of the difference between them, a number of possibilities would spring to mind; for example, we might choose to define the difference m_{ij}, say, as

$$m_{ij} = \sum_{t=1}^{s} |x_{ti} - x_{tj}|,$$

that is, as the sum of the absolute magnitudes of the difference for each species. This is only one among several acceptable definitions for m_{ij}. Another possibility is considered in Section 5.

Having chosen a suitable way of measuring interquadrat difference, the question now is, is it possible to plot points (representing the quadrats) in such a way that the distance between each pair of points is equal to this difference? Gower (1966, 1967b) has shown that it is possible and describes a way of doing it. He calls the process principal coordinate analysis.

What we wish to find are the coordinates of the n points; these coordinates are now *not* equal to the quantities of the species in the quadrats.

Denote by $\mathbf{C}$ the $s \times n$ matrix

$$\mathbf{C} = \{c_{tj}\}, \qquad (t = 1, \ldots, s; j = 1, \ldots, n).$$

The elements of the jth column of $\mathbf{C}$, namely $(c_{1j} c_{2j} \cdots c_{sj})'$ are the coordinates of the jth point; these are the values we shall now find.

The squared distance between the ith and jth points is

$$\sum_{t=1}^{s} (c_{ti} - c_{tj})^2 = \sum_t c_{ti}^2 + \sum_t c_{tj}^2 - 2 \sum_t c_{ti}c_{tj}. \qquad (21.9)$$

This distance^2 is to be *defined* as m_{ij}^2, the square of the difference between the ith and jth quadrats measured in whatever way we choose. These differences are obtained by observation and constitute the data.

Now put

$$\mathbf{C'C = A}.$$

Thus $\mathbf{A}$ is the symmetric $n \times n$ matrix

$$\mathbf{A} = \begin{pmatrix} \sum_t c_{t1}^2 & \sum_t c_{t1}c_{t2} & \cdots & \sum_t c_{t1}c_{tn} \\ \sum_t c_{t2}c_{t1} & \sum_t c_{t2}^2 & \cdots & \sum_t c_{t2}c_{tn} \\ \cdots & \cdots & \cdots & \cdots \\ \sum_t c_{tn}c_{t1} & \sum_t c_{tn}c_{t2} & \cdots & \sum_t c_{tn}^2 \end{pmatrix}$$

$$\equiv \{a_{ij}\}, \text{ say,}$$

where $a_{ij} = \sum_t c_{ti}c_{tj}$. Then, from (21.9)

$$m_{ij}^2 = m_{ji}^2 = a_{ii} + a_{jj} - 2a_{ij}; \qquad m_{ii}^2 = 0. \qquad (21.10)$$

The known values m_{ij}^2 $(i, j = 1, \ldots, n)$ may be written as elements of an $n \times n$ symmetric matrix $\mathbf{M}$. From the given $\mathbf{M}$ we must find $\mathbf{A}$ and then $\mathbf{C}$ whose columns give the coordinates sought. The problem is solved as follows.

Note first that we want the origin of the coordinates to be at the centroid. Therefore we must have $\sum_{j=1}^{n} c_{tj} = 0$ for all t. Then the sum of the elements of the ith row of $\mathbf{C'C = A}$ is

$$\sum_{j=1}^{n} a_{ij} = \sum_{j=1}^{n} \sum_{t=1}^{s} c_{ti}c_{tj} = \sum_{t=1}^{s} c_{ti} \left(\sum_{j=1}^{n} c_{tj} \right) = 0; \qquad (21.11)$$

that is, all the row sums of $\mathbf{A}$ are zero; as also are the column sums, since $\mathbf{A}$ is symmetric.

It will be found that the requirements of (21.10) and (21.11) are fulfilled if we put

$$a_{ij} = -\frac{1}{2} m_{ij}^2 + \frac{1}{2n} \sum_{i=1}^{n} m_{ij}^2 + \frac{1}{2n} \sum_{j=1}^{n} m_{ij}^2 - \frac{1}{2n^2} \sum_i \sum_j m_{ij}^2. \qquad (21.12)$$

Using this formula, $\mathbf{A}$ may be found from $\mathbf{M}$. Now observe that since $\mathbf{A}$ is symmetric we can find an orthogonal matrix $\mathbf{V}$ such that $\mathbf{A} = \mathbf{V}\Lambda\mathbf{V}'$, where Λ is the diagonal matrix whose elements are the latent roots of $\mathbf{A}$, that is, the roots of $|\mathbf{A} - \lambda\mathbf{I}| = 0$.

Now we have that $\mathbf{A} = \mathbf{C}'\mathbf{C} = (\mathbf{V}\Lambda^{1/2})(\Lambda^{1/2}\mathbf{V}')$.

Assume that $s < n$ and recall that $\mathbf{C}'$ and $\mathbf{C}$ have dimensions $n \times s$ and $s \times n$, respectively. Convert them to $n \times n$ matrices Γ' and Γ by adding $n - s$ columns of zeros on the right side of $\mathbf{C}'$ (and, equivalently, $n - s$ rows of zeros at the bottom of $\mathbf{C}$). then

$$\mathbf{C}'\mathbf{C} = \Gamma'\Gamma = (\mathbf{V}\Lambda^{1/2})(\Lambda^{1/2}\mathbf{V}')$$

so that

$$\Gamma = \Lambda^{1/2}\mathbf{V}'.$$

The jth column of Γ is $(c_{1j}\, c_{2j} \cdots c_{sj}\, 0 \cdots 0)'$; these are the coordinates of the jth point in n-space. The jth column of $\mathbf{C}$ is $(c_{1j}\, c_{2j} \cdots c_{sj})'$, which are the coordinates of the points in s-space. Gower (1966) also shows that these coordinates are referred to principal axes; that is, the axes are oriented to meet the requirements listed on page 336.

To demonstrate the process here is a simple numerical example with $s = 2$ and $n = 3$ (see Figure 21.2).

Consider three points, P, Q, and R, in the x_1-x_2 plane. Their coordinates are $(-7, 1)$, $(8, 2)$, and $(-1, -3)$, respectively. The centroid is at the origin. The distances between the points are

$$d_{pq}^2 = 15^2 + 1^2, \qquad d_{pr}^2 = 6^2 + 4^2, \qquad d_{qr}^2 = 9^2 + 5^2,$$
$$d_{pq} = 15.03, \qquad d_{pr} = 7.21, \qquad d_{qr} = 10.30.$$

As a measure of the *differences* between the pairs of points, take the sums of the distances (all treated as positive) parallel with the two coordinate axes. Then

$$m_{pq} = 15 + 1 = 16; \qquad m_{pr} = 6 + 4 = 10; \qquad m_{qr} = 9 + 5 = 14.$$

The matrix $\mathbf{M}$ with elements m^2 is thus

$$\mathbf{M} = \begin{pmatrix} 0 & 256 & 100 \\ 256 & 0 & 196 \\ 100 & 196 & 0 \end{pmatrix}.$$

The elements of $\mathbf{A}$ are found by substituting the elements of $\mathbf{M}$ in (21.12). It will be found that

$$\mathbf{A} = \frac{4}{3}\begin{pmatrix} 43 & -41 & -2 \\ -41 & 67 & -26 \\ -2 & -26 & 28 \end{pmatrix}.$$

Next, we find the latent roots of $\mathbf{A}$ from

$$|\mathbf{A} - \lambda\mathbf{I}| = \lambda(\lambda^2 - 184\lambda + 6400) = 0$$

whence

$$\lambda_1 = 137.4313; \qquad \lambda_2 = 46.5687; \qquad \lambda_3 = 0.$$

Solving $\mathbf{AV} = \mathbf{V}\Lambda$ and normalizing the columns of $\mathbf{V}$ so that $\sum_j v_{ij}^2 = 1$ yields

$$\mathbf{V} = \begin{pmatrix} 0.53788 & 0.61429 & 0 \\ -0.80093 & 0.15867 & 0 \\ 0.26305 & -0.77296 & 0 \end{pmatrix}.$$

The columns of $\mathbf{V}$ are the latent vectors of $\mathbf{A}$. Notice that $\sum_j v_{ij} = 0$; this is a necessary consequence of (21.11). Now put

$$\Gamma = \Lambda^{1/2}\mathbf{V}' = \begin{pmatrix} 6.306 & -9.389 & 3.084 \\ 4.192 & 1.083 & -5.275 \\ 0 & 0 & 0 \end{pmatrix}.$$

The first two elements of the columns of Γ are the coordinates of the new points spaced the desired distances apart. These distances are

$$d'_{pq} = \{[6.306 - (-9.389)]^2 + [4.192 - 1.083]^2\}^{1/2} = 16,$$
$$d'_{pr} = \{[6.306 - 3.084]^2 + [4.192 - (-5.275)]^2\}^{1/2} = 10,$$

and

$$d'_{qr} = \{[-9.389 - 3.084]^2 + [1.083 - (-5.275)]^2\}^{1/2} = 14.$$

Thus, as required,

$$d'_{pq} = m_{pq}, \qquad d'_{pr} = m_{pr}, \qquad \text{and} \qquad d'_{qr} = m_{qr}.$$

The new points are shown in the lower graph of Figure 21.2. The direction of the c_1-axis has been reversed (the values increase from right to left) to make comparison of the graphs easier.

Gower (1968) has shown how to modify the result of a principal coordinate analysis to accommodate a single new data point without doing a complete new analysis. Wilkinson (1970) has demonstrated the procedure.

It is interesting to consider how one might do a principal coordinate analysis if only presence-and-absence data are available; for each quadrat we have merely a list of the species it contains and no observations of the amounts of each species. Let us score 1 for the presence of a species and 0 for its absence; then, if the quadrats are represented by points in an s-dimensional frame, using the scores of the species as coordinates, all the

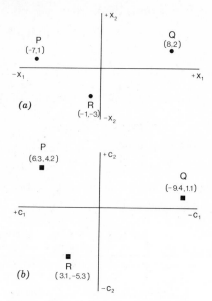

Figure 21.2. (*a*) The quantities of species 1 (x_1-axis) and species 2 (x_2-axis) in three quadrats, *P*, *Q*, and *R*. (*b*) The result of a principal coordinate analysis (the c_2-axis is reversed).

points will be concentrated at the vertices of an *s*-dimensional hypercube. A principal component analysis of the points would be formally possible but not illuminating.

We can, however, use as a measure of the *difference* between any two quadrats the number of species occurring in one or the other but not both; for example, if one quadrat contained species *A*, *B*, *C*, and *D* and another contained *C*, *D*, and *E*, the difference between them would be 3, since the three species *A*, *B*, and *E* occur only once in the pair of quadrats. By a principal coordinate analysis we may plot points, representing the quadrats, so that between every pair the distance is equal to the difference measured in this way.

It is interesting to note that the difference measured, as just described, is equal to the square of the distance between the points when they are plotted at the vertices of a hypercube. This may be shown by an example. Consider two quadrats obtained from vegetation with eight plant species, *A*, ..., *H*. Denoting the presence of a species by a capital letter and its absence by the corresponding lower-case letter, let the quadrats be (*ABcDeFgh*) and (*ABCdEfgh*). If these quadrats were represented in 8-space by points having species-scores as coordinates, the coordinates of the points would be (1, 1, 0, 1, 0, 1, 0, 0) and (1, 1, 1, 0, 1, 0, 0, 0). Obviously the square of the distance between these points is 4, which is the number of species that occur in only one of the quadrats.

5. Nonlinear Data Structures and Catenation

Principal component and principal coordinate analysis and, indeed, nearly all ordination methods used by ecologists until recently, are linear ordinations. They can be relied upon to perform well only if the observed variables (species quantities) have, at least approximately, a linear data structure. Ordination consists in the representation, in some acceptably distortion-free manner in r-space, of a swarm of s-variate data points that in practice can only be *exactly* plotted in s-space ($r \leq s$). We assume that n, the number of points in the swarm, exceeds s. If the data have a linear structure, that is, if the observed data points are scattered throughout an s-dimensional hyperellipse, then linear ordination is appropriate. The axes of the new r-dimensional coordinate frame are straight lines in the old s-dimensional frame. The final step in the ordination, after the directions of these axes have been determined, is the projection of the data points onto the space spanned by the new axes, where they will be scattered throughout an r-dimensional hyperellipse. If it is thought that all "meaningful" (however defined) information in the original data has been retained, and only unwanted "noise" has been discarded, then r can be regarded as the intrinsic dimensionality of the original data.

There is no reason to believe, however, that ecological data sets usually have linear structures. When a set of nonlinear s-variate data, of intrinsic dimensionality r, is plotted in s-space, the swarm of observed points may tend to fall in an r-dimensional hypervolume that is curved and twisted rather than hyperellipsoidal. For such data, linear ordination is inappropriate. An ordination that "flattens out" the data is required. An easily visualized example is that of mapping the *whole* world on a flat sheet of paper. In this case, $s = 3$ and $r = 2$. If a map were constructed by projecting the earth's surface features onto a plane parallel with the plane of the equator, the northern and southern hemispheres would be superimposed and indistinguishable; the resultant map would be uninterpretable without prior knowledge of the true lie of the land. Such a projection is a linear ordination. To present the data clearly, what is required is a method of map projection that, colloquially speaking, opens out and flattens the surface of the globe. Any method that does this can be called nonlinear or curvilinear ordination, or *catenation* (Noy-Meir, 1974).

The salient feature of a catenation is that, starting with a swarm of data points occupying a space of many dimensions, it produces a map of the swarm in a space of fewer dimensions in such a way that the map preserves evidence as to which points were near one another and which distant in the original swarm. To return to the geographical example, a linear ordination of the whole world in the equatorial plane would show Newfoundland and Cape Horn very close to each other; a catenation would not.

Linear ordination is, of course, a special case of catenation. Given a swarm of points of intrinsic dimensionality r, plotted in a space of s dimensions, a catenation will extract (i.e., lead to the derivation of) r new axes, and the points will be found to lie in the space spanned by them. If the original data are nonlinear the r new axes will be curved and the space they span will be a curved hypersurface; if the data are linear the new axes will be straight, the space they span will be flat (or hyperflat) and the catenation in this case will merely be an ordinary linear ordination.

Figure 21.3 shows a simple (artificial) ecological example, with $s = 2$

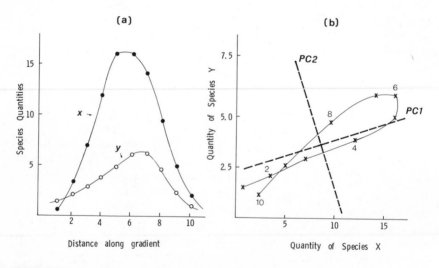

Figure 21.3. (a) The quantities of species X (black dots) and species Y (white dots) in 10 equidistant quadrats along a gradient. (b) The same data replotted. The points representing quadrats 2, 4, 6, 8, and 10 on the gradient are so labeled. The broken lines are the principal component axes $PC1$ and $PC2$. The curve linking the points is the single curved axis (or catena) that a catenation would yield.

and $r = 1$. Envisage a two-species community on an environmental gradient, and suppose observations on quadrats ranged along the gradient yield the results shown in Figure 21.3a; the graph shows the quantities of species X and Y in each of a row of ten equidistant quadrats. The same data have been replotted in Figure 21.3b. This graph is a 2-space containing 10 data points, and we wish to ordinate (or catenate) the data in a space of one dimension, that is, along a straight line. If we supposed (wrongly) that the raw data had a linear structure or, equivalently, that the points in Figure 21.3b should be regarded as scattered throughout a narrow ellipse (as, indeed, they appear to be) we might do a principal component analysis. The result is clear from the figure; the $s = 2$ principal axes extracted by the analysis are the broken lines; ordination in 1-space is obtained by projecting the points onto the first principal axis, PC1. Obviously when this is done, the relative proximity of the original points, in this case proximity on the ground, is lost.

Now suppose (rightly) that the raw data are nonlinear and have intrinsic dimensionality $r = 1$. A catenation (if it can be done) will lead to extraction of the single axis shown as the curve linking the points in Figure 21.3b. The order of the points on the ground is preserved. The fact that their intrinsic dimensionality is one is also clear, from the fact that the points all lie on the curve (recall that this is an artificial example, free of "noise"). Catenation in 1-space is obtained by "straightening out" the curved axis; that is, by redrawing it as a straight line with the data points marked on it in the sequence, and with the spacing, that they had on the curve.

Three comments should now be interpolated before the discussion is resumed.

1. Data with nonlinear structures are common in ecology. They occur whenever the measured variables (species quantities) are nonlinearly dependent on unobserved variables (underlying environmental factors). There is no reason to limit catenation methods to data obtained from habitats encompassing environmental gradients. Data from a gradient are useful for illustrative purposes, as in Figure 21.3, simply because it is often reasonable to assume an approximately linear relationship between the distance of a point (= quadrat) from the start of a gradient and the level of the environmental factor that varies along the gradient. Then curves such as those in Figure 21.3a in effect show the nonlinear relationships between the species and the environmental factor.

2. Noy-Meir (1974), who introduced the methods of catenation to ecologists, proposed the word *catena* for a curved axis obtained by catenation. However, as he points out, a catena in this sense bears no

necessary relation to a soil catena. A catena in Noy-Meir's sense may or may not correspond with an environmental gradient and hence, possibly, with a soil catena.

3. The term linear ordination as here used is to be contrasted with "catenation," that is, nonlinear ordination. It does not mean a one-dimensional, as opposed to a two- or more-dimensional, representation of data.

We now consider how a catenation may be performed. Many methods are possible and Noy-Meir (1974) has compared several of them in ecological contexts. One of the best appears to be that due to Carroll (see Shepard and Carroll, 1966) called, by him, *parametric mapping*. Noy-Meir prefers the name *continuity analysis*. As briefly as possible (which cannot be very brief) the procedure is as follows.

The raw data, which are to be thought of as the coordinates of n data points in s-space, are in the form* of an $s \times n$ matrix $\mathbf{X}$; x_{ij} denotes the amount of species i in quadrat j; equivalently, x_{ij} is the coordinate of the jth point on the x_i-axis. We wish to find an $r \times n$ matrix $\mathbf{Z}$, with $r \le s$, in which z_{gj} is the coordinate of the jth point on the new z_g-axis (which is the gth catena). Here r is either the intrinsic dimensionality of the data or some smaller number adjudged adequate for summarizing them. The elements of $\mathbf{Z}$ are to be such as to maximize the "continuity" of the relation between themselves and the x values which are the raw data.

It remains to consider how continuity is to be measured and then maximized. Carroll (in Shepard and Carroll, 1966) derived a measure of continuity, K, as follows. First put

$$d_{jk}^2 = \sum_{i=1}^{s} (x_{ij} - x_{ik})^2$$

and

$$D_{jk}^2 = \sum_{g=1}^{r} (z_{gj} - z_{gk})^2.$$

Thus d_{jk} is the distance between the jth and kth data points in the old, s-dimensional coordinate frame, and D_{jk} is the distance between these points in the new, r-dimensional frame. Next, put

$$\delta^2(s, r) = \sum_{j \ne k} \frac{d_{jk}^2}{D_{jk}^2} \, w_{jk}. \tag{21.13}$$

The summation is over all $n(n-1)/2$ possible pairs of points; w_{jk} is a

---

*The data matrix discussed here is the transpose of that considered by Shepard and Carroll (1966) and by Noy-Meir (1974). I have transposed their matrix, and changed their symbols, for the sake of uniformity in this chapter.

weighting factor that is a monotonically decreasing function of $|z_{gj} - z_{gk}|$ for every g.

The quantity $\delta^2(s, r)$ is to be interpreted as follows. Imagine that $s = r = 1$. Let the data points be ranked so that $z_{11} \leq z_{12} \leq \cdots \leq z_{1n}$. (It does *not* follow that $x_{11} \leq x_{12} \leq \cdots \leq x_{1n}$.) Put

$$w_{jk} = \begin{cases} 1 & \text{if } k = j+1. \\ 0 & \text{otherwise.} \end{cases} \tag{21.14}$$

That is, $w_{jk} \neq 0$ only for adjacent values of z_1. Then for this case (21.13) becomes

$$\delta^2(1, 1) = \sum_{j=1}^{n-1} \frac{d_{j, j+1}^2}{D_{j, j+1}^2} w_{j, j+1} = \sum_{j=1}^{n-1} \frac{(x_{1, j+1} - x_{1j})^2}{(z_{1, j+1} - z_{1j})^2} . \tag{21.15}$$

Here the denominator of the summand is the distance^2 between the jth and $(j+1)$th points along the z_1-axis or catena (which, since $r = 1$, is the only catena); these points are adjacent on the catena. The numerator is the distance^2 between the corresponding points on the x_1-axis (which, since $s = 1$, is the only x-axis); on this axis, the points are not necessarily adjacent. Equivalently, the numerator is the difference^2 between the amounts of the first (and only) species in the jth and $(j+1)$th quadrats.

In this special case, with $s = r = 1$, we therefore see that $\delta^2(s, r)$ is a sum of squares each of whose (unsquared) terms is the change in x associated with a small change (from one point to the next on the catena) in z. Thus a low value of $\delta^2(1, 1)$ corresponds to a high continuity of the x versus z relationship; $\delta^2(1, 1)$ could be used to measure (inversely) the continuity of this relationship. However, since we seek to maximize continuity by varying the z values, $\delta^2(s, r)$ must be standardized in some way; otherwise we could make its value as small as desired merely by making the z's sufficiently big. Standardization is achieved by replacing $\delta^2(s, r)$ by

$$\frac{\delta^2(s, r)}{\left\{ \sum \frac{1}{D_{jk}^2} \right\}^2} . \tag{21.16}$$

We now return to the general case, with $s > 1$, $r \geq 1$, and $\delta^2(s, r)$ defined as in (21.13); its standardized form is given by (21.16). One final modification is necessary to yield a useful continuity index, since in the general case one cannot define w_{jk} as in (21.14); this is because the notion of adjacency is meaningless for irregularly spaced points in more than one dimension. As an alternative definition for w_{jk}, Carroll recommends (on pragmatic grounds)

$$w_{jk} = \frac{1}{D_{jk}^2} .$$

Thus we finally have, as an inverse measure of continuity in the general case

$$K(s, r) = \frac{\sum\limits_{j \ne k} \dfrac{d_{jk}^2}{D_{jk}^4}}{\left\{ \sum \dfrac{1}{D_{jk}^2} \right\}^2}.$$ (21.17)

To derive the desired values for the elements of $\mathbf{Z}$, that is, to perform the catenation, these elements must be adjusted until $K(s, r)$ is minimized. This is achieved by an iterative process for which Carroll has devised the program PARAMAP. Examples of its use in catenating ecological data have been given by Noy-Meir (1974) who tested it on data from tropical forest, temperate deciduous forest, temperate mixed forest, and semi-arid Australian vegetation. Details are given in his paper.

22

Canonical Variate Analysis and Multiple Discriminant Analysis

1. Introduction

Chapter 21 dealt with ways of simplifying and condensing the raw data obtained from a sample survey of a many-species community of sessile or sedentary organisms. Each quadrat (or other sampling unit) yielded a vector of variate values (the amounts of the several species) and it was shown how, by various techniques, we can find a few combinations of the variates that account for nearly all the variance of the data.

However, observations of the quantities of the different species in a sampling unit from a community may constitute only one of the sets of observations an ecologist makes. He may, in addition, measure several environmental variables; for example, in studying vegetation, he may measure the water-holding capacity of the soil, the quantity of soil organic material, the slope of the ground, the depth of the soil, and various other measurable properties. These environmental data constitute a second set of observations, qualitatively different from the first set. We are naturally led to inquire into the interrelationships among the two *sets* of variates. If the amounts of only one plant species had been recorded and measurements made on only one feature of the environment, it would be straightforward to measure the correlation between them. What we now require is an analogous method appropriate to multivariate data. Canonical variate analysis is such a method.

As in principal component analysis, we are concerned with linear combinations of the observations that have certain desirable properties. Before explaining how these combinations are arrived at, we describe the "input" and "output" of a canonical variate analysis and consider what the analysis achieves. A preview of the output (in symbolic form) will show what is to be gained by examining the relationships among linear combinations of the original variates rather than among the raw data and

351

will provide a motive for the matrix manipulations needed to attain the result.

The input to the analysis is as follows: suppose p environmental variates, measured on each sample unit, yield a vector of observations $(x_1 \, x_2 \cdots x_p)$ and suppose the amounts of q species yield the vector $(y_{p+1} \, y_{p+2} \cdots y_{p+q})$. It is convenient to number the species from $p+1$ to $p+q$ rather than from 1 to q. Also, it is assumed that $p \leq q$; if the number of species is less than the number of measured environmental factors, we need only reverse the labeling of the two sets of data.

Now write $(\mathbf{x}' \, \mathbf{y}')$ for the $(p+q)$-element row vector representing all the observations on any one unit. The $(p+q) \times (p+q)$ covariance matrix of all the data is

$$\mathbf{\Sigma} = \frac{1}{n}\binom{\mathbf{x}}{\mathbf{y}}(\mathbf{x}'\mathbf{y}) = \left(\begin{array}{ccc:ccc}
\sigma_{11} & \cdots & \sigma_{1p} & \sigma_{1,p+1} & \cdots & \sigma_{1,p+q} \\
\cdot & \cdots & \cdot & \cdot & \cdots & \cdot \\
\sigma_{p1} & \cdots & \sigma_{pp} & \sigma_{p,p+1} & \cdots & \sigma_{p,p+q} \\
\hdashline
\sigma_{p+1,1} & \cdots & \sigma_{p+1,p} & \sigma_{p+1,p+1} & \cdots & \sigma_{p+1,p+q} \\
\cdot & \cdots & \cdot & \cdot & \cdots & \cdot \\
\sigma_{p+q,1} & \cdots & \sigma_{p+q,p} & \sigma_{p+q,p+1} & \cdots & \sigma_{p+q,p+q}
\end{array}\right)$$

$$= \begin{pmatrix} \mathbf{\Sigma}_{11} & \mathbf{\Sigma}_{12} \\ \mathbf{\Sigma}_{21} & \mathbf{\Sigma}_{22} \end{pmatrix}. \tag{22.1}$$

We have partitioned $\mathbf{\Sigma}$ into four parts. The $p \times p$ submatrix $\mathbf{\Sigma}_{11}$ is the covariance matrix of the x's (the environmental variates); similarly, the $q \times q$ submatrix $\mathbf{\Sigma}_{22}$ is the covariance matrix of the y's (the species quantities). And the elements of the $p \times q$ submatrix $\mathbf{\Sigma}_{12}$ denote covariances of the form $\text{cov}(x, y)$. Thus $\sigma_{i,p+j}$ (with $i = 1, \ldots, p; j = 1, \ldots, q$) is $\text{cov}(x_i, y_{p+j})$ or the covariance of the ith environmental variate and the quantity of the $(p+j)$th species. Since $\mathbf{\Sigma}$ is symmetric, we see that $\mathbf{\Sigma}_{21}$ is simply the transpose of $\mathbf{\Sigma}_{12}$ or $\mathbf{\Sigma}_{21} = \mathbf{\Sigma}'_{12}$.

Our object now is to find p linear combinations of the x's, say,

$$\xi_i = a_{i1}x_1 + a_{i2}x_2 + \cdots + a_{ip}x_p \qquad (i = 1, \ldots, p),$$

and q linear combinations of the y's, say,

$$\eta_j = b_{j1}y_{p+1} + b_{j2}y_{p+2} + \cdots + b_{jq}y_{p+q} \qquad (j = 1, \ldots, q),$$

such that the $(p+q) \times (p+q)$ correlation matrix of the derived variates (the ξ's and η's) has the form

$$
\mathbf{P} = \binom{\xi}{\eta}(\xi'\eta') = \left\{
\begin{array}{ccccccccccc}
1 & 0 & \cdots & 0 & \rho_1 & 0 & \cdots & 0 & & & \\
0 & 1 & \cdots & 0 & 0 & \rho_2 & \cdots & 0 & & & \\
 & & & & & & & & \mathbf{O}_{p \times (q-p)} & & \\
0 & 0 & \cdots & 1 & 0 & 0 & \cdots & \rho_p & & & \\
\rho_1 & 0 & \cdots & 0 & 1 & 0 & \cdots & 0 & & & \\
0 & \rho_2 & \cdots & 0 & 0 & 1 & \cdots & 0 & & & \\
 & & & & & & & & \mathbf{O}_{p \times (q-p)} & & \\
0 & 0 & \cdots & \rho_p & 0 & 0 & \cdots & 1 & & & \\
 & & & & & & & & 1 & 0 & \cdots & 0 \\
 & \mathbf{O}_{(q-p) \times p} & & & \mathbf{O}_{(q-p) \times p} & & & & 0 & 1 & \cdots & 0 \\
 & & & & & & & & 0 & 0 & \cdots & 1
\end{array}
\right\}
$$

(22.2)

(The subscripts of the null submatrices show their dimensions.)

The variates $\xi_1, \ldots, \xi_p$ and $\eta_1, \ldots, \eta_q$ are known as canonical variates. These variates have the following properties as shown by their correlation matrix $\mathbf{P}$.

1. All the ξ's are uncorrelated with one another.
2. All the η's are uncorrelated with one another.
3. The pair of canonical variates ξ_i and η_i have correlation ρ_i ($i = 1, \ldots, p$) but all other correlations between the ξ's and η's are zero. The ρ's are known as canonical correlations.

It is seen that by using the canonical variates the dependencies between the two original sets of data are reduced to their simplest possible torm. The within-set correlations are all zero. The between-set correlations have been maximized (as shown below) between p pairs of variates, one from each set, and have been reduced to zero between all other pairs.

2.   The Derivation of Canonical Variates

Consider the first pair of canonical variates ξ_1 and η_1. In what follows we shall, for clarity, omit the subscript 1 and write simply

$$\xi_1 = \xi = a_1 x_1 + a_2 x_2 + \cdots + a_p x_p = \mathbf{a}'\mathbf{x}$$

and

$$\eta_1 = \eta = b_1 y_{p+1} + b_2 y_{p+2} + \cdots + b_q y_{p+q} = \mathbf{b}'\mathbf{y}.$$

We now find the correlation between ξ and η, namely,

$$\frac{\mathrm{cov}(\xi, \eta)}{\{\mathrm{var}(\xi)\mathrm{var}(\eta)\}^{1/2}}.$$

First it is necessary to obtain expressions for the variances and covariance. Clearly

$$\mathrm{var}(\xi) = \mathrm{var}(a_1 x_1 + a_2 x_2 + \cdots + a_p x_p)$$

$$= \sum_{i=1}^{p} a_i^2 \, \mathrm{var}(x_i) + \sum_{i \neq j} a_i a_j \, \mathrm{cov}(x_i, x_j).$$

In terms of the symbols in (22.1) we then have

$$\mathrm{var}(\xi) = \sum_{i=1}^{p} \sum_{j=1}^{p} a_i a_j \sigma_{ij}$$

$$= \mathbf{a}'\boldsymbol{\Sigma}_{11}\mathbf{a}.$$

Similarly

$$\mathrm{var}(\eta) = \sum_{i=1}^{q} \sum_{j=1}^{q} b_i b_j \sigma_{p+i,\,p+j} = \mathbf{b}'\boldsymbol{\Sigma}_{22}\mathbf{b}$$

and

$$\mathrm{cov}(\xi, \eta) = \sum_{i=1}^{p} \sum_{j=1}^{q} a_i b_j \sigma_{i,\,p+j} = \mathbf{a}'\boldsymbol{\Sigma}_{12}\mathbf{b} = \mathbf{b}'\boldsymbol{\Sigma}_{12}'\mathbf{a}.$$

The correlation between ξ and η is therefore

$$\rho(\mathbf{a}, \mathbf{b}) = \frac{\mathbf{a}'\boldsymbol{\Sigma}_{12}\mathbf{b}}{\{\mathbf{a}'\boldsymbol{\Sigma}_{11}\mathbf{a} \cdot \mathbf{b}'\boldsymbol{\Sigma}_{22}\mathbf{b}\}^{1/2}}.$$

We have written $\rho(\mathbf{a}, \mathbf{b})$ for the correlation to emphasize that it is a function of $\mathbf{a}$ and $\mathbf{b}$. We now determine what the elements of $\mathbf{a}$ and $\mathbf{b}$ must be for ξ and η to have maximum correlation (positive or negative) and for them to have arbitrary preassigned variances, v_ξ and v_η, say (this is equivalent to choosing scales for ξ and η). In other words, we wish to maximize $|\rho(\mathbf{a},\mathbf{b})|$ subject to the constraints

$$\mathbf{a}'\boldsymbol{\Sigma}_{11}\mathbf{a} = v_\xi \quad \text{and} \quad \mathbf{b}'\boldsymbol{\Sigma}_{22}\mathbf{b} = v_\eta.$$

Write

$$f(\mathbf{a}, \mathbf{b}) = \mathbf{a}'\boldsymbol{\Sigma}_{12}\mathbf{b} - \frac{\lambda}{2}(\mathbf{a}'\boldsymbol{\Sigma}_{11}\mathbf{a} - v_\xi) - \frac{\mu}{2}(\mathbf{b}'\boldsymbol{\Sigma}_{22}\mathbf{b} - v_\eta),$$

where $\lambda/2$ and $\mu/2$ are undetermined (Lagrange) multipliers. Differentiating $f(\mathbf{a}, \mathbf{b})$ with respect to the elements of $\mathbf{a}$ gives

$$\frac{\partial f(\mathbf{a}, \mathbf{b})}{\partial a_i} = \sum_{j=1}^{q} b_j \sigma_{i,\,p+j} - \frac{\lambda}{2} \cdot 2 \sum_{j=1}^{p} a_j \sigma_{ij} \qquad (i = 1, \ldots, p).$$

Similarly, differentiating with respect to the elements of $\mathbf{b}$ gives

$$\frac{\partial f(\mathbf{a}, \mathbf{b})}{\partial b_j} = \sum_{i=1}^{p} a_i \sigma_{i,\,p+j} - \frac{\mu}{2} \cdot 2 \sum_{i=1}^{q} b_i \sigma_{p+i,\,p+j} \qquad (j = 1, \ldots, q).$$

Therefore

$$\frac{\partial f(\mathbf{a}, \mathbf{b})}{\partial \mathbf{a}} = \begin{pmatrix} \dfrac{\partial f}{\partial a_1} \\ \dfrac{\partial f}{\partial a_2} \\ \vdots \\ \dfrac{\partial f}{\partial a_p} \end{pmatrix} = \mathbf{\Sigma}_{12}\mathbf{b} - \lambda\mathbf{\Sigma}_{11}\mathbf{a}$$

and likewise

$$\frac{\partial f(\mathbf{a}, \mathbf{b})}{\partial \mathbf{b}} = \mathbf{\Sigma}'_{12}\mathbf{a} - \mu\mathbf{\Sigma}_{22}\mathbf{b}.$$

To find the values of the a's and b's that will maximize $|\rho(\mathbf{a}, \mathbf{b})|$ we now set these partial derivatives equal to zero and obtain the equations

$$\mathbf{\Sigma}_{12}\mathbf{b} - \lambda\mathbf{\Sigma}_{11}\mathbf{a} = \mathbf{0} \tag{22.3}$$

(both sides are p-element column vectors) and

$$\mathbf{\Sigma}'_{12}\mathbf{a} - \mu\mathbf{\Sigma}_{22}\mathbf{b} = \mathbf{0} \tag{22.4}$$

(both sides are q-element column vectors). Premultiply (22.3) by $\mathbf{a}'$ and (22.4) by $\mathbf{b}'$ to give the equations (in scalars)

$$\begin{aligned} \mathbf{a}'\mathbf{\Sigma}_{12}\mathbf{b} - \lambda\mathbf{a}'\mathbf{\Sigma}_{11}\mathbf{a} &= 0, \\ \mathbf{b}'\mathbf{\Sigma}'_{12}\mathbf{a} - \mu\mathbf{b}'\mathbf{\Sigma}_{22}\mathbf{b} &= 0. \end{aligned} \tag{22.5}$$

If we now scale the variates so that each has unit variance, that is, so that $\mathbf{a}'\mathbf{\Sigma}_{11}\mathbf{a} = \mathbf{b}'\mathbf{\Sigma}_{22}\mathbf{b} = 1$, we see that $\rho(\mathbf{a}, \mathbf{b})$ becomes $\mathbf{a}'\mathbf{\Sigma}_{12}\mathbf{b} = \mathbf{b}'\mathbf{\Sigma}'_{12}\mathbf{a}$ and, from (22.5), that

$$\rho(\mathbf{a}, \mathbf{b}) = \lambda = \mu.$$

Now premultiply the $p \times 1$ matrix equation (22.3) by the $q \times p$ matrix

$\Sigma'_{12}\Sigma_{11}^{-1}$ to give the $q \times 1$ equation

$$\Sigma'_{12}\Sigma_{11}^{-1}\Sigma_{12}\mathbf{b} - \lambda\Sigma'_{12}\mathbf{a} = \mathbf{0}. \tag{22.6}$$

From (22.4) and noting that $\lambda = \mu$ we have

$$\Sigma'_{12}\mathbf{a} = \lambda\Sigma_{22}\mathbf{b}. \tag{22.7}$$

Substituting from (22.7) into (22.6) and premultiplying by Σ_{22}^{-1} gives the $q \times 1$ equation

$$(\Sigma_{22}^{-1}\Sigma'_{12}\Sigma_{11}^{-1}\Sigma_{12} - \lambda^2\mathbf{I})\mathbf{b} = \mathbf{0}. \tag{22.8}$$

Similarly, (22.4) leads to the $p \times 1$ equation

$$(\Sigma_{11}^{-1}\Sigma_{12}\Sigma_{22}^{-1}\Sigma'_{12} - \lambda^2\mathbf{I})\mathbf{a} = \mathbf{0}. \tag{22.9}$$

For (22.8) and (22.9) to have nontrivial solutions, their determinants must vanish; that is, it is necessary that

$$|\Sigma_{22}^{-1}\Sigma'_{12}\Sigma_{11}^{-1}\Sigma_{12} - \lambda^2\mathbf{I}| = 0 \tag{22.10}$$

and

$$|\Sigma_{11}^{-1}\Sigma_{12}\Sigma_{22}^{-1}\Sigma'_{12} - \lambda^2\mathbf{I}| = 0. \tag{22.11}$$

Equation (22.10) is of degree q in λ^2 and (22.11) of degree p in λ^2. It can be shown (e.g., see Kendall and Stuart, 1966) that the nonzero roots of the two equations are identical; they are the eigenvalues of the matrices $\Sigma_{22}^{-1}\Sigma'_{12}\Sigma_{11}^{-1}\Sigma_{12}$ and $\Sigma_{11}^{-1}\Sigma_{12}\Sigma_{22}^{-1}\Sigma'_{12}$. There are p such nonzero roots at most (recall that $p \le q$). Thus (22.10) [or (22.11)] may be solved to yield the roots $\lambda_1^2, \lambda_2^2, \ldots, \lambda_p^2$. Let them be ranked in order of decreasing size so that $\lambda_1^2 > \lambda_2^2 > \cdots > \lambda_p^2$.

Substitution of the numerical value of λ_1^2 into (22.8) and (22.9) enables us to solve these equations to give the required values of the a's and b's. Then the first pair of canonical variates is given by

$$\xi = \sum_{i=1}^{p} a_i x_i, \qquad \eta = \sum_{j=1}^{q} b_j y_{p+j}.$$

The 2nd, $\ldots$, pth pairs of canonical variates are found similarly by substituting $\lambda_2^2, \ldots, \lambda_p^2$, respectively into (22.8) and (22.9).

It may be proved (e.g., see Morrison, 1967) that, when $i \ne j$, $\text{cov}(\xi_i, \eta_j) = 0$. As a result, the correlation matrix of the ξ's and η's is as shown in (22.2).

An alternative derivation of the canonical variate equations has been given by Cramer (1973).

3. A Numerical Example of Canonical Variate Analysis

It is worthwhile to demonstrate the method with a deliberately simplified artificial numerical example.

Imagine that 10 sample units of vegetation are examined, on each of which two environmental factors are measured to give x_1 and x_2 and the quantities of two species are recorded to give y_3 and y_4. Let the results be as shown in the Table 22.1. The left half shows the raw data and the right half the canonical variate values for each unit, which we now derive. [Note that the collection of 10 units is treated as a population, not as a sample, so that we put, for instance, $\text{var}(x_1) = (\frac{1}{10}) \sum_{i=1}^{10} (x_{1i} - \bar{x}_1)^2$.] The covariance matrix of the x's and y's is

$$\begin{pmatrix} \Sigma_{11} & \Sigma_{12} \\ \Sigma'_{12} & \Sigma_{22} \end{pmatrix} = \begin{pmatrix} 82.60 & 23.20 & -41.20 & -48.20 \\ & 7.04 & -12.52 & -13.38 \\ & & 24.01 & 23.24 \\ & & & 30.36 \end{pmatrix}.$$

TABLE 22.1

	Raw Data				Canonical Variate Values		
Environmental Variates		Species Quantities		First Pair		Second Pair	
x_1	x_2	y_3	y_4	ξ_1	η_1	ξ_2	η_2
3	2	17	19	8.22	34.02	-3.88	0.11
4	4	14	18	14.44	30.13	-9.76	-2.00
8	3	18	16	15.83	32.34	-2.32	3.78
12	6	10	18	27.66	26.13	-8.64	-6.00
14	6	12	13	29.66	23.65	-6.64	0.44
16	7	10	12	34.27	20.75	-8.08	-0.67
21	8	7	11	41.88	16.86	-6.52	-2.78
24	8	9	8	44.88	16.17	-3.52	1.89
28	9	3	5	51.49	7.48	-2.96	-1.45
30	11	3	2	58.71	4.79	-7.84	1.22

Notice that x_1 and x_2 are positively correlated, as are y_3 and y_4; but all the between-set correlations (x_1 and y_3; x_1 and y_4; x_2 and y_3; x_2 and y_4) are negative.

Next, evaluate Σ_{11}^{-1} and Σ_{22}^{-1}:

$$\Sigma_{11}^{-1} = \frac{1}{|\Sigma_{11}|} \begin{pmatrix} \sigma_{22} & -\sigma_{21} \\ -\sigma_{12} & \sigma_{11} \end{pmatrix} = \frac{1}{43.264} \begin{pmatrix} 7.04 & -23.20 \\ -23.20 & 82.60 \end{pmatrix}.$$

Likewise,

$$\Sigma_{22}^{-1} = \frac{1}{188.846}\begin{pmatrix} 30.36 & -23.24 \\ -23.24 & 24.01 \end{pmatrix}.$$

Evaluating $\Sigma_{11}^{-1}\Sigma_{12}\Sigma_{22}^{-1}\Sigma_{12}'$ and substituting in (22.11) gives the determinantal equation

$$\begin{vmatrix} 0.7004 - \lambda^2 & 0.1037 \\ 0.9333 & 0.6145 - \lambda^2 \end{vmatrix} = 0$$

or

$$\lambda^4 - 1.3148\lambda^2 + 0.3336 = 0$$

with roots

$$\lambda_1^2 = 0.9714 \qquad \text{and} \qquad \lambda_2^2 = 0.3434.$$

(It will be found that (22.10) has the same roots.) Using the larger root, λ_1^2, and writing a_{11}, a_{12}, b_{11}, and b_{12} for the coefficients of the first pair of canonical variates, (22.9) becomes

$$\begin{pmatrix} -0.2710 & 0.1037 \\ 0.9333 & -0.3569 \end{pmatrix}\begin{pmatrix} a_{11} \\ a_{12} \end{pmatrix} = \mathbf{0}.$$

One of the a's must be chosen arbitrarily. It is convenient to put $a_{11} = 1$. Then $a_{12} = 2.61$. Similarly (22.8) becomes

$$\begin{pmatrix} -0.3152 & 0.3519 \\ 0.2802 & -0.3128 \end{pmatrix}\begin{pmatrix} b_{11} \\ b_{12} \end{pmatrix} = \mathbf{0}$$

with a solution $b_{11} = 1$ and $b_{12} = 0.90$. The first pair of canonical variates is therefore given by

$$\xi_1 = a_{11}x_1 + a_{12}x_2 = x_1 + 2.61x_2,$$
$$\eta_1 = b_{11}y_3 + b_{12}y_4 = y_3 + 0.90y_4. \tag{22.12}$$

The second pair of canonical variates ξ_2 and η_2 is obtained by substituting the smaller root $\lambda_2^2 = 0.3434$ in (22.8) and (22.9). It is found that

$$\xi_2 = a_{21}x_1 + a_{22}x_2 = x_1 - 3.44x_2,$$
$$\eta_2 = b_{21}y_3 + b_{22}y_4 = y_3 - 0.89y_4.$$

Each of the original sample points (x_1, x_2, y_3, y_4) may now be expressed in terms of the canonical variates (ξ_1, ξ_2, η_1, η_2). Their covariance matrix is as follows:

$$\begin{pmatrix} \text{var}(\xi_1) & \text{cov}(\xi_1, \xi_2) & \text{cov}(\xi_1, \eta_1) & \text{cov}(\xi_1, \eta_2) \\ & \text{var}(\xi_2) & \text{cov}(\xi_2, \eta_1) & \text{cov}(\xi_2, \eta_2) \\ & & \text{var}(\eta_1) & \text{cov}(\eta_1, \eta_2) \\ & & & \text{var}(\eta_2) \end{pmatrix}$$

$$= \begin{pmatrix} 251.661 & 0 & -148.354 & 0 \\ & 6.293 & 0 & 3.800 \\ & & 90.030 & 0 \\ & & & 6.683 \end{pmatrix}.$$

The correlation matrix is

$$\begin{pmatrix} 1 & 0 & -0.9856 & 0 \\ & 1 & 0 & 0.5860 \\ & & 1 & 0 \\ & & & 1 \end{pmatrix}.$$

Notice that the correlation between ξ_1 and η_1 is

$$\rho_{\xi_1 \eta_1} = -0.9856, \qquad \text{whence} \qquad \rho_{\xi_1 \eta_1}^2 = \lambda_1^2 = 0.9714.$$

Likewise,

$$\rho_{\xi_2 \eta_2} = 0.5860, \qquad \text{whence} \qquad \rho_{\xi_2 \eta_2}^2 = \lambda_2^2 = 0.3434.$$

These are the *only* nonzero correlations and the purpose of the analysis has been achieved. Figure 22.1 demonstrates the result graphically; Figures 22.1a and b show the relationships between x_1 and x_2, and between y_3 and y_4, respectively; as may be seen, there is fairly strong positive correlation between the x's, and much less pronounced positive correlation between the y's. The first pair of canonical variates, ξ_1 and η_1, for each of the 10 sampling units, are plotted in Figure 22.1c and show very strong negative correlation. The implication is that the weighted sum of the two "environmental variates," x_1 and x_2, weighted in accordance with (22.12), and the weighted sum of the "species quantities," y_3 and y_4, likewise weighted in accordance with (22.12), are the linear combinations of the x's and of the y's that are most strongly correlated.

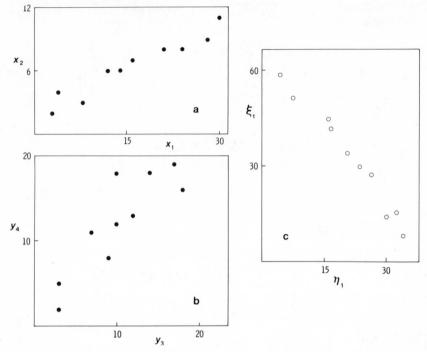

Figure 22.1. To illustrate the numerical example of canonical variate analysis in the text; (*a*) and (*b*) show, respectively, the relationship between x_1 and x_2, and between y_3 and y_4, (*c*) shows the relationship between the first pair of canonical variates ξ_1 and η_1.

4. Possible Ecological Applications of Canonical Analysis

In the foregoing discussion of canonical variate analysis it has been assumed throughout that the observed variate values were to be treated as constituting a *population*, not a sample. The problem of testing the significance of the canonical correlations did not therefore arise. For an account of the sampling theory and for ways of testing for the independence between two sets of variates see, for example, Anderson (1958), Morrison (1967), and Tatsuoka (1971). However, in studies of ecological communities we are generally on safer ground if we treat the data at hand as material to be studied for its own sake. Attempts to make inferences about putative "parent populations" often lead to a welter of statistical difficulties and to the proliferation of unconvincing significance tests.

Canonical variate analysis has not yet been widely used by ecologists; research workers in psychology and education have been much quicker to take advantage of it. Thus educationists use it to study the correlation between the grades and marks achieved by students, first at school and later in a university (Barnett and Lewis, 1963), or between the marks gained by children in a set of tests of their reading ability and another set of their arithmetic ability [Hotelling (1936), quoted in Kendall and Stuart (1966)]. An example of an application in psychology is given by Morrison (1967); sets of test marks designed to test adult intelligence were correlated with another set of variates measuring the subjects' accumulated experience.

Ecologists are often confronted with data that consist of two different sets of observations. Thus, as described above, one data set may consist of lists of the amount of each species in each of a number of sampling units, while the other set consists of records of environmental variables measured in the same units. For example, Barkham and Norris (1970), in a study of ground vegetation in English beech woodlands, obtained the canonical correlations between species quantities (percentage cover for 25 plant species) and soil properties (structure, stoniness, consistence, degree of decomposition of the litter, and the like) in a sample of quadrats arranged in a lattice.

Another ecological context in which knowledge of canonical correlations might give useful insights is that of the interrelationship between species at two different trophic levels in an animal community. One data set would consist of lists of numbers of individuals of each herbivore species in each of many sampling units, and the other data set of corresponding lists of the number of individuals of each primary carnivore species in the same units.

5. Multiple Discriminant Analysis

The foregoing discussion dealt with canonical variate analysis *sensu stricto*, sometimes called (more precisely) canonical correlation analysis. The term "canonical variate analysis" is often applied also to multiple discriminant analysis. The mathematical procedures of the two analyses are identical but the objectives are entirely different. As explained above, the purpose of canonical correlation analysis is to explore the interrelationship between two qualitatively different sets of variates. Multiple discriminant analysis, however, operates on several, say k (with $k \geq 2$) sets of qualitatively similar variates. In ecological contexts, each set usually consists of the amount of each species in each unit of a sample; and each of the k sets comes from a different population, for instance, a different geographical locality. The purpose of multiple discriminant analysis is to find what linear combinations of the variates will give the greatest differences among set means.

This is achieved by performing the same calculations as described above for canonical correlation analysis with the following modification. The environmental variates are replaced by dummy variates that show which of the different sets a given sampling unit comes from. If there are k sets, then $k-1$ dummy variates are required. Call them x_i with $i = 1, \ldots, k-1$. Then for a sampling unit from the jth set (for $j = 1, \ldots, k-1$) the dummy variates are

$$x_j = 1 \qquad \text{and} \qquad x_i = 0 \qquad \text{for } i = 1, \ldots, j-1, j+1, \ldots, k-1.$$

For the kth set, all the dummy variates are set equal to zero. We can now carry out a canonical variate analysis in the way described above. Assume, as before, that q species are observed, and that $q > k-1$. Then the analysis will yield $k-1$ linear combinations of species quantities (the y's) and these are the discriminant functions desired. (The linear combinations of the dummy variates that might also be obtained are of no interest in the present context.)

A demonstration that the discriminant functions represent those linear combinations of the measured species quantities that will maximize the ratio of the between-set to the within-set sum of squares of the quantities is given by Tatsuoka (1971). He also describes how to test which, if any, of the discriminant functions derived in a particular analysis signify true differences among the data sets. The tests presuppose that the species quantities have a q-variate normal distribution, and also that the sampling units have been drawn at random from some specified, homogeneous parent population. When, as is so often the case in ecological work, these assumptions are unwarranted, significance testing may be misleading. It is informative, however, to use plots of the set-means of the discriminant functions to demonstrate graphically how the set-means differ from one another.

An example has been described by Buzas (1967), who compared foraminifera populations in sediment samples collected from the continental shelf in the Gulf of Mexico. One of his analyses is of data on 45 species of living forams whose shells were found in samples collected from five areas (labeled $h_4, \ldots, h_8$ by Buzas). The number of sampling stations in each area ranged from 25 to 31. Figure 22.2 shows a three-dimensional plot of the means, for these five areas, of the first three discriminant functions; it had been found that these three functions accounted for almost 95% of the total variability. From inspection of the graph it seems reasonable to hypothesize that areas h_4 and h_6 form a

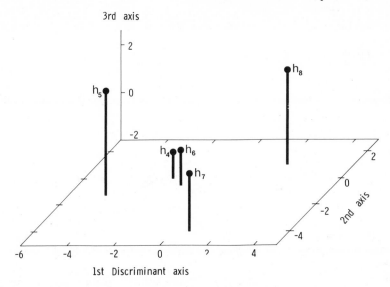

Figure 22.2. A plot of the means of the first three discriminant functions ("canonical variates") from five populations of foraminifera. (Adapted from Buzas, 1967.)

similar pair from which, and from one another, areas h_5, h_7, and h_8 differ markedly.

It will be noticed that canonical variate analysis and multiple discriminant analysis are designed for data that have a linear structure (cf. page 345). The analysis done in the text example (see page 357 and Figure 22.1) would not have been satisfactory if the relationship between x_1 and x_2, or between y_3 and y_4 (or between both sets) had been curvilinear. We should then have wished to find canonical variates by some catenation method such as continuity analysis.

Research on this topic remains to be done.

The same can truly be said of a host of topics that challenge ecologists. I hope this book has pointed to some of them.

Bibliography

Anderson, T. W. (1958). *An Introduction to Multivariate Statistical Analysis.* Wiley, New York.

Anscombe, F. J. (1950). Sampling theory of the negative binomial and logarithmic series distributions. *Biometrika* **37**: 358–382.

Bailey, N. T. J. (1964). *The Elements of Stochastic Processes with Applications to the Natural Sciences.* Wiley, New York.

Barkham, J. P., and J. M. Norris (1970). Multivariate procedures in an investigation of vegetation and soil relations of two beech woodlands, Cotswold Hills, England. *Ecology* **51**: 630–639.

Barnard, G. A. (1947). Significance tests for 2×2 tables. *Biometrika* **34**: 123–138.

Barnett, V. D. (1962). The Monte Carlo solution of a competing species problem. *Biometrics* **18**: 76–103.

Barnett, V. D., and T. Lewis (1963). A study of the relation between G. C. E. and degree results. *J. R. Statist. Soc.* **A126**: 187–226.

Bartlett, M. S. (1960). *Stochastic Population Models in Ecology and Epidemiology.* Methuen, London.

Bartlett, M. S. (1974). The statistical analysis of spatial pattern. *Adv. Appl. Probab.* **6**: 336–358.

Barton, D. E., and F. N. David (1959). The dispersion of a number of species. *J. R. Statist. Soc.* **B21**: 190–194.

Basharin, G. P. (1959). On a statistical estimate for the entropy of a sequence of independent random variables. *Theory Probab. Appl.* **4**: 333–336.

Batcheler, C. L. (1971). Estimate of density from a sample of joint point and nearest neighbor distances. *Ecology* **52**: 703–709.

Béland, P. (1974). On predicting the yield from brook trout populations. *Trans. Amer. Fish. Soc.* **103**: 353–355.

Bennett, B. M., and C. Horst (1966). *Supplement to Tables for Testing Significance in a 2×2 Contingency Table.* Cambridge University Press, Cambridge.

Berthét. P., and G. Gérard (1965). A statistical study of microdistribution of Oribatei (Acari), Part I. The distribution pattern. *Oikos* **16**: 214–227.

Birch, L. C. (1948). The intrinsic rate of natural increase of an insect population. *J. Anim. Ecol.* **17**: 15–26.

Bliss, C. I. (1965). An analysis of some insect trap records. In "Classical and Contagious Discrete Distributions" (G. P. Patil, Ed.) Statistical Publishing Society, Calcutta.

Bliss, C. I., and R. A. Fisher (1953). Fitting the negative binomial distribution to biological data and a note on the efficient fitting of the negative binomial. *Biometrics* **9**: 176–200.

Boswell, M. T., and G. P. Patil (1971). Chance mechanisms generating the logarithmic series distribution used in the analysis of number of species and individuals. In "Statistical Ecology," Vol. 1 (G. P. Patil, E. C. Pielou, and W. E. Waters, Eds.), Pennsylvania State University Press.

Boulton, D. M., and C. S. Wallace (1970). A program for numerical classification. *Comput. J.* **13**: 63–69.

Brauer, F., and D. A. Sanchez (1975). Constant rate population harvesting. *Theor. Pop. Biol.* **8**: 12–30.

Bray, J. R. (1956). A study of mutual occurrence of plant species. *Ecology* **37**: 21–28.

Brian, M. V. (1953). Species frequencies in random samples from animal populations. *J. Anim. Ecol.* **22**: 57–64.

Brillouin, L. (1962). *Science and Information Theory.* 2nd ed. Academic, New York.

Broadbent, S. R., and D. G. Kendall (1953). The random walk of *Trichostrongylus retortaeformis. Biometrics* **9**: 460–466.

Bulmer, M. G. (1974). On fitting the Poisson lognormal distribution to species-abundance data. *Biometrics* **30**: 101–110.

Bulmer, M. G. (1975). Phase relations in the ten-year cycle. *J. Anim. Ecol.* **44**: 609–621.

Buzas, M. A. (1967). An application of canonical analysis as a method for comparing faunal areas. *J. Anim. Ecol.* **36**: 563–577.

Caughley, G. (1966). Mortality patterns in mammals. *Ecology* **47**: 906–918.

Caughley, G. (1967). Parameters for seasonally breeding populations. *Ecology* **48**: 834–839.

Chiang, C. L. (1954). Competition and other interactions between species. In "Statistics and Mathematics in Biology" (O. Kempthorne et al., Eds.), Iowa State College Press, Ames.

Chiang, C. L. (1968). *Introduction to Stochastic Processes in Biostatistics.* Wiley, New York.

Clark, P. J., and F. C. Evans (1954). Distance to nearest neighbor as a measure of spatial relationships in populations. *Ecology* **35**: 445–453.

Clark, P. J., and F. C. Evans (1955). On some aspects of spatial pattern in biological populations. *Science* **121**: 397–398.

Cochran, W. G. (1954). Some methods for strengthening the common χ^2-tests. *Biometrics* **10**: 417–451.

Cohen, J. E. (1968). Alternative derivations of a species-abundance relation. *Amer. Natur.* **102**: 165–172.

Cole, L. C. (1949). The measurement of interspecific association. *Ecology* **30**: 411–424.

Conover, W. J. (1971). *Practical Nonparametric Statistics.* Wiley, New York.

Cooper, C. F. (1961). Pattern in Ponderosa pine forests. *Ecology* **42**: 493–499.

Cormack, R. M. (1971). A review of classification. *J. R. Statist. Soc.* **A134**: 321–367.

Cramer, E. M. (1973). A simple derivation of the canonical correlation equations. *Biometrics* **29**: 379–380.

Dagnelie, P. (1966). A propos des différentes méthodes de classification numerique. *Rev. Stat. Appl.* **14**: 55–75.

Darwin, J. H., and R. W. Williams (1964). The effect of time of hunting on the size of a rabbit population. *N. Z. J. Sci.* **7**: 341–352.

David, F. N., and N. L. Johnson (1948). The probability integral transformation when parameters are estimated from the sample. *Biometrika* **35**: 182–190.

David, F. N., and P. G. Moore (1954). Notes on contagious distributions in plant populations. *Ann. Bot. Lond. N. S.* **18**: 47–53.

David, H. A. (1970). *Order Statistics*. Wiley, New York.

DeAngelis, D. L., R. A. Goldstein, and R. V. O'Neill (1975). A model for trophic interaction. *Ecology* **56**: 881–892.

DeBach, P. (1966). The competitive displacement and coexistence principles. *Ann. Rev. Ent.* **11**: 183–212.

Deevey, E. S. (1947). Life tables for natural populations of animals. *Quart. Rev. Biol.* **22**: 283–314.

Dice, L. R. (1945). Measures of the amount of ecologic association between species. *Ecology* **26**: 297–302.

Dublin, L. I., and A. J. Lotka (1925). On the true rate of natural increase. *J. Amer. Stat. Ass.* **20**: 305–339.

Edwards, A. W. F., and L. L. Cavalli-Sforza (1965). A method for cluster analysis. *Biometrics* **21**: 362–375.

Engen, S. (1974). On species frequency models. *Biometrika* **61**: 263–270.

Feller, W. (1943). On a general class of contagious distributions. *Ann. Math. Stat.* **14**: 389–400.

Feller, W. (1968). *An Introduction to Probability Theory and Its Applications. Volume I.* 3rd ed., Wiley, New York.

Field, J. G. (1971). A numerical analysis of changes in the soft-bottom fauna along a transect across False Bay, South Africa. *J. Exp. Mar. Biol. Ecol.* **7**: 215–253.

Finney, D. J., R. Latscha, B. M. Bennett, and P. Hsu (1963). *Tables for Testing Significance in a 2 × 2 Contingency Table.* Cambridge University Press, Cambridge.

Fisher, R. A., A. S. Corbet, and C. B. Williams (1943). The relation between the number of species and the number of individuals in a random sample of an animal population. *J. Anim. Ecol.* **12**: 42–58.

Frank, P. W. (1968). Life histories and community stability. *Ecology* **49**: 355–357.

Gallopin, G. C. (1971). A generalized model of a resource-population system. I: General properties; II: Stability analysis. *Oecologia* **7**: 382–413; **7**: 414–432.

Gani, J. (1973). Stochastic formulations for life tables, age distributions and mortality curves. In "The Mathematical Theory of the Dynamics of Biological Populations" (M. S. Bartlett and R. W. Hiorns, Eds.), Academic, New York.

Gause, G. F. (1934). *The Struggle for Existence*. Hafner, New York.

Gehrs, C. W., and A. Robertson (1975). Use of life tables in analysing the dynamics of copepod populations. *Ecology* **56**: 665–672.

Gittins, R. (1965). Multivariate approaches to a limestone grassland community: I. A stand ordination. *J. Ecol.* **53**: 385–401.

Goldman, S. (1953). Some fundamentals of information theory. In "Information Theory in Biology" (H. Quastler, Ed.), University of Illinois Press.

Good, I. J. (1950). *Probability and the Weighing of Evidence.* Griffin, London.

Good, I. J. (1953). The population frequencies of species and the estimation of population parameters. *Biometrika* **40**: 237–264.

Goodall, D. W. (1954). Objective methods for the classification of vegetation: III. An essay in the use of factor analysis. *Aust. J. Bot.* **2**: 304–324.

Goodall, D. W. (1963). Pattern analysis and minimal area: some further comments. *J. Ecol.* **51**: 705–710.

Goodall, D. W. (1966). Hypothesis testing in classification. *Nature* **211**: 329–330.

Goodall, D. W. (1973a). Sample similarity and species correlation. In "Handbook of Vegetation Science," (R. H. Whittaker, Ed.) Junk, The Hague.

Goodall, D. W. (1973b). Numerical classification. In "Handbook of Vegetation Science," (R. H. Whittaker, Ed.) Junk, The Hague.

Goodall, D. W. (1974). A new method for the analysis of spatial pattern by random pairing of quadrats. *Vegetatio* **29**: 135–146.

Goodman, L. A. (1969). The analysis of population growth when the birth rates and death rates depend upon several factors. *Biometrics* **25**: 659–681.

Gower, J. C. (1966). Some distance properties of latent root and vector methods used in multivariate analysis. *Biometrika* **53**: 325–338.

Gower, J. C. (1967a). A comparison of some methods of cluster analysis. *Biometrics* **23**: 623–637.

Gower, J. C. (1967b). Multivariate analysis and multidimensional geometry. *The Statistician* **17**: 13–28.

Gower, J. C. (1968). Adding a point to vector diagrams in multivariate analysis. *Biometrika* **55**: 582–585.

Greig-Smith, P. (1952). The use of random and contiguous quadrats in the study of the structure of plant communities. *Ann. Bot. Lond. N. S.* **16**: 293–316.

Greig-Smith, P. (1964). *Quantitative Plant Ecology.* 2nd ed. Butterworths, London.

Greig-Smith, P., M. P. Austin, and T. C. Whitmore (1967). The application of quantitative methods to vegetation survey: I. Association analysis and principal component ordination of rain forest. *J. Ecol.* **55**: 483–503.

Greville, T. N. E., and N. Keyfitz (1974). Backward projection by a generalized inverse. *Theor. Pop. Biol.* **6**: 135–142.

Grundy, P. M. (1951). The expected frequencies in a sample of an animal population in which the abundances of the species are log-normally distributed. Part I. *Biometrika* **38**: 427–434.

Hairston, N. G. (1967). Studies on the limitation of a natural population of *Paramecium aurelia. Ecology* **48**: 904–909.

Heatwole, H., and R. Levins (1972). Trophic structure stability and faunal change during recolonization. *Ecology* **53**: 531–534.

Herdan, G. (1958). The mathematical relation between Greenberg's index of linguistic diversity and Yule's characteristic, *Biometrika* **45**: 268–270.

Heyer, W. R., and K. A. Berven (1973). Species diversities of herpetofaunal samples from similar microhabitats at two tropical sites. *Ecology* **54**: 642–645.

Hoel, P. (1954). *Introduction to Mathematical Statistics*. Wiley, New York.

Holgate, P. (1969). Species frequency distributions. *Biometrika* **56**: 651–660.

Holling, C. S. (1965). The functional response of predators to prey density and its role in mimicry and population regulation. *Mem. Entomol. Soc. Can.* **45**: 3–60.

Holling, C. S. (1973). Resilience and stability of ecological systems. *Ann. Rev. Ecol. Syst.* **4**: 1–23.

Hopkins, B. (1955). The species area relations of plant communities. *J. Ecol.* **43**: 409–426.

Hopkins, B., and J. G. Skellam (1954). A new method for determining the type of distribution of plant individuals. *Ann. Bot. Lond. N. S.* **18**: 213–227.

Horn, H., and R. H. MacArthur (1972). Competition among fugitive species in a harlequin environment. *Ecology* **53**: 749–752.

Howard, R. A. (1960). *Dynamic Programming and Markov Processes*. Wiley, New York.

Hurlbert, S. H. (1971). The nonconcept of species diversity: a critique and alternative parameters. *Ecology* **52**: 577–586.

Iwao, S. (1970). Analysis of contagiousness in the action of mortality factors on the Western Tent Caterpillar population by using the $\overset{*}{m} - m$ relationship. *Res. Pop. Ecol.* **12**: 100–110.

Iwao, S., and E. Kuno (1971). An approach to the analysis of aggregation patterns in biological populations. In "Spatial Patterns and Statistical Distributions" (G. P. Patil et al. Eds.), Pennsylvania State University Press.

Jeffries, C. (1974). Qualitative stability and digraphs in model ecosystems. *Ecology* **55**: 1415–1419.

Kemeny, J. G., and J. L. Snell (1960). *Finite Markov Chains*. van Nostrand, New York.

Kendall, M. G. (1957). *A Course in Multivariate Analysis*. Griffin, London.

Kendall, M. G., and A. Stuart (1966). *The Advanced Theory of Statistics*. Vol. 3. Griffin, London.

Kendall, M. G., and A. Stuart (1967). *The Advanced Theory of Statistics*. Vol. 2. 2nd. ed. Griffin, London.

Kershaw, K. A. (1960). The detection of pattern and association. *J. Ecol.* **48**: 233–242.

Kershaw, K. A. (1961). Association and covariance analysis of plant communities. *J. Ecol.* **49**: 643–654.

Keyfitz, N. (1968). *Introduction to the Mathematics of Population*. Addison-Wesley, New York.

Khinchin, A. I. (1957). *Mathematical Foundations of Information Theory*. Dover, New York.

Kilburn, P. D. (1966). Analysis of the species area relation. *Ecology* **47**: 831–843.

Knight, W. (1974). A run-like statistic for ecological transects. *Biometrics* **30**: 553–555.

Krishna Iyer, P. V. (1949). The first and second moments of some probability distributions arising from points on a lattice and their application. *Biometrika* **36**: 135–141.

Krishna Iyer, P. V. (1952). Factorial moments and cumulants of distributions arising in Markoff chains. *J. Ind. Soc. Agr. Stat.* **4**: 113–123.

Lambert, J. M., and W. T. Williams (1966). Multivariate methods in plant ecology. VI. Comparison of information analysis and association analysis. *J. Ecol.* **54**: 635–664.

Lancaster, P. (1969). *Theory of Matrices.* Academic, New York.

Lance, G. N., and W. T. Williams (1968). Note on a new information-statistic classificatory program. *Comput. J.* **11**: 195.

Langford, A. N., and M. F. Buell (1969). Integration, identity and stability in the plant association. *Adv. Ecol. Res.* **6**: 83–135.

Laughlin, R. (1965). Capacity for increase: a useful population statistic. *J. Anim. Ecol.* **34**: 77–91.

Lawley, D. N., and A. E. Maxwell (1963). *Factor Analysis as a Statistical Method.* Butterworths, London.

Lefkovitch, L. P. (1965). The study of population growth in organisms grouped by stages. *Biometrics* **21**: 1–18.

Lehmann, E. L. (1974). *Nonparametrics: Statistical Methods based on Ranks.* Holden-Day, San Francisco.

León, J. A., and D. B. Tumpson (1975). Competition between two species for two complementary or substitutable resources. *J. Theor. Biol.* **50**: 185–201.

Leslie, P. H. (1945). The use of matrices in certain population mathematics. *Biometrika* **33**: 183–212.

Leslie, P. H. (1948). Some further notes on the use of matrices in population mathematics. *Biometrika* **35**: 213–245.

Leslie, P. H. (1958). A stochastic model for studying the properties of certain biological systems by numerical methods. *Biometrika* **45**: 16–31.

Leslie, P. H. (1959). The properties of a certain lag type of population growth and the influence of an external random factor on a number of such populations. *Physiol. Zool.* **32**: 151–159.

Leslie, P. H. (1966). The intrinsic rate of increase and the overlap of successive generations in a population of guillemots (*Uria aalge* Pont.) *J. Anim. Ecol.* **35**: 291–301.

Leslie, P. H., and J. C. Gower (1960). The properties of a stochastic model for the predator-prey type of interaction between two species. *Biometrika* **47**: 219–234.

Leslie, P. H., and T. Park (1949). The intrinsic rate of natural increase of *Tribolium castaneum* Herbst. *Ecology* **30**: 469–477.

Leslie, P. H., and R. M. Ranson (1940). The mortality, fertility and rate of natural increase of the vole (*Microtus agrestis*) as observed in the laboratory. *J. Anim. Ecol.* **9**: 27–52.

Levin, S. (1974). Dispersion and population interactions. *Amer. Nat.* **108**: 207–228.

Levins, R. (1968). *Evolution in Changing Environments*. Princeton University Press, Princeton.

Levins, R. (1974). The qualitative analysis of partially specified systems. *Ann. N. Y. Acad. Sci.* **231**: 123–138.

Lewis, E. G. (1942). On the generation and growth of a population. *Sankhyā* **6**: 93–96.

Lloyd, M. (1967). Mean crowding. *J. Anim. Ecol.* **36**: 1–30.

Lotka, A. J. (1925). *Elements of Physical Biology*. Williams and Wilkins, Baltimore.

MacArthur, R. H. (1957). On the relative abundance of bird species. *Proc. Nat. Acad. Sci. Wash.* **43**: 293–295.

Macdonald, P. D. M., and L. Cheng (1970). A method of testing for synchronization between host and parasite populations. *J. Anim. Ecol.* **39**: 321–331.

Macnaughton-Smith, P., W. T. Williams, M. B. Dale and L. G. Mockett (1964). Dissimilarity analysis: a new technique of hierarchical subdivision. *Nature* **202**: 1634–1635.

Margalef, D. R. (1958). Information theory in ecology. *General Systems* **3**: 36–71.

Matérn, B. (1960). Spatial variation. Stochastic models and their application to some problems in forest surveys and other sampling investigations. *Medd. fran Statens Skogsforskningsinstitut.* **49**: 1–144.

May, R. M. (1971). Stability in one-species community models. *Math. Biosci.* **12**: 59–79.

May, R. M. (1972). Limit cycles in predator-prey communities. *Science* **177**: 900–902.

May, R. M. (1973). *Stability and Complexity in Model Ecosystems*. Princeton University Press, Princeton.

May, R. M. (1974). Ecosystem patterns in randomly fluctuating environments. *Prog. Theor. Biol.* **3**: 1–48.

May, R. M. (1975). Biological populations obeying difference equations: stable points, stable cycles and chaos. *J. Theor. Biol.* **51**: 511–524.

Maynard Smith, J. (1974). *Models in Ecology* Cambridge University Press, Cambridge.

McIntosh, R. P. (1967). The continuum concept of vegetation. *Bot. Rev.* **33**: 130–187.

Mead, R. (1974). A test for spatial pattern at several scales using data from a grid of contiguous quadrats. *Biometrics* **30**: 295–308.

Miller, C. A. (1963). The spruce budworm. In "The Dynamics of Epidemic Spruce Budworm Populations" (R. F. Morris, Ed.), *Mem. Entomol. Soc. Can.*, No. 31.

Moore, P. G. (1954). Spacing in plant populations. *Ecology* **35**: 222–227.

Morisita, M. (1954). Estimation of population density by spacing method. *Mem. Fac. Sci. Kyushu U. Series E (Biol.)* **1**: 187–197.

Morisita, M. (1959). Measuring the dispersion of individuals and analysis of the distributional patterns. *Mem. Fac. Sci. Kyushu U. Series E (Biol.)* **2**: 215–235.

Morrison, D. F. (1967). *Multivariate Statistical Methods*. Wiley, New York.

Mountford, M. D. (1961). On E. C. Pielou's index of non-randomness. *J. Ecol.* **49**: 271–275.

Murdie, G., and M. P. Hassell (1973). Food distribution, searching success and predator-prey models. In "The Mathematical Theory of the Dynamics of Biological Populations" (M. S. Bartlett and R. W. Hiorns, Eds.), Academic, New York.

Murray, M. D., and G. Gordon (1969). Ecology of lice on sheep. *Aust. J. Zool.* **17**: 179–186.

Neill, W. E. (1974). The community matrix and interdependence of the competition coefficients. *Amer. Nat.* **108**: 399–408.

Nelson, W. C., and H. A. David (1967). The logarithmic distribution: a review. *Virginia Journal of Sci.* **18**: 95–102.

Neyman, J., T. Park, and E. L. Scott (1956). Struggle for existence. The *Tribolium* model. *Proc. 3rd Berkeley Symposium on Math. Stat. and Prob.* **3**: 41–79.

Noy-Meir, I. (1974). Catenation: quantitative methods for the definition of coenoclines. *Vegetatio* **29**: 89–99.

Odum, E. P. (1959). *Fundamentals of Ecology*, 2nd ed. Saunders, Philadelphia.

Orloci, L. (1966). Geometric models in ecology: I. The theory and application of some ordination methods. *J. Ecol.* **54**: 193–215.

Orloci, L. (1967a). An agglomerative method for classification of plant communities. *J. Ecol.* **55**: 193–205.

Orloci, L. (1967b). Data centering: a review and evaluation with reference to component analysis. *Syst. Zool.* **16**: 208–212.

Orloci, L. (1968). Information analysis in phytosociology: partition, classification and prediction. *J. Theor. Biol.* **20**: 271–284.

Orloci, L. (1969). Information analysis of structure in biological collections. *Nature Lond.* **223**: 483–484.

Orloci, L. (1975). *Multivariate Analysis in Vegetation Research*. Junk, The Hague.

Oster, G. F. (1975). Review of "Ecological Stability" (M. B. Usher and M. H. Williamson, Eds.), *Ecology* **56**: 1462.

Park, T. (1954). Experimental studies on interspecies competition: II. Temperature, humidity and competition in two species of *Tribolium*. *Physiol. Zool.* **27**: 177–238.

Pearson, E. S. (1947). The choice of statistical tests illustrated on the interpretation of data classed in a 2×2 table. *Biometrika* **34**: 139–167.

Pennycuick, L. (1969). A computer model of the Oxford great tit population. *J. Theor. Biol.* **22**: 381–400.

Phillips, M. E. (1953). Studies on the quantitative morphology and ecology of *Eriophorum angustifolium* Roth.: I. The rhizome system. *J. Ecol.* **41**: 295–318.

Pielou, D. P., and W. G. Matthewman (1966). The fauna of *Fomes fomentarius* (Linnaeus ex Fries) Kickx. growing on dead birch in Gatineau Park, Quebec. *Can. Entomol.* **98**: 1308–1312.

Pielou, D. P., and E. C. Pielou (1967). The detection of different degrees of coexistence. *J. Theor. Biol.* **16**: 427–437.

Pielou, D. P., and E. C. Pielou (1968). Association among species of infrequent occurrence: the insect and spider fauna of *Polyporus betulinus* (Bulliard) Fries. *J. Theor. Biol.* **21**: 202–216.

Pielou, E. C. (1959). The use of point-to-plant distances in the study of the pattern of plant populations. *J. Ecol.* **47**: 607–613.

Pielou, E. C. (1961). Segregation and symmetry in two-species populations as studied by nearest neighbor relations. *J. Ecol.* **49**: 255–269.

Pielou, E. C. (1962a). The use of plant-to-neighbor distances for the detection of competition. *J. Ecol.* **50**: 357–367.

Pielou, E. C. (1962b). Runs of one species with respect to another in transects through plant populations. *Biometrics* **18**: 579–593.

Pielou, E. C. (1963a). The distribution of diseased trees with respect to healthy ones in a patchily infected forest. *Biometrics* **19**: 450–459.

Pielou, E. C. (1963b). Runs of healthy and diseased trees in transects through an infected forest. *Biometrics* **19**: 603–614.

Pielou, E. C. (1964). The spatial pattern of two-phase patchworks of vegetation. *Biometrics* **20**: 156–167.

Pielou, E. C. (1965a). The concept of randomness in the patterns of mosaics. *Biometrics.* **21**: 908–920.

Pielou, E. C. (1965b). The concept of segregation pattern in ecology: some discrete distributions applicable to the run lengths of plants in narrow transects. In "Classical and Contagious Discrete Distributions" (G. P. Patil, Ed.). Statistical Publishing Society, Calcutta.

Pielou, E. C. (1966a). Species-diversity and pattern-diversity in the study of ecological succession. *J. Theor. Biol.* **10**: 370–383.

Pielou, E. C. (1966b). Shannon's formula as a measure of specific diveristy: its use and misuse. *Amer. Nat.* **100**: 463–465.

Pielou, E. C. (1966c). The measurement of diversity in different types of biological collections. *J. Theor. Biol.* **13**: 131–144.

Pielou, E. C. (1967a). The use of information theory in the study of the diversity of biological populations. *Proc. 5th Berkeley Symposium on Math. Stat. and Prob.* **4**: 163–177.

Pielou, E. C. (1967b). A test for random mingling of the phases of a mosaic. *Biometrics* **23**: 657–670.

Pielou, E. C. (1972). Niche width and niche overlap: a method for measuring them. *Ecology* **53**: 687–692.

Pielou, E. C. (1974a). Competition on an environmental gradient. In "Mathematical Problems in Biology" (P. van den Driessche, Ed.), Springer-Verlag, Berlin.

Pielou, E. C. (1974b). Vegetation zones: repetition of zones on a monotonic environmental gradient. *J. Theor. Biol.* **47**: 485–489.

Pielou, E. C. (1974c). *Population and Community Ecology: Principles and Methods.* Gordon and Breach, New York.

Pielou, E. C. (1975a). *Ecological Diversity.* Wiley, New York.

Pielou, E. C. (1975b). Ecological models on an environmental gradient. In "Applied Statistics" (R. P. Gupta, Ed.), North Holland Publishing Co., Amsterdam.

Pielou, E. C. (1976). The factual background of ecological models: tapping some unused resources *Proc. Conf. Environmental Modeling & Simulation*: 668–672. U.S. Environmental Protection Agency.

Pielou, E. C., and R. E. Foster (1962). A test to compare the incidence of disease in isolated and crowded trees. *Can. J. Botany* **40**: 1176–1179.

Pielou, E. C., and R. D. Routledge (1976). Salt marsh vegetation. Latitudinal variation in the zonation patterns *Oecologia* **24**: 311–321.

Pollard, J. H. (1966). On the use of the direct matrix product in analysing certain stochastic population models. *Biometrika* **53**: 397–415.

Pollard, J. H. (1973). *Mathematical Models for the Growth of Human Populations.* Cambridge University Press, Cambridge.

Poole, R. W. (1976a). Empirical multivariate autoregressive equation predictors of the fluctuations of interacting species. *Math. Biosci.* **28**: 81–97.

Poole R. W. (1967b). An autoregressive model with level-dependent error terms for predicting fluctuations in animal or plant abundance. *Biometrics* In Press

Preston, F. W. (1948). The commonness and rarity of species. *Ecology* **29**: 254–283.

Quirk, J. P., and R. Ruppert (1965). Qualitative economics and the stability of equilibrium. *Rev. Econ. Stud.* **32**: 311–326.

Rapport, D. J., and J. E. Turner (1975). Predator-prey interactions in natural communities. *J. Theor. Biol.* **51**: 169–180.

Renyi, A. (1961). On measures of entropy and information. *Proc. 4th. Berkeley Symposium on Math. Stat. and Prob.* **1**: 547–561.

Rescigno, A. (1968). The struggle for life: II. Three competitors. *Bull. Math. Biophys.* **30**: 291–298.

Rescigno, A., and I. W. Richardson (1967). The struggle for life: I. Two species. *Bull. Math. Biophys.* **29**: 377–388.

Sanchez, D. A. (1968). *Ordinary Differential Equations and Stability Theory: An Introduction.* Freeman, San Francisco.

Sarukhan, J., and M. Gadgil (1974). Studies on plant demography; *Ranunculus repers* L., *R. bulbosus* L. and *R. acris* L. *J. Ecol.* **62**: 921–936.

Scott, A. J., and M. J. Symons (1971). On the Edwards and Cavalli-Sforza method of cluster analysis. *Biometrics* **27**: 217–219.

Scudo, F. M. (1971). Vito Volterra and theoretical ecology. *Theor. Pop. Biol.* **2**: 1–23.

Shannon, C. E., and W. Weaver (1949). *The Mathematical Theory of Communication.* University of Illinois Press, Urbana.

Shepard, R. N., and J. D. Carroll (1966). Parametric representation of nonlinear data structures. In "Multivariate Analysis" (P. R. Krishnaiah, Ed.) Academic, New York.

Siegel, S. (1956). *Nonparametric Statistics for the Behavioral Sciences.* McGraw-Hill, New York.

Simpson, E. H. (1949). Measurement of diversity. *Nature* **163**: 688.

Skellam, J. G. (1951). Random dispersal in theoretical populations. *Biometrika* **38**: 196–218.

Skellam, J. G. (1952). Studies in statistical ecology: I. Spatial pattern. *Biometrika* **39**: 346–362.

Skellam, J. G. (1967). Seasonal periodicity in theoretical population ecology. *Proc. 5th Berkeley Symposium on Math. Stat. and Prob.* **4**: 179–205.

Smith, F. E. (1963). Population dynamics in *Daphnia magna. Ecology* **44**: 651–663.

Smith, R. H., and R. Mead (1974). Age structure and stability in models of prey-predator systems. *Theor. Pop. Biol.* **6**: 308–322.

Sokal, R. R., and P. H. A. Sneath (1963). *Principles of Numerical Taxonomy.* Freeman, San Francisco.

Strobeck, C. (1973). *N* species competition. *Ecology* **54**: 650–654.

Sutherland, J. P. (1974). Multiple stable points in natural communities. *Amer. Nat.* **108**: 859–870.

Swed, F. S., and C. Eisenhart (1943). Tables for testing randomness of grouping in a sequence of alternatives. *Ann. Math. Stat.* **14**: 66–85.

Switzer, P. (1965). A random set process in the plane with a Markovian property. *Ann. Math. Stat.* **36**: 1859–1863.

Switzer, P. (1967). Reconstructing patterns from sample data. *Ann. Math. Stat.* **38**: 138–154.

Sykes, Z. M. (1969a). On discrete stable population theory. *Biometrics* **25**: 285–293.

Sykes, Z. M. (1969b). Some stochastic versions of the matrix model for population dynamics. *Jour. Amer. Stat. Assn.* **64**: 111–130.

Tanner, J. T. (1975). The stability and the intrinsic growth rates of prey and predator populations. *Ecology* **56**: 855–867.

Tatsuoka, M. M. (1971). *Multivariate Analysis: Techniques for Educational and Psychological Research.* Wiley, New York.

Thompson, H. R. (1956). Distribution of distance to *n*th neighbor in a population of randomly distributed individuals. *Ecology* **37**: 391–394.

Usher, M. B. (1972). Developments in the Leslie matrix model. In "Mathematical Models in Ecology" (J. N. R. Jeffers, Ed.), Blackwell, Oxford.

Utida, S. (1957). Cyclic fluctuations of population density intrinsic to the host-parasite system. *Ecology* **38**: 442–449.

Vandermeer, J. H. (1975). Interspecific competition: a new approach to the classical theory. *Science* **188**: 253–255.

Volterra, V. (1931). *Leçons sur la Théorie Mathematique de la Lutte pour la Vie.* Gauthier-Villars, Paris.

Wallace, C. S., and D. M. Boulton (1968). An information measure for classification. *Comput. J.* **11**: 185–194.

Waters, W. E. (1959). A quantitative measure of aggregation in insects. *J. Econ. Ent.* **52**: 1180–1184.

Watson, G. N. (1944). *Theory of Bessel Functions.* Cambridge University Press, Cambridge.

Watt, A. S. (1947). Pattern and process in the plant community. *J. Ecol.* **35**: 1–22.

Watterson, G. A. (1974). Models for the logarithmic species abundance distributions. *Theor. Pop. Biol.* **6**: 217–250.

Whitworth, W. A. (1934). *Choice and Chance.* Steichert, New York.

Wilkinson, C. (1970). Adding a point to a principal coordinate analysis. *Syst. Zool.* **19**: 258–263.

Wilkinson, D. H. (1952). The random element in bird "navigation." *J. Exp. Biol.* **29**: 532–560.

Williams, C. B. (1964). *Patterns in the Balance of Nature.* Academic, New York.

Williams, E. J. (1961). The distribution of larvae of randomly moving insects. *Austr. J. Biol. Sci.* **14**: 598–604.

Williams, W. T. (1971). Principles of clustering. *Ann.Rev. Ecol. Syst.* **2**: 303–326.

Williams, W. T., and J. M. Lambert (1959). Multivariate methods in plant ecology: I. Association analysis in plant communities. *J. Ecol.* **47**: 83–101.

Williams, W. T., and J. M. Lambert (1960). Multivariate methods in plant ecology: II. The use of an electronic digital computer for association analysis. *J. Ecol.* **48**: 689–710.

Williams, W. T., and J. M. Lambert (1966). Multivariate methods in plant ecology: V. Similarity analyses and information analysis. *J. Ecol.* **54**: 427–445.

Williamson, M. H. (1959). Some extensions of the use of matrices in population theory. *Bull. Math. Biophys.* **21**: 13–17.

Williamson, M. H. (1967). Introducing students to the concepts of population dynamics. In "The Teaching of Ecology" (J. M. Lambert, Ed.) Blackwell, Oxford.

Williamson, M. H. (1972). *The Analysis of Biological Populations.* Edward Arnold, London.

Wylie, C. R. Jr. (1951). *Advanced Engineering Mathematics.* McGraw-Hill, New York.

Yule, G. U. (1912). On the methods of measuring association between two attributes. *J. R. Stat. Soc.* **75**: 579–642.

Yule, G. U. (1924). A mathematical theory of evolution based on the conclusions of Dr. J. C. Willis F.R.S. *Phil. Trans. R. Soc. Lond.* **B213**: 21–87.

Zahl, S. (1974). Application of the S-method to the analysis of spatial pattern. *Biometrics* **30**: 513–524.

Author Index

376

Subject Index

Index of Organisms